Africa: What It Gave Me, What It Took from Me

Africa: What It Gave Me, What It Took from Me

Remembrances from My Life as a German Settler in South West Africa

Margarethe von Eckenbrecher

Translated, edited, and introduced by
David P. Crandall, Hans-Wilhelm Kelling,
and Paul E. Kerry

LEHIGH UNIVERSITY PRESS
Bethlehem

Published by Lehigh University Press
Copublished by The Rowman & Littlefield Publishing Group, Inc.
4501 Forbes Boulevard, Suite 200, Lanham, Maryland 20706
www.rowman.com

Unit A, Whitacre Mews, 26-34 Stannary Street, London SE11 4AB

British Library Cataloguing in Publication Information Available

Library of Congress Cataloging-in-Publication Data

Eckenbrecher, Margarethe von, 1875–1955.
[Was Afrika mir gab und nahm. English]
Africa: What it gave me, what it took from me : remembrances from my life as a German settler in South West Africa / Margarethe von Eckenbrecher ; Translated, edited, and introduced by David P. Crandall, Hans-Wilhelm Kelling and Paul E. Kerry.
pages cm
Includes bibliographical references and index.
ISBN 978-1-61146-150-3 (cloth : alkaline paper) — ISBN 978-1-61146-198-5 (paperback : alkaline paper) — ISBN 978-1-61146-151-0 (electronic) 1. Eckenbrecher, Margarethe von, 1875–1955. 2. Germans—Namibia—Biography. 3. Women pioneers—Namibia—Biography. 4. Farmers—Namibia—Biography. 5. Frontier and pioneer life—Namibia. 6. Namibia—History—1884–1915—Biography. 7. Namibia—History—1915–1946—Biography. 8. Namibia—Social life and customs—20th century. 9. Namibia—Ethnic relations—History—20th century. 10. Okombahe (Namibia)—Biography. I. Crandall, David P., 1960– II. Kelling, Hans-Wilhelm. III. Kerry, Paul E. IV. Title. V. Title: Africa.
DT1608.E25A313 2015
968.8103—dc23

2015004497

∞™ The paper used in this publication meets the minimum requirements of American National Standard for Information Sciences—Permanence of Paper for Printed Library Materials, ANSI/NISO Z39.48-1992.

Printed in the United States of America

Contents

Acknowledgments

It is a satisfying duty to give thanks to those who have supported us on this journey. Much is written about interdisciplinary work, and in undertaking this project we have encountered its benefits, but also its difficulties. Combining the insights of an anthropologist, Germanist, and historian to translate and contextualize Margarethe von Eckenbrecher's memoir sounded like a fine idea over a decade ago, and along the way we have learned from each other, even when we had contrasting ideas over the translation or the content and direction of the scholarly Introduction. Often, we were on different continents (Africa, Europe, United States), and so we had to rely on the Internet to keep the project moving. Yet a warm cooperation prevailed as we laid out our perspectives openly and then agreed on how to move forward. We have become better colleagues for having worked on this project together.

Our efforts were helped enormously by the peer review provided by Lehigh University Press, a process that improved the manuscript in important ways. We are also particularly grateful to Professor Jens-Uwe Guettel in the Department of History at Penn State University who read our Introduction and made several valuable suggestions, one of which led us to alter the translation of the title. We thank Professor Jesse S. Crisler in the Department of English at Brigham Young University for copyediting our Introduction—it was hard work to unify stylistically three different voices. Mr. Hans-Henning von Eckenbrecher kindly agreed to meet Professor Crandall in Cape Town and share his remembrance of his grandmother.

We acknowledge the institutional support of Brigham Young University and specifically the departments of Anthropology, German, and History, as well as the David M. Kennedy Center for International Studies. We express deep appreciation to Professor Monica Najar, former director of Lehigh University Press, who found merit in our labor and guided us patiently through the publication process. And we have had the pleasure of working with the new director, Professor Kate Crassons, and a former director, Professor Scott Paul Gordon. In addition, Tommy J. Moore at the Press has assisted us, and we happily recognize him.

Biographical Entries

David P. Crandall is an Oxford University–trained anthropologist who has lived and worked extensively among the Himba and explored the varied cultural and natural landscapes of southwestern Africa. He teaches in the Department of Anthropology at Brigham Young University and has served as chair. He is the author of *The Place of Stunted Ironwood Trees: A Year in the Lives of the Cattle-herding Himba of Namibia*.

Hans-Wilhelm Kelling was born and raised in Germany and received his PhD from Stanford University. He has taught in the Department of German at Brigham Young University for many years and has served as chair of the department as well as director of the European Studies program. He is perhaps best known for his popular textbook, *Deutsche Kulturgeschichte*, which has been reprinted several times.

Paul E. Kerry is an associate dean of undergraduate education at Brigham Young University and teaches in the Department of History. He took his DPhil from the University of Oxford and is a Fellow of the Royal Historical Society. He has published on Goethe, Franklin, and Carlyle, among others. He is currently an associate visiting research fellow at The Rothermere American Institute at the University of Oxford.

A Reader's Introduction

I

To introduce the first full-length English-language translation of Margarethe von Eckenbrecher's memoir, *Africa: What It Gave Me, What It Took from Me, Remembrances from My Life as a German Settler in South West Africa*, to a new audience some one hundred years after it first appeared in the bookshops of Berlin is not without its difficulties.[1]

There is an increase in scholarly interest and writing about the German colonial period in Africa during the past several decades that correlates with a revived curiosity among a cross-section of the general public in Europe, but particularly in Germany, as well as in the United States about the history and legacy of the short-lived German Colonial Empire, including female voices.[2] Thriving, present-day German-Namibian communities with their architecture, restaurants, markets, bakeries, butcheries, and breweries, featuring a marked German influence, and German-language local newspapers in what some might consider a rather isolated corner of the world, often astonish the hundreds of thousands of European, British, and American tourists who visit Namibia annually. Indeed, evidence of late nineteenth- and early twentieth-century German settlement and occupation appears throughout coastal, central, and southern Namibia, while descendants of early German settlers continue to live in the erstwhile German colony that finally achieved independent status in 1990.

Both reflecting and generating this renewed interest is the publication of books on German colonialism, not only in Europe and the United States but in Namibia as well, particularly the publication of locally authored books through the National Archives, Namibian Scientific Society, and international publishing houses in partnership with Namibian printers. Peter's Antiques, an eclectic shop in Swakopmund, for example, for several decades has reproduced a variety of German colonial-era books, ranging from governors' reports to settler histories and memoirs. Indeed, von Eckenbrecher's *Was Afrika mir gab und nahm: Erlebnisse einer deutschen Ansiedlerfrau in Südwestafrika* is available in softcover, hardcover, and occasionally in special leather-bound editions through the auspices of this shop.

This first English-language edition of her book was a translation of the original 1907 edition (Book I) along with new materials contained in the 1936 edition (Book II), both published by Ernst Siegfried Mittler & Sohn, Berlin. The 1936 edition, actually published in 1940, was typed by Irma Schimansky Hübner,[3] who passed by the von Eckenbrecher home in Klein-Windhoek after having played a few sets of tennis one Sunday afternoon. Von Eckenbrecher called to her from the window, bidding her to come inside her house. A little over twenty at the time, Hübner had been a secondary school pupil of von Eckenbrecher, whom she later described as a "kindly though rather formidable woman, a woman whose words and injunctions were taken seriously and acted upon. Her whole bearing was nothing if not proper and strict. Strangely, Frau von Eckenbrecher's nose—which could only be described as *eine lustige Nase* (a comical nose)—seemed quite out of place with her otherwise stern demeanor and look." Having recently completed an updated edition of her memoir in longhand, von Eckenbrecher pressed Hübner into typing the entire manuscript, a not unpleasant task, according to Hübner. Several weeks later, the typescript was on its way to Berlin for publication.

Although Book I of von Eckenbrecher's memoir is now over one hundred years old, particular thoughts, feelings, and experiences she chronicled seem to transcend the borders of time and culture, immediately resonating with a remarkable familiarity. Yet for readers who live in a different century and in places far removed from the book's setting, much of the actions, intentions, statements, sentiments, and daily congress of persons she describes, indeed the whole milieu of her early twentieth-century life in southwestern Africa, may today seem a surreal vestige of an alien world. Thus, an introduction essay to the Germany she willingly left behind is useful. In this vein, we attempt to throw light on von Eckenbrecher herself—her inner life, her fundamental beliefs, her understandings of the nature of the world and human beings, her moral universe, and the character traits she embodies in her day-to-day life—and provide some context for that life.

It is difficult to capture the course of any life, and to reduce it by interpreting it through a lens is an inevitable act of scholarly structuring that some might call an act of violence but others an interpretative focus. All theory requires reductions, and in some ways it seems that Margarethe von Eckenbrecher's life can be, as with any life, flattened to fit a thesis. But in placing her life in the context of Namibian history, her memoir becomes an important historical source and one that could contribute to the ongoing memory work of contemporary Namibia and Germany; in other words, a working through a briefly overlapping past.

II

When von Eckenbrecher was born on November 30, 1875, Germany as a unified nation-state, the *Kaiserreich* or German Empire, had been in existence for a mere four years. Before 1871, the year in which Otto von Bismarck, the "Iron Chancellor," had forged an empire with the help of the Prussian army by defeating France, "Germany" consisted of almost three dozen kingdoms, duchies, principalities, and free cities, loosely allied by linguistic, cultural, economic, but not strong political ties. The two most powerful states, Prussia in the north and the Austrian Empire in the south, had battled for political supremacy for years, the last time in 1866. Prussia decisively won that short war, and from the moment of Austria's defeat, no observer doubted that Prussia was now the dominant force among German states, especially those located in the northern and central parts of German territory. After 1871, Prussia also dominated the southern states as well, with the exception of what had become the Austro-Hungarian Empire.

Since the Industrial Revolution with its concomitant infusion of huge amounts of money into the economy, the division between rich and poor in Germany had become more evident. The contrast between rich landowners and poor farm workers, opulent industrialists and destitute workers, attracted the attention of social critics. As in other industrializing nations, the proletariat was oppressed, denied basic human dignity and rights, and was forced to labor twelve-to-sixteen-hour days in often debilitating circumstances and for a pauper's wage. Bankers and industrialists ruled economic life, while some Junkers—wealthy, aristocratic landowners in Prussia and Mecklenburg—treated their farm workers like sharecroppers. Clearly, property led to social status, as did education. Professors and educators, although not part of the richest class, were nonetheless highly respected.

Political parties in Germany had come into existence just after the middle of the nineteenth century, including the oldest labor party in the world, but they were initially relatively ineffective. A majority of army officers, landowners, ministers, and high officials belonged to conservative parties. These conservatives were proagriculture and anti-industrialization; they also supported the idea of divine right of kingship, a powerful military, and the supreme influence of Prussia in German affairs. Some were anti-Semitic and proclaimed that capitalism was exploited by Jewish people.

The Catholic Party became a strong minority after the exclusion of Austria from the Empire. Bismarck, who was supported by the National Liberals, engaged in a struggle with the Catholics, a *Kulturkampf*, to wrest power away from the Church and elevate the State. He ordered the dissolution of the Jesuit order, made civil marriage obligatory, passed legislation against Church education, and increased state control over bishops and the Church. Even so, in the end Bismarck was forced to rescind most of these laws, and the Catholic Party ultimately emerged strengthened. The conflict did, however, lead to a further division between Protestants and Catholics.

Socialism was also regarded as an enemy of the state. Karl Marx, Friedrich Engels, Ferdinand Lassalle (founder in 1863 of the ADAV or General German Workers' Association), August Bebel, and Wilhelm Liebknecht were dominant forces in the battle for workers' rights. Besides radical ideas about the abolition of the state and all class structures and privileges, they struggled for full economic and political emancipation of the working class, universal suffrage for men, and decent labor laws. Although for decades the government suppressed the activities of the labor movement, its influence and power grew steadily until it eventually became the largest party in Parliament.

Economically, Germany benefited substantially from a huge influx of capital owing to reparation payments by France following the Franco-Prussian War, and by the turn of the twentieth century the country had transformed from a largely agrarian nation into a powerful industrial state dominating Europe and outproducing and challenging both France and England in world markets. The coal and steel industries, captained by the Krupp, Thyssen, and Stump-Halberg dynasties, contributed to German economic leadership, as did the construction of heavy weapons and in particular the creation of the second largest naval fleet to Great Britain's. In the minds of political leaders, a way of thinking similar to that in the United States emerged; namely, that a large merchant fleet was needed to accommodate a growing export trade, and battleships and heavy cruisers thus had to be launched to protect the merchant fleet itself.

Germany also witnessed a steady increase in population from twenty-four to more than sixty million by the end of the century. A substantial increase in per capita income accompanied this growth in population. Many argued that Germany needed foreign colonies to supply raw materials, to purchase German goods, and to signal Germany's international prestige. Foreign colonies would also lighten the overpopulation in cities caused by an influx of farm laborers from the country, enticing people to move abroad and yet remain German, in a German sphere of influence. Immigration to the American continent and other areas of the world had significantly lessened with political stabilization in Germany. Yet migration from farms and proportionately less farming activity contributed to a significant drop in agricultural production. Indeed, by the end of the nineteenth century Germany was no longer agriculturally self-sufficient, depending instead on substantial imports to feed its growing population. The rush to urban centers also led to large increases in urban poverty. At the turn of the century, Berlin, for example, had to sustain a population of more than two million with insufficient housing and social benefits.

The Hopfers, von Eckenbrecher's natal family, lived in Bernburg,[4] capital of the small Duchy of Anhalt, which had been under Prussian hegemony for some time and for all practical purposes had become an integral part of its much larger neighbor, although Anhalt's dukes reigned semiautocratically until 1918 when Germany lost World War I and became a democratic republic for the first time in its long, colorful history.

The Hopfer family was conservative, supporting the existing political situation, including the monarchy, and celebrating the country's unification under a powerful Prussian emperor. As a well-to-do middle-class family, the Hopfers had doubtless gained an influential position befitting their social standing in both the community and local government.[5] Some of their relatives, the von Eckenbrechers, even enjoyed noble rank. When in 1902 Margarethe married her cousin, Themistokles von Eckenbrecher II, she increased her own social position.

Owing to the comparatively secure economic situation disposition of the Hopfer family, Margarethe did not suffer the many wants of the working class in Germany. Undoubtedly, her thinking was partially informed by the conservative economic and political views of her parents and relatives. She enjoyed a fine education, one at that time not available to most women. When she was twelve years old, her family sent her to a boarding school for "women of nobility," which further reinforced her conservative views. The school's curriculum included philosophy, mathematics, sciences, English, and French. She later spent three years in England, perfecting her English language skills. Back in Berlin, the new imperial capital, she attended lectures in philosophy and theology by highly respected professors at the renowned Humboldt University.[6] Religion does not play a central role in her memoir, yet spirituality does, and her mind drew on these influential sources of modern theology, though she managed to avoid being dominated by them. Instead, she framed her own view of the world and her own sense of morality, in both of which she appears to have felt quite secure.

As an educated woman, then, one who had lived in England for three years and for an extended period in Berlin, she was exposed to a rich intellectual life in *fin-de-siècle* Germany. Even so, von Eckenbrecher's memoir provides surprisingly little detail about the intellectual influences that informed her life and thinking. She never mentions any of the emerging modernist literature, music, or philosophy that was quickly gaining favor and popularity in her native Germany and would come to dominate its intellectual scene during the Weimar Republic. As a teacher, she appears to have been at least partially a representative of the *Bildungsbürger* or educated middle class in Germany, with its taste for Goethe and a broad range of German classical and Romantic thought, as borne out by her grandson's recollection of her personal library, which contained the foundational sources to which she returns again and again throughout her memoir. In some ways, although never explicitly stated, she seems to have been a self-made woman, and, aside from an acknowledged homage to Goethe, she includes few direct clues about those whom she admired or who might have molded her. Rather, in many ways, her memoir is about how life in Africa tutored her.

As for the major social changes and political upheaval in Germany, she also fails to mention either in her narrative, though she certainly witnessed both in Berlin and elsewhere, but it was in South West Africa that she saw opportunity to lead an agrarian existence, something that was becoming an increasingly remote possibility in the new Germany. If her memoir appears at times to idealize life in the Südwest she certainly has no qualms about providing details—which are decidedly not complaints—about the incredibly hard work that such a life demanded. She seems to have had a sensibility uniquely attuned to the life she chose to lead there and never conveys the impression that she would have preferred to have lived anywhere else.

On the last day of April 1902, von Eckenbrecher sailed from Hamburg to German South West Africa with her husband of only a few days. They intended to build a farm from the ground up, to create a comfortable rural existence for themselves, and to contribute to the colonial enterprise so desired by a new German state. Within a year or so of arriving in the tiny settlement of Okombahe, she gave birth to a son. When, early in 1904, the native Nama and Herero peoples began an uprising against German colonial rule (that the colonists brutally put down), von Eckenbrecher and her son returned to the safety of Germany; her husband followed not long thereafter. In May 1914, now divorced with two young sons, she returned to South West Africa, took up residence in Windhoek, the capital city, and taught secondary school, a career she continued until she turned seventy-four, after which she remained on her own in Windhoek for another five years. Then as old age overcame her, she moved to Cape Town, South Africa, where her elder son and his family cared for her until she died on July 18, 1955, her body dispatched back to South West Africa for burial beside her younger son in the Windhoek City Cemetery.

III

When it first appeared in 1907, von Eckenbrecher's memoir sold well in Germany. She traveled the country, lecturing on her life in South West Africa and encouraging others, particularly women, to earnestly consider participating in building the colony. To some degree her memoir follows the trends of German colonial literature, both in fiction and personal narratives, in which German female writers played a substantial role. It was a literature that became enormously popular through the end of the nineteenth century up to the beginning of World War I.[7] Though in the later edition of her memoir von Eckenbrecher asserted that it was not a work of propaganda designed to bring her compatriots to South West Africa,[8] she made this comment only after many years of living in South West Africa, her identity by then perhaps more Südwest than German. By 1936, when she completed the expanded version of her memoir, Germany had in any case experienced violent transitions as it convulsed from the *Kaiserreich* to the Weimar Republic and then to the Third Reich. But in 1906 her intention is clear:

> I wish to recruit women for the colony and plead with you: German women, do not let yourselves be held back by pedantic doubts. If you are confronted with the decision to follow the man you love to Africa, do not hesitate, follow him! Certainly, you will give up particular things from your life in Germany, but you will gain so much more over there. You are young, strong, and German. The colony can only blossom and prosper again with the help of strong women. Help your Fatherland to achieve this grand and beautiful mission. (p. 44)

This statement and others like it, of course, provide ready grist for the mill to those who argue that she is merely a propagandist for colonial emigration. What is not usually noted, however, is that she was in fact convinced that what she wrote was true, since she had experienced it herself: German women, she believed, would indeed find opportunities for sacrifice, for growth, and for supporting the cause of their homeland. In 1906 her aim was external, a part of the "beautiful mission" of the "Fatherland" to establish an outpost of German civilization. But between the early patriotic sentiments and the conclusion of her memoir three decades later, von Eckenbrecher had matured in many ways. The tempestuous political conditions in both South West Africa and a Germany that no longer possessed—indeed had been stripped by international mandate of its colonies—made her initial mission impossible to carry out.

As will be seen later, some scholars argue that von Eckenbrecher focused on that which she could control; namely, the domestic sphere. Although this is certainly the case, it is possible to detect other processes subtly at work.

She chose to live and rear her children in the Südwest even after it was no longer a German colony. One might imagine, given the displacement of political power and her subsequent double minority status vis-à-vis the new white authority and natives, that one could discern the dawning emergence of a postnational identity, but that would be too strong a formulation.[9] Similarly, to suggest that von Eckenbrecher developed some form of hybrid identity is to be anachronistic; nevertheless, it would be just as inaccurate to claim that she was merely a German nationalist.

Her identity, at least as manifested in her memoir, shows signs of having developed into something altogether different, the sources of which lie in her reverence for the particular kind of Germanity promulgated by Goethe, whose formulation of that identity remained steadfast even after Napoleonic aggression and the later loss of German political sovereignty, a circumstance that von Eckenbrecher possibly saw as parallel to her own. Goethe acknowledged German culture, true, but a German culture that embraced *Weltkultur* (world culture), and at its center it was not a nation-state but a humane entity ("das Menschliche"), not the nurturing of nationalism, but the fostering of a flourishing individuality through the process of *Bildung* or self-cultivation.[10] On the latter, von Eckenbrecher seems to have laid most emphasis throughout her life—an openness to learning that was no doubt also practical, if not pragmatic, given her status in a country that saw a quick succession of political regimes.

The memoir opens with a bleak description of the port city of Hamburg, standing as a symbol for all of Germany and reflecting von Eckenbrecher's depressed emotional state. She thus sets South West Africa up as a place not only of natural beauty, where the industrialization and urbanization that had gripped Imperial Germany were absent, but also of existential egress, where she could release her personal powers as she developed and used her skills and abilities on behalf of the colonial mission. To this end the von Eckenbrechers joined a small but committed German community of about two thousand people, working a farm until the Herero uprising in 1904.[11] The family of three, including her first son who was born in South West Africa, eventually moved to Weimar, the home of German Classicism.[12] Her memoir is thus an important document not only of her life in a remote African colony but also of the way she toiled and thrived as a European woman who had inserted herself into a new society.

The colonial sphere offered—in some instances demanded—a fresh way of viewing traditional gender roles. Indeed, the German Colonial Society attempted to attract a certain type of German to South West Africa. Immigration was critical to the survival of the colony on a number of levels, but many felt that a particular kind of immigrant would adapt best to colonial conditions and subsequently contribute most to the homeland. Daniel Walther argues that the vision of local authorities was preindustrial: they attempted to attract "a hard-working, parsimonious, Protestant agrarian class filled with staunch nationalist values and devotion to the emperor, with the 'traditional' German family at the core of society."[13] Further, as Roger Chickering shows, German women were drawn to South West Africa through the appeal of tradition as well as of innovation. They followed German soldiers who had crushed the natives in the uprising. While they were to provide spheres of traditional domesticity, they also must represent a new kind of German woman, one who possessed "a physical, mental and moral toughness" and would be "versed in agricultural techniques, the fabrication of clothing and utensils and in a broad range of medical skills; they required robustness to withstand both the African climes and the isolation that life there entailed."[14]

Von Eckenbrecher, not without sympathy, observes that her husband, who had nearly succumbed to fever in Africa, was physically exhausted and mentally broken from his exertions in the colony and, after settling his family in Weimar, refused to return. But von Eckenbrecher herself did return with her sons in 1914, even though she had not secured any kind of employment in advance—an action that speaks not only to her sense of independence but also to her hopeful and optimistic outlook. In Africa she would either succeed or fail. On the eve of a war that would change Germany and all of Europe, she chose South West Africa as the place where she would create a new life for herself and her family, guided through the years not by the values of the new republic or the later dictatorship, but by the compass of her own moral sense. For although some ideas in her memoir can seem modern in many ways, her Germanity, not merely her sense of propriety but her deepest personal values, appear to have been formed, at least to some small degree, by middle-class German expectations.

Since the last decade of the twentieth century, Germans, separated by a century from their colonial past, have become increasingly interested in South West Africa. Several historians have also produced work concerning German colonial imagination and the interplay of race and gender.[15] It is easy to misunderstand von Eckenbrecher's life if one approaches it only through a theory that it is meant to exemplify. For example, colonialist narratives provide easy targets for spotting instances of racism, and von Eckenbrecher's work displays racism as well as other prejudices of her day and position as an imperialist on a civilizing mission.[16] If one leaves it at that, there are fundamental aspects her life that will go unnoticed, including her complex relationship with the native peoples.

Scholars have productively analyzed her life on many levels. Her narrative has been seen to be creating a "conservative utopia" as well as mapping out her "autobiographical self" as she "interweaves the constant threats to the civilization she helps build with the successes she can celebrate in domesticating both the harsh land and its inhabitants."[17] She is held to have a "gossipy" narrative voice in which the "imperial I" retreats to the background of her self-construction.[18] Furthermore, her life is said to embody a "displaced sort of feminism that avoided direct demands for equality."[19] The often hazardous realities of colonial life led to "parallel rather than separate" male and female spheres.[20] In some ways she was simply doing what she had previously done in Germany: enshrining German domestic values such as cleanliness, order, and *Gemütlichkeit* (a sensation of charming coziness, a feeling of well-being), this time, however, in her South West African home, using natives and their way of living as negative contrasts.[21]

Von Eckenbrecher's candid account challenges artificial dichotomies that are sometimes forced on complex ideas when examining European colonialism. She lived in Africa and therefore interacted with others in ways that are too easily lost when one examines merely the metanarratives of nineteenth-century geopolitics, as one example illustrates. She was, as were many settlers, often asked to attend ailing natives:

> It seemed that whenever I thought I had a moment to rest, I was called to visit the sick. I was not a volunteer nurse or anything of that sort; people simply came and fetched me. [. . .] People came to me with the most unbelievable illnesses and requests. My main drugs were quinine, calomel, and opium, and with them, one can cure most diseases. (p. 164)

She describes how she dressed wounds, delivered babies, diagnosed internal illnesses, and treated fevers, "which any white person can treat over here." She continues: "It was most unpleasant for me to treat people who had festering sores or who had smeared the universal remedy—cow dung—all over their wounds to prevent air from 'disturbing' them. Countless flies and maggots infested these wounds and often it required all my emotional strength to simply touch them." [p. 164] Ironically, she seems willing to assist the sick, but at the same time folk remedies repulse her. Although she casts her thoughts in obvious racial registers, what she says here reflects more than racism. When she writes "white" in this instance, it is in connection with advances in European medicine—she speaks particularly of "antiseptics" for which Central European and in particular German scientists had performed leading research. In her mind, native remedies are simply insufficient and even damaging. Her unpretentious and frank narrative presents the reader not only with deeply complex historical questions in connection with colonial interactions, but also with vexing present-day concerns surrounding "development" and aiding delivery such as intervening with medical vaccinations to fight disease in developing nations or organizing social, political, and economic pressure on communities that traffic in human trade, practice female genital mutilation and honor killings, and abuse women, children, and even men in ways inconsistent with Western standards.[22]

"I was deeply distressed about the superstition and outright stupidity of the people," von Eckenbrecher interjects, recounting a story of being asked to help a five-year-old burn victim. Through quite extraordinary efforts, she aided the native girl, half of whose body had been badly burned, by cleaning and dressing her wounds. Afterward, she conferred with Captain Victor Franke, commander of German troops at Omaruru, offering her ox cart and oxen to transport the child to the doctor stationed there. At first the mother consented, only later to withdraw that consent when the cart was ready for the journey. "Miraculously," however, the girl's condition improved, and her wounds began to heal. Making daily visits, von Eckenbrecher hoped for a full recovery. Unfortunately, the day came when she could no longer treat the girl because her mother had decided to rely on traditional methods. But as the girl's condition steadily deteriorated, a day or so later von Eckenbrecher was summoned to the family's hut in the middle of the night. Against her better judgment, she gathered her things and journeyed to the hut. A traditional healer was present, and as von Eckenbrecher surveyed the scene, she noticed that all the bandages she had carefully made were in tatters and the girl's skin was smeared with cow dung, rancid butter, and mashed roots. The makeshift bed von Eckenbrecher had brought for the child had also been replaced by "magical sticks" and herbs, while the food she had left for the patient had fed others. Not surprisingly, within a few hours the girl died. Von Eckenbrecher states,

> I had cared for the sick child and now it was up to me to provide a coffin. Normally, the body is sewn into animal skins and buried, but in Okombahe such a burial would have been disgraceful. "After all, we are Christians, and like Christians everywhere, we must be buried in a coffin." Wooden planks were a rarity in Okombahe, so the family came to me and said, "Just give us a wooden box instead of a casket, and provide us with linen for burial cloth." I did.
>
> Like a pauper's bell, the church bell tolled pitifully shrill, and the little creature was returned to earth. Soon, the mourners came to me again. "Madam, you took care of the child," they said. "You furnished the casket and the linen, now give us coffee and sugar for the funeral repast." On occasions like these, one can well imagine that often I reached the end of my tether. (p. 167)

This account, with its skewering of what von Eckenbrecher accepted as superstition and the deadly consequences of its practice, echoes the genre of "reports from the countryside" that Voltaire and Franklin wrote. Historians make much of her desire to flee industrial Europe, but the heart of her personal philosophy, one of two subjects she had studied—indeed she continued to read Kant, Schopenhauer, and Nietzsche in South West Africa (p. 158)—is grounded, it seems, in a commitment to the German Enlightenment, mediated by Goethe, which included above all in its Kantian formulation the freedom and courage to reason for oneself. Her intellectual independence is the hallmark of her personality as it comes through in her memoir. Her descriptions of life in Africa may at times border on the Romantic, but far more prosaic descriptions dominate the narrative: growing potatoes (something that carried a historical resonance, given that Frederick the Great planted them in his own lands), avoiding baboons, paying taxes on mining claims, cooking chicken dinners, purchasing land, and combating snakes, all in all the day-to-day life of settling a new land.

But she also ranges farther afield, criticizing religious narrowness; for example, such as the "utter nonsense" of soothsaying (p. 203) or the refusal by Christian converts to allow "heathens" to bury their dead in coffins. Von Eckenbrecher had heard lectures in theology at the University of Berlin from a pair of the greatest Protestant theologians since Friedrich Schleiermacher—Adolf von Harnack and Reinhold Seeberg, who later trained Dietrich Bonhoeffer—both of whom advanced forms of rationalistic theology. She thus observed, with understanding, that some natives questioned the teachings of their Christian missionaries by asking about God's justice or the notion of being equal in the eyes of God when they recognized the inequalities that they suffered in the present:

> Here in South West—like everywhere else in the world—people wrestle with the attempt to understand the workings and motivations of God. I was frequently asked these and similar questions. Why do we maintain that after students have received such limited instruction at the missionary station they are capable of carrying Christianity to the poor heathens on the dark continent? The heathens also think; not all of them are dull-witted! (p. 163)

Here her phrase *not all of them* stings the contemporary ear, yet its existence implies, however grudgingly, an acknowledgment that "natives" cannot be solidly lumped together; that is, in her view some are as capable as Europeans. Her critique of Christian missionaries is also inescapable.

Of course, one may also point to von Eckenbrecher's sometimes sentimental and even sympathetic moments when describing natives and their ways, scattered amid her general belief in European superiority.[23] Yet she exhibits more than sentimentality in her occasional moments of genuine admiration for natives—even during the uprising—when she comments on the expert use to which the Herero put the poison of the euphorbia plant (189), and, like

Rudyard Kipling in "Fuzzy Wuzzy" who praises the East African Hadendoa for breaking a British infantry square at the battles of Tamai and Abu Klea, she pays respect to natives who bested whites at warfare despite European technological advances in arms (p. 270). These comments notwithstanding, von Eckenbrecher remained skeptical as to whether natives deserved "equal rights," without ever setting out in detail her reasons. She seems to have believed that Africans and Europeans lived in separate cultural worlds, although her very existence in Africa featured moments when bridges between those worlds were of necessity built. Despite these attempts or perhaps from her view because of them, her underlying belief appears to have been that until such time as natives left behind their own world or at least embraced Germanity, any sort of equality would be largely chimerical.

Von Eckenbrecher's status as a woman still led her to occupy, even in South West Africa, an intermediate social position. After castigating natives in her narrative for forbidding her access to the injured child, she later describes how a white family whose child lay ill also refused her offer of assistance, opting to await the arrival of a male doctor. The child died before the doctor arrived. Later, after the Union of South Africa took control of South West Africa and her own status as a German settler altered into a disadvantaged one, she observes with keen interest the relationship between the new, non-German white leadership and native populations. Moreover, she experiences the sudden change in how natives interacted with the remaining German settlers who were no longer their masters: "it was we who had to give way to them" (p. 270). She further describes how some natives attempted to warn German settlers when their lives were endangered, and occasionally points out, after a specific experience with natives, that they actually preferred German colonial government to South African rule. Her life and observations, especially given ever-changing geopolitical circumstances in South West Africa, thus provide unique insights into difficult questions about the meaning and lingering legacy of colonialism, some of which are only now being reevaluated.[24]

Another reason her memoir remains compelling is that it challenges the reader's own assumptions. Her narrative does not dwell on theoretical implications but on a life lived. Yet she supplies no grand summary of that life: try to pigeonhole her in any one category and her life provides evidence that subverts such an attempt. Instead, readers must work their way through her individual experiences. She rarely articulates any overt principles by which one should live a life, although one may derive many implicit principles from her actions. But in one instance she does insert something of a life's motto, perhaps the closest one can come to her own self-fashioned view; namely, when she casts herself in the role of Goethe's Faust. *Faust* (1808, 1832) was a favorite. She not only quotes it at the beginning and near the end of her memoir, but she also notes that she packed it away on a long journey she and her husband undertook to deliver livestock to a guano company (p. 140).

In her preface, she quotes a key passage from the climactic scene: "Only that man earns freedom and life / Who must reconquer both in constant daily strife." But a look at a number of lines preceding this passage is even more revealing:

> A swamp surrounds the mountains' base;
> It poisons all I have achieved till now.
> I'll drain it too; that rotten place
> Shall be my last great project. I see how
> To give those millions a new living-space:
> They'll not be safe, but active, free at least.
> I see green fields, so fertile: man and beast
> At once shall settle that new pleasant earth,
> Bastioned by great embankments that will rise
> About them, by bold labour brought to birth.
> Here there shall be an inland paradise:
> Outside, the sea, as high as it can reach,
> May rage and gnaw; and yet a common will,
> Should it intrude, will act to close the breach.
> Yes! to this vision I am wedded still,
> And this as wisdom's final word I teach:
> Only that man earns freedom, merits life,
> Who must reconquer both in constant daily strife. (lines 11,560–11,576)[25]

Here Faust emerges as a colonizer, a position that did not escape von Eckenbrecher's notice, as she applies its meaning to settling South West Africa. She calls Goethe "our great master" [p. 44], and by mimicking the actions of Faust who believed he was conquering the elements to win and cultivate land on which future generations could create a thriving community, she suggests that she views her own life in Africa as purposive, not merely in a patriotic sense, but in a way that ultimately gives it profound meaning. Although Faust's vision occurs in his mind's eye—he is physically blind and does not realize that the "clash of spades" (line 11539) he hears from the workmen is actually the work of the minions of Mephistopheles preparing his own grave—poets often associate the trope of blindness with an ability to see through an inner, spiritual eye. Even as Faust seems to be deceived, the Deceiver himself is actually deceived since he is not able to claim Faust's soul when he appears to lose their bargain. Faust's vision of creating a mechanically formed safe haven for the multitudes, as beneficent as it is, was not a saving grace but, rather, a realization that self-restraint and self-cultivation, the harmonious colonization of his passions, was essential for real human freedom.

As von Eckenbrecher's memoir progresses, one narrative arc that emerges is that the very challenge of hewing a life out of the vastly different circumstances and rugged landscape of South West Africa mobilized her most personal, deeply rooted, and vital character traits, including her desire to bring order to what she perceived as chaos, through the microcosm of her own domesticating projects and her belief that she was contributing to a civilizing effort that would serve a similar purpose in the macrocosm. But completely changed political circumstances made this outwardly directed "clash of spades" impossible. As does Faust, she recognized that the way of self-cultivation contained the more pressing personal imperative. Her perceived gifts to Africa, like the technology of Faust's dikes and dams, are matched and exceeded by what Africa offers her. Her self-described life shows a work in progress of virtue, in the Aristotelian sense of reason governing the passions. Her own sense of Germanity, confirmed at first through the colonial enterprise, did not diminish when any hope of a recrudescence of a German colonial return ended; rather, it revealed itself to be more about her own human flourishing, in the sense of *Bildung*, particularly in its Goethean resonances of moral or cultural education.

She also carried on the work of *Bildung* in another way. Her vocation as a teacher—one of the few strands that bind together her life through different times and disparate places—provided an avenue for cultivating the next generation. Thus, she forsook the cultivation of fruits, vegetables, and livestock for the cultivation of young minds as a schoolteacher, providing them a mixture of practical skills and moral and spiritual teachings to carry on the work of *Bildung* in their own lives.

Still, one may well ask what von Eckenbrecher's exceptional memoir reveals about her as a person in both ordinary and extraordinary settings. What does it disclose about her intellectual and practical gifts? How does she employ these gifts as a colonist, a single working mother in a man's world, a European in Africa, often irritated and bewildered on the one hand by the thoughts and actions of the colony's native peoples while on the other never quite settled regarding natives' proper role in the order of things? In the opening pages of her narrative she informs her reader that from her earliest youth under her mother's influence, she cultivated a desire to learn and work so that she might become independent.

But independent from what—the needs of her extended family, male authority, the industrializing Germany's increasingly dismal atmosphere, perhaps even certain German middle-class conventions? Margarethe von Eckenbrecher deliberately chooses to remain single after her divorce from

Themistockles von Eckenbrecher II. She does not elaborate on the causes or the repercussions of this action, who initiates the divorce, or its final consequences. It appears that Themistockles never plays any role in his sons' lives—though the elder son visits the von Eckenbrecher estate in Germany on several occasions and is always warmly received. Frau von Eckenbrecher very much enjoys the company of "good" men, and has strong acquaintance with certain aristocratic men who immigrated to South West Africa, as well as those in the higher professions. That she is the only female teacher in her school for a good long time bothers her not at all (other than wondering initially whether or not she would be accepted). That there is an acute shortage of females in the colony, and that she could remarry if she so desired, is old news to her. Men she describes positively in her memoir are men of character, backbone, and resolve, men who get on with things, often men who enjoy higher intellectual and cultural life, and men who are honest and successful—whether it be in farming, the military, the professions, even in merchandising. During the 1904 uprising, von Eckenbrecher mentions a "burley" sergeant who cares for her ailing infant son with such tenderness and exceptional competence that it touches her deeply—not to mention how it much relieves her burden. He is greatly admired. This sergeant, like Margarethe von Eckenbrecher herself, is someone who crosses the boundaries of established gender roles without fear. That her growing sons need good male role models does not escape her notice, and she finds trustworthy men who teach her sons "manly" skills and take them on excursions into the countryside.

As much as von Eckenbrecher claimed a deep, strong, emotional, intellectual, and cultural bond with the Fatherland, she clearly was far happier *not* living in Germany. Indeed, what she found so appealing about German South West Africa was that it was not densely populated, not green and wooded, but raw and largely untamed, yet with hard work, the land could be converted and made fruitful, even prosperous. Life in South West Africa was an adventure, a place where she could be German, where she was not rigidly bound by the conventions of German society, where she was freer to be whatever it was she imagined for herself. For instance, she thought nothing of hiding weapons and smuggling German documents, with considerable risk to herself and her sons, from Windhoek to the coast during the South African occupation in World War I, while simultaneously a magnificent vista, a place of spectacular natural beauty in South West, reduced her to tears and the rare employment of rapturous language. In sum, von Eckenbrecher experienced a sense of personal freedom in South West Africa that she did not in Germany, and neither all of the political upheaval, nor the challenges she faced personally, would ever cause her to leave and risk losing that personal freedom.

As a writer of unadorned prose, von Eckenbrecher's voice is strong, without much regret, and founded in deep conviction. Occasionally, she casts herself as a victim of dishonesty (theft of small stock and provisions, etc.), but more generally she is the heroine, the person who rises to the occasion and does what few others have the ability, the courage, the tenacity, or the endurance to do (administering medical care to the Damara and Herero of Okombahe, organizing the rescue of people stranded by the floods that rage through Klein-Windhoek, etc.). Quite naturally, then, she filters everything she describes through her own moral imagination.

She had received a sound, even a superior, education for her day and time, and she continued to read German classics throughout her life. On a lazy weekday, early afternoon on the shady veranda of her home in Okombahe with the midday meal happily behind her, for example, while others nap she searches her "well-stocked library" for something to read. "Kant and Schopenhauer" she deems "too challenging at this hour, and Nietzsche simply create[s] pessimism" [p. 158], indicating that she had more than just a passing acquaintance with German philosophers. The degree to which she actually understood Kant, Schopenhauer, Nietzsche, and others is an unanswered question, but she certainly enjoyed her forays into higher intellectual life. Yet the bulk of her memoir suggests that she was not a true intellectual in the strict academic sense and that her greatest intellectual gifts tended more toward the practical, toward the everyday, toward resolving matters, toward creating a well-devised plan and executing it with as much precision as possible, amending it when needed but always bringing things to a proper conclusion. Indeed, it was in the practical skills of diurnal living that von Eckenbrecher truly excelled.

Her thoughts, decisions, and actions were thus not arbitrary but deeply rooted in a core of fundamental values and understanding of the nature of the world that she deliberately chose to accept as true. What grounds them, both consciously and unconsciously, more than anything else was her idiosyncratic version of *Deutschtum*, her own personal, fungible version of Germanity—the ideas, beliefs, sentiments, habits, practices, and more, that arise and develop within Germanic culture—that she personally accepted as true and authoritative. Like all intellectual worlds, Germanity is ethereal and malleable to some extent to idiosyncratic taste. For von Eckenbrecher, it was deeply held, deeply personal; it stirred her emotions and stimulated her mind, and she used an ever-evolving conception of Germanity to define the content and limits of her own world. Having begun a life in Okombahe, she became all the more aware of the goodness and correctness of that conception when setting against the backdrop of Damara and Herero culture, ideas, and practices. The necessity of creating order and schedules, coupled with hard work, diligence, German manners, organization, honesty, cleanliness of self and place, as well as certain Christian virtues, allowed her to see in sharp relief the goodness—as she would have it—of her own world juxtaposed against the chaos and primitiveness of the native world.

Yet von Eckenbrecher's personal Germanity remains a tangle of ideas, practices, sentiments, taste, conduct, habits, fashions, emotions, dispositions, even class-consciousness, all of which are open to situational fine-tuning and change. It manifests itself in everything she does, from the satisfaction she derives from preparing delicious meals to the creation of Gemütlichkeit in her own home, something a reclusive and rather pessimistic distant neighbor, "Old Ertmann," savors, much to her own delight. Her Germanity is evident in the painting of her new home, the murals her husband creates, the fireplace he and she design and construct, the painstaking care she gives her livestock and poultry, the parties and gatherings she hosts, the small, yet delightful, flower garden she cultivates in front of her house, and the beauty she finds in the homegrown vegetables she dries. These are indications of a practical Germanity—everything well and properly done—while the satisfaction, the aesthetic pleasure, she derives from the successful completion of mundane projects suggests instances of a more ethereal Germanity that penetrates her personal world. Yet in contrast to the practical, when circumstances nudge her to explain a small piece of her ideological Germanity, when a native asks her a question relating to her core beliefs about the world, when she must translate the "ethereal" into the everyday, she exhibits great difficulty in clearly articulating her understanding to another person.

Still, both the fervency of her writing and actions reveal that she does not doubt that high, classical German culture, particularly German philosophy and literature—especially the works of Goethe—combined with what she sees as Germanic traits of innovation, industry, efficiency, sanitation, and medical advancements make German culture superior to any other country. In her opinion, Germany, not France or Britain, will be the bearer of culture to Africa; nevertheless, it is curious that in her memoir Frau von Eckenbrecher rarely, if ever, mentions music and Germany's illustrious musical heritage.

Another seeming contradiction concerns the strength or weakness of von Eckenbrecher's religious convictions. Though she was Protestant and mentions lectures by famous theologians she heard in Berlin, it would not be inaccurate to state that religion does not appear to inform her Germanity significantly, even though religious pageantry was something she thoroughly enjoyed. When her husband and other German settlers volunteer to join German Schutztruppe in an effort to suppress a native insurrection, she looks on as they receive the Eucharist in an outdoor ceremony, administered by missionaries standing at an altar table set between two huge fires blazing against the night sky the evening prior to their riding to battle. The following morning when Friedrich Bernsmann, a missionary she much admired, dressed in flowing robes, his hands extended upward, blesses the "weather-beaten warriors" [p. 213], this spectacular confluence of Protestant pageantry (Germanity personified?) and the departure of her husband and others whom she admires into the unknown reduces her to tears.

Conversely, von Eckenbrecher discloses a lack of fondness for the missionaries stationed at Okombahe, nor did she find most of the missionaries in the colony as a whole to be people of any significance. The several genuinely devout converts whom she met among the Damara of Okombahe she respected, and she was also happy to discuss theology with them, though always from a position of assumed superiority. She also clearly relished the chance to disabuse a local clergyman or missionary of his misconceptions, his naïveté, and his turning a blind eye to the corruption of the locals. The doctrines of Protestantism may be important to her, but she kept them very private, not mentioning them in her memoir to any extent. Indeed, in Part II she notes a sermon preached by a German clergyman in Windhoek shortly after South Africa's occupation of South West Africa at which she takes umbrage, not over a point of doctrine but over her own perception of the clergyman's acquiescence to the occupiers. Perhaps more telling than anything else is the fact that affixed to the inside of the oxcart she and her husband initially used as transportation was an image of Saint Christopher, but in place of a Holy Bible was a copy of Goethe's *Faust* lying beneath the image of the good Saint whenever they traveled.

Though von Eckenbrecher herself was unable to take up farming upon her return to South West Africa, she retained to her death a clear and distinct bias in favor of that occupation. Farming was a form of creation, and she believed it a virtuous activity, a worthwhile and deeply satisfying means of making a livelihood. She contrasted it with treasure hunting for diamonds, gold, and other so-called valuables in which she found no real virtue. That her elder son became involved in prospecting for a time was an unhappy circumstance. When he married and settled on a farm, she worried about his success in this endeavor, no doubt wondering if he would follow in his father's footsteps. Perhaps, given her talent for the practical, her ability to strive for and achieve an ideal, together with a lingering belief in a slightly romanticized form of feudalism (for she deliberately continued to nurture the bearing of an aristocrat and to think of herself as such), she harbored some regret, possibly even disappointment, at not having been able to pursue the rural ideal, yet she derived great satisfaction in teaching. She taught various levels and subjects for a time but then settled into the practical and spent many years teaching what might be called "home economics."

This should come as no surprise, for German South West Africa as a place, as a physical presence, is tangible, thorny, often unforgiving, and difficult, yet to von Eckenbrecher's mind, that harsh and capricious reality joined another one both romantic and purposeful, that is a South West that was an animated, though ethereal, entity, possessing both purpose and will. When upon receiving notice that the South African administration had granted her permanent residency, she exclaims, Africa "needed me!" [p. 276], she does so with all the emotional fervor of her heart, as though some Being embodied as "Africa" controls the affairs of the land and she has been chosen to play a significant role in its destiny. This form of expression also occurs elsewhere in her memoir, placing her within a greater, cosmic scheme of things. Tellingly, for example, when her younger son succumbs to an outbreak of typhus, she simply observed, "Africa had taken Büdi from me" [p. 323]. In her mind she has a calling, a vocation, and so does he.

When von Eckenbrecher first became interested in Africa she does not disclose, though she writes that her brother and her cousin (her eventual husband)—both of whom were militarily involved in German South West Africa—piqued her curiosity and fired her enthusiasm to become a settler. What she might have read about Africa, whether fictional or true, is also almost entirely unknown. Of all the Germans who wrote of Africa, the only one she cites is Wilhelm Busch, a nineteenth-century caricaturist, painter, and humorist, probably best known for his satirical children's picture stories of Max and Moritz. As she recounts the details of her first voyage to South West Africa, recalling naked, dark-skinned African children she encountered at different ports along the way, she employs a line tinged with the colonialist racism of the day from a poem by Busch to explain their lack of necessity for clothing: "one is black and that's enough" [p. 58].

What von Eckenbrecher's brother and husband told her in their "glowing colors" [p. 43] of South West regarding the Protectorate's natives forms no part of her narrative. Yet well before she set foot on African soil she seemed very well assured that her education and cultural background made her superior to natives, and in her actual day-to-day dealings with them as their employer, neighbor, *ersatz* medical doctor, or any of the other roles she perforce assumes, she always placed herself in a higher position.

To be sure, a few local natives (Damara, Herero) von Eckenbrecher genuinely admired for various reasons (her first son's wet nurse, an earnest seeker after God's truth, a pious man who disciplines his children in an odd manner), though she never attempted to cultivate an actual friendship with any of them. It is as if she belonged to one world, the natives to another, and that these worlds were either incommensurable or must be so considered. This raises the interesting question of whether von Eckenbrecher considered natives physically, biologically, or mentally deficient. Did they actually possess an inferior intellectual capacity? But on this topic von Eckenbrecher gives no clear, unambiguous answer, though the preponderance of evidence suggests that she viewed the gulf between these two worlds—and the peoples who inhabit them—in terms of culture: Germany's was a vastly superior culture, the Herero's and Damara's distinctly inferior. She carries this view of German superiority even further in her memoir in which she is rather democratic in her contempt, castigating not only natives but also Germans who do not measure up to her standards, English-speaking South Africans, and perhaps to her mind the most inferior of all whites, Afrikaans-speaking Trek-Boers, resettled from Angola, who quickly ran down farms appropriated by the South African government from their erstwhile German owners who had built them up from nothing. Von Eckenbrecher never ascribes a Rousseauian innocence, a native simplicity, or an ecological wisdom that European societies have long since shed, to her African neighbors. Instead, her comments reveal that she thought these neighbors could only improve their lot in life by accepting German culture, habits, and ingenuity.

No European farmer in South West Africa could make a decent living without inexpensive labor, and as native lands were acquired by treaty or fiat, the need for labor followed, which brought together the settlers of Okombahe and laborers who wished to earn money. They lived side by side, yet trust between black Africans and white settlers was troubled at best. When insurrection began, von Eckenbrecher learned that she could not take native loyalty for granted, even in those she has known for nearly two years. Yet trust works both ways, and she relates an interesting story about a former German soldier who, presumably after his discharge, became Herero. He learned the language, lived as a Herero, dressed exactly as a Herero man, married a Herero woman, and had children with her. Quite simply, this German-cum-Herero was no longer German and had nothing to do with white settlers. But when the uprising commenced, he learned precisely how fragile his new Herero identity was. Once the insurrection intensified, the white Herero was executed by his "fellow" Herero because his loyalty, as they saw it, could not be assured. The moral to this story, contends von Eckenbrecher, is that there is a white world, there is a black world, and there is no substantial bridge between them. This was, for her, a rather practical philosophy she followed to the end of her days.

In Part II of her memoir, von Eckenbrecher refers to "new racial view" attitudes [p. 245], by which she may mean preapartheid South African attitudes, but perhaps also Nazi racial theories, both of which appear not too distant from her own ideas.[26] Occasionally through her teaching, she came into contact with mixed-race children, "coffee with cream," as she tells us they are called [p. 245]. She writes that she sorrowed for such children because their mixed-race origin is not of their own doing, yet when they grow into adulthood they will not be part of white society and may not be accepted into black society, either. From her memoir, it appears that she never met a person of color she considered her equal. Coloureds (mixed race) had already formed their own society, or subsociety (Rehoboth Basters) long before German settlers had arrived, but without doubt, German soldiers and other German men in the fledgling Protectorate contributed to the Coloured population (little is known about how German women may have contributed to this community).

Von Eckenbrecher's belief in the superiority of German culture and the clear inferiority of African cultures continued, perhaps even strengthened, during the course of her life. She viewed Europeans and their culture as essential for the improvement of Africa. Thus, native cries such as "Africa for Africans" [267, give adjusted pages for both quotations] not only threatened the white population but potentially denied, on her view, Black Africans a better life. Even so, von Eckenbrecher never actually came to terms with what a Germanized African society might look like, and what role—beyond servitude—Germanized Black Africans might play in such a society. Such a future seems quite beyond her immediate imagination.

With all her markers and boundaries of culture, class, and race, her intellectual motivations, her competence and taste, her classical strivings, her occasional displays of bravado, her strength of devotion to an ideal, her powerful idiosyncratic sense of life and how one ought to comport oneself, her cognizance of her own vulnerabilities, von Eckenbrecher must have understood and sensed acutely how different she was from most of her contemporaries. It is this difference that ultimately makes her memoir a remarkable read.

IV

The von Eckenbrechers sailed from Germany on the last day of 1902 and landed several weeks later in Swakopmund, German South West Africa, in mid-January, 1903. They immediately began the process of making a new life for themselves in the colony. A year later, having established themselves in Okombahe with a new house, cattle, sheep, goats, and large vegetable gardens, even a newborn son, they surely thought that the various threads of their ideal rural life were coming together. What they did not know, however, was that their new life would soon coincide with a native uprising, a revolt against German rule and German colonial settlement. Indeed, the Herero-Nama uprisings, partially chronicled in a very personal manner in Book I of von Eckenbrecher's memoir, cannot be fully appreciated unless they are contextualized within the greater colonial and social setting of German South West Africa.

"Largely desolate and bleak" is a common but not inaccurate description of Namibia's shoreline and may explain why European seafarers of the fifteenth, sixteenth, and seventeenth centuries never sought to make the few natural harbors along its coast into revictualization stations. Two exploratory expeditions set out from Cape Town, itself, founded in 1652 as a supply station for the Dutch East India Company, in 1670 and 1677, respectively, to investigate the Namibian coast for the purpose of determining the feasibility of establishing profitable trading relations with peoples living along the coast. Upon their return, both expeditions lodged negative assessments.[27]

More than a century later in 1791, an overland expedition headed by Willem van Reenen was sent to prospect for valuable minerals,[28] making its way almost as far north as present-day Windhoek. Van Reenen returned to the Cape with samples of ore that upon analysis were determined not to contain gold "but copper of a very high quality."[29] Copper in far-flung Namibia, however, was of insufficient interest at that time to motivate a company to undertake the creation of a mine. But rumors of gold mines persisted, and the occasional prospector-of-fortune made his way to that area, though none succeeded in finding gold. The next ripple of Europeans to enter Namibia did so with the desire of establishing permanent settlements—missionary settlements. Two missionaries, sent in 1806 by the London Mission Society, were successful in erecting two rather humble mission stations, one in Warmbad, the other in Heirachabis, in the southern part of the territory.[30]

Reliable sources of information about the life of native peoples in turn-of-the-nineteenth-century Namibia are scarce. In central Namibia, where the cattle-herding Herero had settled, and in central/south-central Namibia where the slightly less pastoral Nama lived, conflicts between the two peoples arose over grazing lands—conflicts that, according to Heinrich Vedder, himself a missionary as well as a controversial historian of Namibia, had as their primary objective the elimination of their opponents and the confiscation of their livestock and grazing lands. Vedder presents a remembrance from a very old Herero man who describes pre-nineteenth-century life as vastly different from early- and mid-nineteenth-century life. According to him, life was characterized by a lack of competition for grazing; when grass was depleted, people simply trekked their livestock to another area where grazing was fresh and used all the lands of central and northwestern Namibia without conflict.[31] Considering that other Christian missionaries of the early nineteenth century describe the Herero and Nama/Orlams as frequently in conflict, one can reasonably inquire after the origins of these conflicts.

Jon M. Bridgman answers by pointing to the larger geographical sphere of southern Africa, writing of the Herero and Nama as further examples of pastoral peoples of that time and place engaging in range wars rooted in competition for grazing lands and water. Yet in the specific context of Namibia, Bridgman suggests that the causes of conflicts were twofold: the consequences of ivory hunting and the arrival between 1815 and 1830 of the Orlams from South Africa.[32] During the second decade of the nineteenth century, white hunters in search of ivory crossed into Namibia. "Because the hunters' only interest was in the ivory and the Namas only wanted meat," Bridgman asserts, "the two were willing to cooperate."[33] White hunters also introduced firearms to the Nama, which, according to Bridgman, changed the way the Nama were able to pursue game. Whereas they had previously relied on bows, arrows, spears, and other weapons, which unintentionally served to keep an ecological balance, with guns they could take down animals with an ease and frequency never before imagined. Great herds of elephants and other game animals were thereby decimated in rather short order. This increase in Nama food supplies in turn led to a corresponding increase in the Nama population. But the longer-term result of a more accurate and efficient method of hunting was that game animals in Nama lands became virtually extinct.[34]

The migration of the Orlam people from the Cape to Namibia between 1815 and 1830 was the second contributing factor to warfare between Herero and Nama. Both Nama and Orlam are of Khoikhoi or Hottentot (meaning "men of men") stock, more or less indistinguishable from San or Bushmen. Bridgman observes that in "contrast to the Nama and Herero, who were primarily pacific and fought only when provoked, the Orlam were fighters who only reluctantly pursued a peaceful way of life."[35] When the drought of 1829 to 1830 reduced Herero pasturelands to vast barren stretches, they trekked their starving herds ever farther south into Nama grazing lands. Unhappy over this situation, the Nama sought and enlisted the great Orlam leader and raider, Jonker Afrikaner, to organize and lead them against the groups of Herero who were now in Nama lands. Thus, Nama and Orlam joined together to fight—though the fighting was actually little more than a series of cattle raids and Herero counterraids. Despite the frequency, the fighting resulted in little bloodshed and loss of life, but it continued from 1830 to 1842 when warring parties finally signed a brokered peace.[36]

Thus, Europeans did not as such bring war to Namibia, rather, "the cycle of wars began more than a decade before the first traders and missionaries arrived and [at least] fifty years before the Germans proclaimed their protectorate."[37] Though Bridgman is correct in his analysis that by the time missionaries and traders arrived in Namibia it was hardly in a state of "primitive paradise," one must also understand that these vanguards of European civilization—wittingly or unwittingly—brought with them implements of both their intellectual and material cultures, some of which had profound and unintended consequences for local populations.

Missionaries from various European and American mission societies were especially drawn to southern Africa from the early nineteenth century onward. To Namibia came both Catholic and Protestant missionaries from sponsoring societies headquartered in England, Germany, and Finland, among others. From the beginning, missionary work was not simply about Christianizing local peoples but about wholesale culture change as well. Often missionaries also promoted economic change as they tried to introduce or encourage intensive horticulture so that transhumant cattle herders might switch from a roaming existence to a more settled one, making for more stable congregations.

In certain ways missionaries were ambiguous persons. Mostly whites from northern Europe, they learned local languages (indeed, often compiling the first grammars and dictionaries of native tongues), gradually acquired a good deal of knowledge about local cultures and customs,[38] and frequently engaged in trade with natives in order to earn their own keep, since official stipends and maintenance allowances were often erratic, sometimes nonexistent, thus introducing locals to new products ranging from food and clothing to tools and weapons, items that often become luxury goods whose possession gave the native possessing them a higher social standing than that of his fellows. Because many looked upon missionaries as agents of European colonial governments, they were often instrumental in settling disputes among natives, communicating colonial government policies, brokering peace treaties, and engaging in other pursuits that on the surface had little to do with proselytizing. Von Eckenbrecher writes a very interesting critique of the missionaries at Okombahe: their activities, the state of their converts, and the usefulness of their work. Hence, whether missionaries actually saw themselves as the front line of colonialism, they sometimes functioned as its unofficial emissaries—and this is certainly true for Namibia.

Because some missionaries were clearly more devout and genuine in their work than others, and because the world in which they lived and performed their deeds appeared in a different light to them than it does to a contemporary reader, missionaries are easy targets of criticism. Yet the sincerity of many of them can hardly be disputed. Consider the following passage in a letter dated February 4, 1844, from Emma Sarah Hahn, wife of one of Namibia's best-known missionaries, Carl Hugo Hahn, to her mother:

> I shall be very glad when the time arrives for us to go among them [the Nama]. At first, I doubt not, many will be the difficulties and trials to be encountered in our intercourse with this nation, as is the case in all heathen countries during the first years of a missionary's residence among them. They have long

> been willing captives to Satan, and he will not easily bear the spoiling of his kingdom, but we know that the Lord will make out of the heathen a people for His name, and we know that He who is for us, is greater than those who are against us, therefore we will not fear. Pray much for us, my dearest Mother, pray for my dear Hugo, that he may be made a great blessing among a people who are now sitting in darkness and that he may preach the Gospel in season and out of season, that many may be turned from darkness unto light and from the power of Satan unto God. Pray especially that we may be enabled to learn the language, enlightening the eyes of our understanding to that effect, that we may be more useful among them, for I am persuaded it is the only way of being really and extensively a benefit to them.[39]

Such convictions, however sincere, also exhibit of a self-evident understanding that native peoples would benefit not only from Christianity but also from adopting the template of northern European life. Missionaries were not simply one-dimensional, single-minded persons. Their diaries reveal that their lives of necessity were full and busy; that they had their moments of despair, sorrow, and joy; that they were engaged in life as husbands and wives, fathers and mothers, breadwinners and householders, leaders and followers; and that they possessed multiple—often situationally conflicting—loyalties, perhaps most acutely felt when difficult matters arose concerning the natives among whom they carried on their labors.

As missionaries entered Namibia on a more or less permanent basis, traders were not far from their wake. Traders represented quite a different brand of white man in Namibia; local peoples were important to them as well, but mostly to the extent that they could make a profit from them. Still, traders had complex and temperamental relationships with both their black clientele and white missionaries and, like the latter, frequently entered the fray. In this regard, Charles Andersson, a Swedish-born, Cape Town–based trader operating in central South West Africa with significant dealings with both the Orlam/Nama and the Herero, knew from personal experience the difficulties involved in trading among these peoples:

> *5 November 1864* The Rehobothers have traded away all their cattle and sheep. Abraham dissatisfied about the horses that died whilst coming down to Otjimbingwe. . . . Frederick told me this evening that he could not eat my ewe sheep. They were so poor. I eat them and feel thankful. But it is always so with the needy and improvident; the less they have, the greater their carelessness and demands. . . . How stingy Mr. Hahn is! Not a sheep has he yet given me for slaughter, and yet how many has he not received from me? Did I not feed him and his party at the Bay at great cost to myself, to say nothing of many other gratuitous services rendered him and the Mission.

> *7 November 1864* Poor Smith! A trader's life amongst such ruffians as the Hottentots and the [Herero] must be something fearful, he is cuffed and kicked worse than a dog. One moment they tell the poor fellow that he can go, and the next they threaten him with loss of property if he attempts to move! When he asks old Willem: "Can I go?" he receives the answer: "Yes, as far as I am concerned you may trek," and yet it is well-known that all along they have made up their mind to stop the poor fellow. I regret to say they are countenanced in these their proceedings by Mr. Hahn![40]

Mid-nineteenth century economic activity in the central interior of Namibia was quite steady and often very profitable. Traders were interested in ostrich feathers but even more so in livestock that could be transported to and sold in cattle markets of Cape Town. Traders initially came to central Namibia heavily laden with the kinds of goods (the variety of which steadily increased over time) that proved attractive to local populations, and when their goods were exhausted, they left with what they had acquired through trade and returned the following season. Some traders/merchants also eventually established permanent trading posts and sold firearms, ammunition, alcohol, cloth, foods of various sorts, cheap clothing, lamps, and cooking wares. Yet whether itinerant or settled, traders competed with missionaries in some instances, as certain concentrations of natives had come to settle near missionary compounds where churches, stores, and, occasionally, schools operated. Indeed, among the Nama having a resident missionary attached to a particular grouping of Nama headed by a successful local leader conferred a greater impression of importance on that leader.

The presence of stores and trading posts coupled with the growing availability of goods directed at natives led to unforeseen changes in local economies. By mid-nineteenth century, Brigitte Lau asserts, a dramatic shift had occurred among the Nama and Orlam peoples as a largely self-sufficient pastoral economy gave way to one based on commando-style raids in order to acquire through plunder the means of purchasing goods from the traders.[41] The size of Nama and Orlam herds and flocks dwindled to alarmingly low numbers, and people actually began to lose, indeed, eventually lost—to the proportion that cattle raiding became the dominant feature of their economy—the very skills that had allowed them to make a living from the land. As the Orlam and Nama needed cattle, sheep, and goats to make purchases and trades, they became significantly dependent on both the labor of the Herero and the goods traders had to offer them. In order to acquire the latter, they concentrated their raiding skills on the former. Around 1850, Jonker Afrikaaner explained that "he would and could not tolerate missionaries among the Herero [who were not already under his domination]: the Herero were his servants whom he needed for war and whom he could not allow to sit around with the missionaries to listen to subversive teachings."[42] Year after year, Nama and Orlam stole and traded thousands upon thousands of Herero cattle that considerably depleted Herero livestock and endangered Herero livelihoods.

In another letter, dated September 25, 1865, Hahn describes the tension and anxiety the Herero endured as they anticipated these cattle raids against them:

> My dear Matilda,
>
> In great haste a few lines which will serve as an answer to yours and sequel to dear Rosy's, to whom I think I wrote that we were daily expecting hostilities. . . . On the 27th of last month we received a few hasty lines from Mr. Brincker, the missionary now living on New Barmen, announcing that a large commando of Namaquas was in their neighbourhood, that horsemen had the previous night been on the place and announced its approach. Very few people were on Otjimbingwe, but Maharero, the Chief, had immediately on hearing Mr. Brincker's letter called on his people and cattle posts from the mountains and grazing-places where they were staying, and each evening fresh fires on the heights around us shewed that new arrivals were continually taking place.
>
> On Sunday morning, before it was light, Hugo and I were simultaneously awakened by someone knocking at the door of the sitting-room . . . [Hugo] went to the door and found two of his catechumens who at once told him the Namaquas were near, about 3 hours distant, and begged for ammunition. The news soon spread like wildfire through the place and the men and youths all betook themselves to Maharero's [domicile to] prepare for the expected attack. Still they did not come . . . Hugo walked over quietly to Maharero's werft and found only about 200 men gathered there. . . . But we had heard that a very large company had been continually traveling in a parallel line with the Namaquas, but unobserved by them, ever since they had left New Barmen, and from among them we were, or rather Maharero and the others were, always being informed of the enemy's maneuvers.[43]

Around 1860, Andersson enlarged his business by selling firearms in great numbers to the Herero with the result that for the first time a more or less level playing field came to exist, at least as far as firearms were concerned, between Herero and Nama. It was a lucrative trade for Andersson, who died in 1867, but in 1870, German missionaries, not to be outdone, banded together to form the Missions-Aktiengesellschaft, resulting by the end of the 1880s in a virtual monopoly of firearms trade in the area.[44]

The availability of firearms and ammunition to both Herero and Nama did nothing to promote peace, and in 1863 another "Herero War of Freedom" began in earnest. Whether the fighting was more intense this time or just better known among the white population, the conflict and cattle raiding had reached a point that white officers actually led the Herero into battle on several occasions. Some of the white residents of Namibia petitioned both the Cape and German governments for help in bringing an end to these ongoing "tribal wars."[45]

A regional Herero leader called Maharero emerged in 1870 after the negotiated end of this conflict, known as the Peace of Okahandja, in a very powerful position, indeed as a kind of "obvious" leader of people and as a statesman that Herero, Nama, and whites felt would be capable of brokering and fostering decent relations among their communities. It was, of course, an unenviable and uncomfortable position, and Maharero asked the Cape government to dispatch a commissioner to Namibia to act in this capacity. William Coates Palgrave was sent. While Maharero knew him and trusted him, the Nama were unimpressed by the choice. Palgrave, to his credit, worked diligently in this capacity, yet failed because of a simple lack of confidence and trust among all parties. When he withdrew, he left behind a power vacuum.[46]

But the Peace of Okahandja held for ten years before cattle raiding started anew. Bridgman notes that, unlike previous tribal wars that had stimulated the trading economy, this one depressed it, and exports from Namibia to the Cape "dropped from 138,886 German marks in 1880 to 52,032 German marks" in 1882.[47] The result was that many missionary and nonmissionary trading enterprises went bankrupt. With ongoing skirmishes between Herero and Nama, economic decline, and unwillingness by any European nation, even the Cape government, to bring a measure of order to Namibia, the situation—from the perspective of missionaries, traders, settlers, and perhaps even the Herero and Nama—seemed nearly hopeless.

Germany as a unified nation was then in its infancy, and its chancellor, Otto von Bismarck, initially expressed no interest whatsoever in acquiring colonies. However, in 1882 when a tobacco merchant from Bremen, Adolph Lüderitz, spoke with Bismarck about the possibility of being granted protection from the *Kaiserreich* over the lands he hoped to claim in Namibia, Bismarck answered affirmatively, though with two absolute preconditions: 1) that Lüderitz acquire a harbor, and 2) that he have clear and exclusive title to any lands that should come into his possession. Lüderitz's agent, Heinrich Vogelsang, did acquire a harbor along with several vast tracts of land adjacent to it on behalf of Lüderitz, hoisting the German flag over them in mid 1883. Bismarck officially committed his government to the protection of this territory on April 24, 1884, thereafter sending Dr. Gustav Nachtigal, Imperial Consul General and Commissioner for West Africa, where Germany would acquire two more colonies, Togo and Cameroon, to German South West Africa with the assignment to broker treaties with "the tribes of the interior" such that they agreed to place themselves under the protection of the Kaiser.[48] In this he was but partially successful. Minor players signed treaties, but Maharero and his Nama counterpart, Hendrik Witbooi, along with a number of other native leaders, were unwilling to sign such an agreement.[49]

In April 1885, Dr. Heinrich Ernst Göring, Imperial Commissioner for German South West Africa, arrived in the protectorate, charged with gathering the remaining signatures to a treaty guaranteeing German protection.[50] His main task was to persuade both Maharero and Witbooi to sign the document. While Göring was on an official visit to Maharero at the latter's homestead outside of Okahandja, Hendrik Witbooi arrived. The two visited for a time as Göring looked on, though in the course of events, gunfire was heard, and a short gunfight ensued in which several men were wounded. Witbooi and his company withdrew, and Göring is reported to have helped dress the wounds of several Herero men. On October 21, 1885, Maharero signed the treaty guaranteeing him and his people protection. However, cattle raiding began once again, and by 1888 a frustrated Maharero had yet to receive German "protection" in any form—whether from men, equipment, or arms. Having determined that German "protection" was entirely worthless, he renounced not only the treaty but also various agreements he had signed deeding mineral rights to German companies as well.[51]

Conditions in German South West Africa continued to deteriorate, prompting Bismarck to send Curt von François with twenty-one soldiers in mid 1889 in a bid to impose order and assert Germany's rightful authority over the land. Maharero was not the only Herero leader doubting the practical value of the German treaty. Von François records the content of a letter, dated June 14, 1891, that Manasse Tjiseseta, another Herero leader who lived near Omaruru, had sent him in which Manasse respectfully complained of the complete absence of the promised help, particularly the munitions that were to have been shipped by wagon to his people: "What type of help," he asks, "do you mean when you say that you have brought it to us? Nothing has been brought to us."[52] Then, almost in the form of a parable, Manasse recounts a supposed incident in which a

> Herero ventures to Windhoek, to a place in Windhoek where a German lives, and is refused both water and food, and must remain outside in the rain even though a German's house stands before him. There he loiters until he is chased away like an enemy! . . . According to the announcement that you will, from time to time, send [German] soldiers our way, I say, certainly, this is fine. We have also committed to do this in our treaty with Kaiser Wilhelm. But friend, if this which I have just mentioned above, and which I hold to be a falsehood, should turn out to be true—that a German would treat a Herero in such a manner—then please, do not send us any soldiers.[53]

The clear intention of Manasse's letter was to question both the value and the honesty of the German treaty of protection.

Von François, rather quickly determining Hendrik Witbooi to be his principle adversary, laid plans to bring him to heel and end the fourth Herero War. Witbooi had become a Christian at the age of twenty-five; as a student of Old Testament stories (note that the Dutch Reformed Church of South Africa was very much an Old Testament Christianity), he saw in their storylines and imagery a means of understanding his own role and destiny in life; namely, that he should unite the various Nama factions and lead them to a land of promise, a land that happened to be the grazing lands of the Herero.[54]

In November 1892, Witbooi signed a treaty with Maharero's son, Samuel Maharero, that brought the final war between the Nama and the Herero to a close.[55] It appears that Witbooi had long understood the German subterfuge of "protection" for what it was: a legal means of acquiring authority that would eventually lead to dispossession and complete subordination of a native people. Lau writes that Witbooi's "realisation that the African conflicts were secondary to a threat of entirely new, immense proportions, namely conquest and colonization by Germany which could only be staved off by African unity, makes him one of the wisest pre-colonial rulers in Africa."[56] Indeed, three months before the treaty was signed, Witbooi wrote a long letter, dated August 4, 1892, to the British magistrate in Walvis Bay in which he expressed what was now clear to him: Germans, not black nor red tribes, were the true enemy; not only that, Germans were cruel and brutal and would stop at nothing to impose their rule.[57]

Captain von François sought and received an additional two hundred and fifty troopers, and on April 12, 1893, he directed a raid with the object of "annihilating the Witbooi tribe" at Witbooi's stronghold in Hornkranz. Von François succeeded in killing around one hundred and fifty of Witbooi's people, seventy-eight of them women and children, but he failed in capturing or killing Witbooi himself, who escaped, gathered his surviving followers, and staged what can only be termed as an audacious raid on a German military installation in which he succeeded in capturing nearly all of the German military's horses. This was one of several equally daring raids (another of which saw Witbooi attack and carry off thousands of head of livestock from a German experimental agricultural station). News of the massacre at Hornkranz coupled with news of Witbooi's successful raids against German installations generated a good deal of uncomfortable publicity in the German press, and Berlin quietly dispatched the eventual replacement for von François: Major Theodor Leutwein.[58]

Leutwein, arriving in Windhoek in January 1894, took some time to acquaint himself with the situation before deciding how best to achieve the objectives of the task before him: 1) to subdue the non-European peoples of German South West Africa and maintain that subordination so that German settlers would have a secure political environment in which to farm, and 2) to make available the huge tracts of land necessary to farm cattle and sheep in an arid land. Because of Witbooi's continuing raids on Herero cattle, Leutwein was at least partially successful in persuading Herero leaders that Witbooi was the greater danger to them—as well as to the Germans. Leutwein also successfully planned and executed displays of his military might among other groups of Nama and Orlam, which had as their design the further isolation of Witbooi. With far better armaments and the construction of fortresses in strategic locations around the colony, Leutwein decided to pummel Witbooi into surrender. When Witbooi resisted, Leutwein began a two-week campaign of demonstrating the effectiveness of his repeating rifles and artillery. Finally, on September 15, 1894, Witbooi signed a peace treaty with Leutwein. In exchange for the cessation of raiding and the promise of abiding by a general peace, Leutwein allowed Witbooi and his followers to retain their guns, their horses, and their territory.[59]

Relieved that the natives had been at least temporarily subdued, Leutwein began to work on his second objective: the acquisition of land and livestock for German settlers. He was convinced that a substantial rural population of German settlers in the colony was key to creating a lasting peace. Given the fact that South West Africa was semiarid to arid, securing the best possible grasslands for incoming settlers was a priority. The Nama and Orlam had settled on, for the greater part, thirstier lands than the Herero, who possessed the very best pasturelands in the colony—land the Herero would have to share with German settlers. To begin the process of acquiring Herero land, Leutwein elected to create an official southern boundary to Hereroland and establish an east-west parallel that began in Gobabis in the far east of the colony and ran westward through Otjimbingwe (located north and slightly west of Windhoek) to the farthest westerly stretches of Hereroland. Part of this agreement—an obvious attempt to lay hold of cattle for next to nothing—was a stipulation that any Herero cattle found grazing south of the boundary would be deemed officially without owner and become property of the German government, which then would have the right to sell these "ownerless" cattle, though half of the proceeds would be given to the local Herero leader at Okahandja, Samuel Maharero.[60]

For his part, Maharero was paid two thousand German marks to sign this agreement (a sum to be given him annually), though he had no authority to do so on behalf of the Herero associated with his leadership, let alone the vast majority of Herero who had no political, ritual, nor economic ties to him whatsoever. Maharero had also sold the land south of the boundary that his people had previously used as well as a huge tract of land to the east over which he had neither control nor right to control. When Germans began to confiscate some of the cattle that had drifted south of the line, and the full impact of what had been done became clear to the Herero dwelling in the southeastern portions of Hereroland, resistance began. Though Maharero certainly profited from these cattle confiscations, he was clearly the junior partner and often fell victim to both his own avarice and German deception.[61]

In November 1895, writes Gerhard Pool, a German patrol found Herero cattle grazing south of the line near Heusis. By local Herero calculations, between four thousand and five thousand cattle were grazing in the area. Major Mueller demanded two hundred head of cattle as the 5 percent penalty for pasturing south of the line, though he ended up confiscating all of the cattle from one of the cattle posts since Leutwein had increased the penalty by an additional 5 percent to help offset the "costs of patrolling the border."[62] Once the corresponding number of cattle had been removed, Maharero received the value of half of the remaining cattle—far less than he had expected. Complaining, he claimed he understood the agreement to mean that every person who had cattle grazing south of the boundary line had to pay a fine of 5 percent of his total herd and that the Germans would be responsible for collecting these fines. The official response was that the patrols did not have sufficient time to collect fines in this manner, but that he himself ought to make these collections. Of course, Maharero understood very well that it was one thing for German soldiers to collect cattle fines and quite another for him to attempt to do so: he was, after all, Herero. He did receive 1401.25 German marks, but considering that he owed 13,200.00 German marks to just one of his several creditors, Herr Ludwig Conradt, his prospects for deriving handsome revenues from confiscated cattle were dashed.[63]

An emergency conference that took place in January 1896 in Okahandja ultimately failed to settle the matter of cattle confiscation in a satisfactory manner. Unrest began that was crushed within weeks, and by April 1896 the proper conditions were in place for white settlement to begin. In 1897 a devastating rinderpest struck central South West Africa resulting in Herero cattle losses estimated at 80 to 90 percent of their herds: Herero cattle had not been inoculated as had the settlers' cattle. When the Swakopmund-Windhoek narrow-gage railroad was finally completed in 1903, the European population in the colony had grown to 4,640, and the majority of whom farmed cattle.[64]

From 1896 onward, Herero not only continued to lose cattle but also steadily lost land that was surreptitiously sold or through various other means fell into the hands of German merchants, land syndicates, and settlers. John H. Wellington, citing a passage from Kurd Schwabe's *Mit Schwert und Pflug* (1904), writes that a primary means of acquiring cattle was through trade:

> The Herero brings the oxen he wishes to sell. "How much do you want for the oxen?" says the trader. "Fifty pounds sterling," replies the Herero. "Good!" says the trader, "here is a coat valued at 20 pounds, trousers worth 10 pounds, and coffee and tobacco worth 20 pounds, that is in all 50 pounds." The Herero is satisfied: he knows that according to the custom of the traders he cannot expect more for his cattle. He may probably exchange the coat for a blanket and get some sugar in lieu of tobacco, and he will also (as is customary) by begging get a little extra; if, however, he does not succeed, the transaction is closed.[65]

Worse still were traders' credit systems, which meant that Herero or Nama who had no livestock at their immediate disposal could purchase items on credit at exorbitant interest rates (calculated in livestock) and thus become perpetually indebted to traders. Many natives fell victim to a weakness for certain European goods such as alcohol and clothing, the conspicuous use and display of which served to elevate them in the eyes of their fellows, as well described by von Eckenbrecher.

Many Herero were not oblivious to what was happening. Neither was Leutwein ignorant of the injustice of the credit system. Furthermore, he knew that natives "were mad with rage at the robbery of their cattle, with no hope of redress, and that there was a possibility of an armed revolt against the situation,"[66] a revolt he and his small force could not possibly suppress. In January 1899, Leutwein introduced an ordinance that forbade granting credit to natives, something missionaries in Namibia had been advocating for years. When settlers whose livelihoods were connected to this practice—either traders themselves or farmers who purchased cattle from them—bitterly complained and threatened to take the matter to the German Parliament, Leutwein repealed the ordinance but continued lobbying against the practice both within the colony and with the Imperial government, which did finally abolish it. But by then the die had been cast.[67]

Perhaps worse than the credit system was the mostly spurious acquisition of land. While Maharero had been unwilling to sell land—indeed, the whole idea that land could be bought and sold was certainly not part of Herero tradition—when approached by Europeans, particularly missionaries, he did grant permission for missionaries to settle on a piece of land on an open-ended basis, but such temporary occupancy was definitely not ownership. This contrasts significantly with Witbooi practice, as "[b]etween 1894 and 1904 the Witboois sold almost a third of their land to the settlers who gradually came."[68] Samuel Maharero, however, once he assumed the position of "paramount chief" (again, a European-inspired political position), broke with longstanding Herero practice and began to sell huge parcels of land simply because he needed to finance a lifestyle that necessitated the purchase and consumption of imported goods offered by traders. By September 1901,

Maharero owed the trading company Wecke and Voigts 17,154.5 German marks,[69] money he simply did not have. Consequently, tract after tract of land was clandestinely sold to traders who either resold it or lived on it themselves. "It was primarily because of Samuel Maharero's crushing debts," writes Pool, "that the twin problems of trade and land possession became so inextricably linked at the end of the nineteenth century."[70]

An excerpt from a 1986 interview with two elderly Herero on the subject of Samuel Maharero and his land sales provides a very interesting explanation of how a man now so greatly revered among the Herero could possibly have dispossessed his fellows with such recklessness: "They gave Samuel strong drink, and when they saw that he was drunk, they asked for land to live [on]: 'We want to buy such and such a place.' They offered him money and boxes of brandy and bags of sugar, mealie meal, rice and coffee: all these [things] they gave him for nothing. For they had given him drink, and now that he was drunk, they were asking for land to live [on]."[71]

In 1888, Germany enacted an ordinance that land sales between native and German be abolished *unless* transacted through the administrator of South West Africa as intermediary. Difficulties arose as tracts of land, well north of the southern boundary line of Hereroland established by Leutwein, were "sold" to and then settled by white farmers. In August 1901, German missionary Herr Lang wrote a letter on behalf of distressed Herero near Okatumba who complained that white settlers were building houses on Herero land, land granted to the Herero in the name of the Kaiser himself. Missionaries were distraught over the loss of land among the Herero and had repeatedly petitioned that something be done, a view with which Leutwein sympathized.[72]

The majority of settlers, however, did not share this view. Wellington cites an article in a local newspaper, the *Deutsche Südwestafrikanische Zeitung*, from January 22, 1901, that well articulates the sentiments of the local white settler community: "That the land, of course, must be transferred from the hands of the natives to those of the Whites, is the object of colonization in the territory. The land shall be settled by Whites. So the Natives must give way and either become servants of the Whites or withdraw to the reserves allotted to them."[73] Leutwein realized that settlers needed the land in order to farm successfully, but he also felt that natives had to be accommodated and not left landless.

Leutwein was hardly without his critics, who felt he was coddling natives, favoring them above settlers and thus impeding German colonization of Namibia.[74] Indeed, Leutwein, soldier and governor of German South West Africa for eleven years, is not an unsympathetic man in the midst of an unfolding drama that was far greater than he. Leutwein clearly recognized the goodness, even the greatness, of persons who were declared enemies of the state. In this regard, his summary of the life of Hendrik Witbooi, killed by German military forces in 1905, is a revealing addendum to this historical overview and illustrates well the complex nature of the relationships between people and peoples during the colonial encounter:

> Had he [Hendrik Witbooi] fallen into our hands we could no longer have taken for granted his loyal service to us; he would have had to have paid the penalty. I therefore regard the German bullet that finally struck him as a mercy for both him and for us. It gave him an honorable warrior's death and spared us a painful dilemma. The name of this little Captain will, however, remain engraved upon the history of South West Africa forever. His stubborn resistance against the mighty Germany Empire as the head of a small warlike band, ragged and poor; his ten years of loyalty to our cause; and finally his daring in leading another uprising against us: these have inseparably linked his name with the history of the Protectorate. I still see him before me, the little Captain, ten years my loyal brother-in-arms. Modest though self-possessed, loyal yet not without political cunning, never deviating from what he considered his duty or his right, fully understanding the superior culture of the European, while by no means always in love with those who purveyed it, a born leader and ruler: this was Hendrik Witbooi, who without question would have become immortalized in world history had not the fates chosen him to be born to an insignificant African throne. He was the last national hero of a race doomed to destruction.[75]

Leutwein's loyalty to the German program of colonization was certainly not in doubt, but his ability to recognize Witbooi, a friend-cum-enemy, as a great leader and "ruler" worthy of admiration and praise demonstrates his capacity to look, at least occasionally, beyond color and race in assessing a person's worth.

Conversely, Paul Rohrbach, head of the Settlement Commission in the colony and a leading proponent of land transfers and of native tribes taking their "proper positions" within the new order of things, explains his own stance regarding the land issue:

> The decision to colonize . . . means nothing more than that the native tribes must withdraw from the lands on which they have grazed their cattle in order that the white man may graze his cattle on those very same lands. Should the moral correctness of this position be brought into question, the answer is that for people at the cultural level of the natives of southern Africa, the loss of their state of barbarism and their subsequent development into a class of laborers who will be in the service of, and fully dependent upon, the whites is nothing more than a basic law of human existence. For a people, as well as for an individual, one's very existence seems to be justified to the degree that that existence proves useful in promoting a general progress in the development of civilization. No argument in the world can prove that the perpetuation of any degree of the national

> independence, national prosperity, and political organization of the natives of South West Africa would be of greater or even equal advantage in the development of humankind in general, or to the people of Germany in particular, than that these races should develop skills to serve the whites, so that the whites may more fully enjoy the lands once possessed by the natives."[76]

As land transfers continued, often through the instrumentality of Maharero, anger and dissention among the Herero was rife, and Maharero found his safety and well-being under threat. In 1903, largely because of him, three-and-a-half million hectares of the twelve million comprising Hereroland were now in the hands of settlers. Consequently, he left his homestead in Okahandja to take refuge in Osana—or in Windhoek when he truly felt himself to be in danger. Just as Leutwein, in consequence of Witbooi having signed a peace treaty, deeded the Witbooi people a reserve of 120,000 hectares, he also hoped that the creation of a reserve for the Herero might pacify them. There was some support for this among the Herero, though the accompanying fears that they would be allocated too little land or that the land would be barren and hence useless for grazing cattle was strong.[77]

In the middle of 1903, the year the von Eckenbrechers arrived in South West Africa, the German chancellor decreed that the credit system existing among traders and natives would be abolished, but that traders had one year in which to collect all debts owed to them; at the end of that year the government would no longer legally recognize the existence of such debts. The result was that traders collecting debts fell like predators upon the Herero and devoured them. It proved to be, as Leutwein lamented, one of the key sparks that finally ignited the well-filled powder keg that led to full-scale revolt. Von Eckenbrecher and her ailing baby son sailed for Germany on March 7, 1904, a month or two after the revolt began, her husband following several months later. It would be ten years before she returned to her former home.[78]

Kaiser Wilhelm eventually replaced Leutwein with Lieutenant-General Lothar von Trotha, whose proposed solution to the Herero/Nama revolt against German rule was expulsion or annihilation. The resulting genocide, complete with concentration camps, reduced the Herero from about eighty thousand to about fifteen thousand; the Nama suffered over a 50 percent casualty rate: ten thousand died; nine thousand survived.[79]

V

In October 2007, by invitation of Herero chiefs from six of Herero royal houses, some of Lothar von Trotha's descendants traveled to the small town of Omaruru to apologize formally for their ancestor's role in Herero genocide.

Family spokesman Wolf-Thilo von Trotha said, "We, the von Trotha family, are deeply ashamed of the terrible events that took place one hundred years ago. Human rights were grossly abused at that time. We say sorry, since we bear the name of General Lothar von Trotha. We, however, do not only want to look back, but also look to the future." As to the call for German reparations, another descendant, Ulrich von Trotha, responded, "Our family cannot become involved in the demand for reparations from a government."[80]

An estimated three hundred Herero and Nama skulls were eventually transported to Germany after the revolt had been crushed, arriving sometime between 1909 and 1914. The skulls were studied and eventually used by Nazi race theorists to promote Aryan racial superiority. Twenty of these skulls were more recently discovered at the Charité Hospital in Berlin as well as at the University of Freiburg. German researchers believe that eleven are Nama and the other nine Herero. Four are female, sixteen are male; one of the latter that of a boy. They were returned to Namibia in October 2011 amid continuing calls for German reparations.

Interestingly, every Herero Day—celebrated each year on the Sunday closest to August 23—Herero from all over Namibia converge on the town of Okahandja. En masse they proceed to the graveyard where several great Herero chiefs are buried. They dress the graves and commemorate the chiefs as well as other fallen heroes. They don their ethnic costumes, which for women is ironically a long Victorian-era dress, precisely the attire worn by German missionaries' wives, with a folded cloth headdress that imitates the large horns of a prized ox. Herero men wear military uniforms patterned, also ironically, almost exactly after the uniforms of German colonial troops and engage in what is called the *Truppenspiele*—an act of bravery and defiance against German colonial rule. Although the von Eckenbrechers had returned to Germany before these tragic events occurred, it is highly unlikely that Margarethe could have been unaware of them.

In May 1914, von Eckenbrecher returned to German South West Africa to begin a new life in Windhoek, eventually settling into a large property in the suburb of Klein-Windhoek. In time, she began teaching practical arts, continuing to do so until her retirement decades later. As Europe anticipated a new war, von Eckenbrecher and her sons enjoyed a simple meal of bread, jam, honey, and apple cake, all the while extolling the grand beauty of their adopted homeland on a lazy afternoon, when dogs begin to bark as a black courier made his furious way to their open door, a one-page special edition of a local German-language newspaper in his hand: World War I had begun.

Eventually, the South Africa army engaged German colonial forces, but the latter were greatly outnumbered and, in comparison with the South Africans, poorly outfitted as well. They soon capitulated, and the fate of German South

West Africa hung in the balance until after the war ended. With full German surrender in Europe the League of Nations eventually declared South West Africa a Mandate, ceding its governance to South Africa. Von Eckenbrecher viewed her new situation as uncertain since many German settlers were being sent back to Germany. But when her file came before the "British" review board, it accepted her appeal, granting her residency: South West Africa had truly become her place under the sun.

With South African rule came an influx of white South African settlers—often very undesirable settlers in von Eckenbrecher's estimation. Other changes in administration, governance, and sentiment made her realize that Germans and South Africans often did not see eye to eye.[81] So it was that toward the end of her memoir, as Hitler was rising to power and calling for the return of previously German territories, that on occasion von Eckenbrecher and her pupils listened to the Führer's speeches via radio transmission. Indeed, the final words of her narrative form a kind of prayer, a lamentation with great hope that South West Africa will eventually be returned to the Fatherland.

In 1966, the United Nations, successor of the League of Nations, terminated the South African Mandate over South West Africa, an act largely ignored by South Africa, and a war of liberation began between the South African Defense Force and the military arm of the South West Africa Peoples Organization (SWAPO), greatly aided by Cuba. In 1988, a peace agreement was reached through United Nations Resolution 435, calling for free elections and an independent Namibia. On March 21, 1990, South West Africa officially became the independent nation of Namibia.

In 2011, the World Bank estimated Namibia's population at 2,324,004. The number of Namibians of primarily European origin ranges from 124,000 to 150,000. Sixty percent of these speak Afrikaans as their first language, 32 percent speak German, 7 percent English, and 1 percent Portuguese.[82] While English is Namibia's official language, nearly everyone speaks at least two, if not three or four, other languages, and language rights are protected. In addition to German-speaking Namibians of European origin, some fifteen thousand African Namibians, many of them children at the time, lived for some years in the former German Democratic Republic, attended German schools, and returned to Namibia after the cessation of combat between SWAPO and the South African Defense Force. The extent to which these persons identify themselves with Germany is unclear. Windhoek (in German, *Windhuk*), Swakopmund, Lüderitz, Otjiwarongo, Omaruru, and other towns have, proportionately, fair-sized German-speaking communities with German language schools. In central and southern Namibia, many towns and hamlets have German names. A daily newspaper—*Die Allgemeine Zeitung*—has been in existence for more than a hundred years. Although it is not unusual for German Namibians to attend university in Germany, many of them return to Namibia, or, if not to Namibia, then to South Africa, rather than settle in Europe or Anglophone New World countries.

Given the facts that Namibia was declared a German Protectorate in 1884, that the first German troops arrived in 1889, that on July 9, 1915, German colonial troops surrendered unconditionally to the South African Union, and that South Africa effectively governed and developed Namibia from that time until 1990, it might seem audacious even to suggest the existence of some enduring positive legacies or inheritances of German colonization. Even so, German Namibians have had an influence on the country as a whole that greatly exceeds their rather small percentage of the total population. These inheritances—mostly having to do with ordinary life—cannot and do not in any manner compensate for the brutalities of German colonization, yet some of them have become so much a part of the fabric of Namibian life that they seem practically transparent. Architecture, cuisine, beer, and education are perhaps the most prominent of these inheritances.

First-time visitors to Namibia are often surprised by the presence of so many structures—residential, commercial, educational, religious, and public—built in Jugendstil (Art Nouveau), traditional, and what has come to be called "German colonial" styles. Steep roofs and oriel windows bedeck houses, villas, and small castles in Windhoek, Swakopmund, and Lüderitz, as do domes, towers, turrets, sculptures of Atlas supporting the earth, beautiful gables, jutting bay windows, and crenellation, all suggesting northern European edifices awaiting heavy winter snows. With South African governance, new architectural styles appeared, some of which were poor attempts at modern concrete buildings (hospitals, schools, even government administration buildings). Many of these have since been razed because of defective design or workmanship. Perhaps the most distinctive of the South African styles found in Namibia is Cape Dutch. Ubiquitous in South Africa, Cape Dutch has a much smaller representation in Namibia. For large shopping malls, whether indoors or out, as well as arcades and smaller retail strip malls, versions of German colonial style have become the current norm. Perhaps German colonial style is Namibia's "comfort" architecture.

Namibian cuisine has, for many, many decades, shown a pronounced German sensibility, perhaps even an older German sensibility than would be currently found in Germany itself. German restaurants are common in all cities and larger towns. Preparing meat in Schnitzel fashion with traditional breading and seasonings is ubiquitous in the country, as are grilled, smoked, and prepared meats, such as bratwurst, mettwurst, hackepeter, and cold cuts. German bakeries are found throughout southern, central, and north central Namibia and have made the common Brötchen (a light, crusty bread roll) a staple throughout the country. The commercial dairy industry also produces distinctive German commodities such as quark (smooth cottage cheese) and a range of yogurts and thick milk products. The fact that foods such as these have grown in popularity and sales indicates a market that stretches far beyond ethnic Germans and the larger white population. Interestingly, German haute cuisine in Namibia is currently undergoing a welcome change. Now infused with local plants, wild plants, fungi, and flavorings known to indigenous peoples, traditional German cooking is being transformed into a unique, interesting, and subtle Namibian cuisine, a culinary fusion of the two cultures.

True German-style beer is brewed in Namibia and exported to other countries in southern Africa. Following precisely the Reinheitsgebot (the German law of purity) of 1516, beer is made from malted barley, hops, and water. Brewing of beer in Namibia is not particularly noteworthy, but doing so according to the Reinheitsgebot promotes a kind of practice for which Germans are well known: high quality and exactitude. As the Namibian government has not yet been very successful in its attempts to attract enduring manufacturing industries to the country, the kinds of industries that could employ substantial numbers of people at livable wages, the principles of the Reinheitsgebot, the work ethic attached to it, and even the brewery that continues to be successful could become guidelines for labor and management practices.

Perhaps the most important inheritance from German colonial times is the education system established then. Small as it was, it formed the beginning of what is now a massive effort toward universal education. German schools in Namibia today are certainly among the very best in the country. In earlier times they taught manual arts along with an academic curriculum. Today, German language high schools are highly competitive, attracting students from all races, language groups, and (with financial assistance) income levels. Beyond formal schooling, Germans were instrumental in creating clubs or associations based on intellectual interests. In 1925, for instance, the South West Africa Scientific Society was established by and for persons interested in learning and investigating the flora, fauna, geography, history, geology, peoples, and more of Namibia. Most of the founders were of German origin, and eighty-eight years later it is still a thriving enterprise.

It is difficult to overstate the challenges of bringing together a country of distinct peoples—peoples separated and bounded by race, language, geography, history, and culture, peoples who were largely encouraged, if not forced under South African administration, to nurture and promote these very differences and to live as separate, almost sequestered, peoples.[83] Though this remains a legitimate concern, the policies of founding president Sam Nujoma and his successor, Hifikepunye Pohamba, to some extent have promoted peace, reconciliation, and stability in Namibia and among Namibians with a degree of success. President Pohamba, nearing the end of his final term in office, recently designated Hage Geinkop as the SWAPO candidate for president in the next election. Given that SWAPO draws its largest support from Ovambos, and that Geinkop is Damara, not Ovambo, this designation suggests signs of a growing political maturity. Despite the difficulties of enlarging an economy; providing quality education; moving people out of poverty; overcoming entrenched cultural ideas about gender inequity and violence; combating the plagues of graft, corruption, and presumed entitlement; overcoming the burden of unsustainable debt; reorganizing a bloated and inefficient civil service; and a fairly long list of other difficulties, in certain ways the country is working and people are succeeding.

Curiously, in the years since independence, some of the greatest battles waged among the various communities of people that comprise Namibia have been over the naming and renaming of streets—whether in hamlets, towns, or cities. A fair number of streets in the Windhoek suburb of Klein-Windhoek have been renamed, yet von Eckenbrecher Strasse, named for Margarethe von Eckenbrecher, has not.

D. P. Crandall
Hans-Wilhelm Kelling
Paul E. Kerry

Book I

PREFACE

During my childhood I saw much of the world, and from my earliest years onward my young mind absorbed a myriad of diverse impressions about life. At the age of twelve, I was admitted to a convent school for girls from aristocratic families; it was there that I changed from a happy, carefree child to a ponderous and serious-minded teenager whose only touch of beauty were her long, blond braids.

Because of difficult family circumstances and a strict, serious education, my life became anything but happy and carefree. Under my mother's influence, I quickly developed a single desire: to learn and to work so that I could become fully independent. Against this background, at the age of eighteen I passed my examinations and began teaching at an advanced secondary school for girls in a western suburb of Berlin. There I taught for four years before deciding to travel to England to continue my studies. Upon returning to Germany, I began to attend—with passionate interest—lectures in philosophy and theology at the University of Berlin.

It was also during my youth that a deep love of Africa began to weave itself like a crimson thread through the fabric of my being. My older brother, Hans, and my cousin, Themis (now my husband), described Africa in glowing colors, and it became the land of my dreams. During our holidays, as the three of us met together at the Cadet Room in Berlin, sitting close together in the gathering twilight, we often tried to best one another with new and fantastic tales of the dark continent. After each holiday, before we went our separate ways, we promised one another that someday all three of us would go to Africa.

A few years later, Themis left Germany in order to start a farm in South West Africa. Hans withdrew from the army and also went to Africa—German East Africa—where he endured many hardships. After seven long, difficult years, he had finally begun to achieve some success, only to be killed by rebellious natives in August of last year (1905).

Themis returned in 1900. Because of lingering differences, we decided to break our engagement and go our separate ways—he went off to Scotland to learn sheep farming while I remained in Berlin and continued my studies. In February of 1902, we reinstated our engagement. It had become clear to us that we much preferred to be together and lead a life of want and privation than to be separated again. Six weeks later, we married. Several days following the wedding, we set sail for South West Africa.

A greater contrast can hardly be imagined: Berlin, the epicenter of German culture and science, and a lonely farm in Damaraland. And yet, if I had to choose again, I would make the same decision. The constant struggle and the difficulty of wrestling for a simple justification of one's existence possesses something so intensely satisfying that I would not want to trade places with all those who sit within their four walls not daring to venture out into the air, into that greater freedom. Our great master, Goethe, is correct when he states:

> Nur der verdient sich Freiheit wie das Leben,
> Der täglich sie erobern muss. (Faust II, 11575, 76)
>
> (Only he deserves both freedom and life
> Who daily must conquer them.)

I hope the time soon comes when we may prepare for our return to South West. I remember the promise I made to Lieutenant Colonel von Estorff in Omaruru a couple of years ago, when he admonished me, "The colony needs women, come back!"

His words, "We need women in the colony!" continue to echo in my mind. Thus I wish to recruit women for the colony and plead with you: German women, do not let yourselves be held back by pedantic doubts. If you are confronted with the decision to follow the man you love to Africa, do not hesitate, follow him! Certainly you will give up particular things from your life in Germany, but you will gain so much more over there. You are young, strong, and German. The colony can only blossom and prosper again with the help of strong women. Help your Fatherland to achieve this grand and beautiful mission!

To this end have I written the following pages. May they thus be read, and may they be convincing. Yet one more thing: I must declare that I come from Africa and cannot write beautiful phrases but unadorned ones—only the way

I think and feel. I ask forgiveness if now and then I should write something of which the reader may not approve. I only want to present the truth, not to cause anyone hurt or grief.

Berlin, Summer of 1906
Margarethe von Eckenbrecher, née Hopfer

LEAVING HOME

1. The First Ten Days

It was the last day of April in the year 1902. Thick, damp fog blanketed the old Hanseatic city of Hamburg. From the gray sky fell a soft but relentless drizzle. To experience Hamburg in such weather leaves one with a rather dreary impression of it. In the Ratskeller, however, we tried—though without much success—to counteract the somber mood. The hours seemed interminable as we waited to board our steamer, scheduled to depart at three o'clock in the afternoon. At last we drove to the harbor and saw dense fog hovering above the Elbe River; the thousands of masts and chimneys from sailing vessels and steamers were barely visible.

The small boat that would transfer us to our ship, the *Eduard Bohlen*, lay anchored at the wharf, at the Amerika-Kai. A few well wishers had assembled in the waiting room, ready to bid farewell to the handful of passengers now preparing to board the *Eduard*. I glanced at them curiously; some had interesting features, to say the least. Africa, the dark continent, the continent of privation and struggle, had already marked some of those returning there. For others, the hope of a bright, golden future seemed to beautify their countenances and press back the bitter pain of leaving behind their loved ones. I doubt that any of the departing men and women foresaw how cruelly fate would destroy all their plans. Even now, as I write this, various faces appear in memory. Two-thirds of them already lie buried under the tropical sun, while the rest, broken in body and soul, struggle on in what seems like a futile existence.

The boat's tiny bell rang for the passengers to board. A last handshake, a kiss, many tears, and slowly the boat left the wharf. White handkerchiefs fluttered as icy wind, rain, and fog began to separate the ones remaining behind from the ones departing. A quarter of an hour later we reached the *Eduard*. People started pushing and shoving, and it was with difficulty that we climbed the ship's wet, steep, slippery ladder and boarded the *Eduard Bohlen*, one of the older vessels of the Woermann Line. It was an old-fashioned ship devoid of any luxurious fittings and had fully seen its day (though it has been completely refurbished and now sails between Cape Town and Swakopmund).

The head steward led us to the cabin that had been reserved under our name. A slight feeling of unease befell me when I realized that we would have to pass four to five weeks in these cramped quarters. On one side, two bunks were attached to the wall, one above the other. Separated by a tiny area of open space, a sofa of sorts stood on the other side of the cabin. At its end was a narrow wardrobe with a lock and key. Between the bunks and the sofa, a retractable box-like washbasin was affixed to the wall. A net stretched above each bunk—like those in railroad cars—provided additional storage. The most beautiful thing about the room were the two large portholes that afforded a view of the water, but only if one stood on the sofa or lay on the top bunk. To access the upper bunk, one had to be either a bit of a gymnast or to climb a small, polished mahogany ladder—which required some skill when the sea was rough. Once on top, one had the pleasant assurance of being able to climb down far more rapidly than actually intended!

We immediately took to making ourselves at home in the cabin by stowing both of our trunks beneath the beds and tucking various boxes and a crate filled with oranges on the floor beneath the sofa. Our formal wear and black coats were hung in the wardrobe, books and toiletries placed in the nets, and weapons and money deposited with the captain. As soon as we had finished these arrangements, we headed to the dining hall. A bottle of champagne helped us overcome our feelings of sadness. Loud voices and laughter greeted us as we approached, for the less melancholy were already boisterously celebrating the voyage with drinking, singing, and laughing. Amid all this noise, the ship's siren roared, the anchors moaned as they were hoisted up, and at last the *Eduard* began its voyage.

At dinner the guests not only met the captain and the officers, and introduced themselves to one another, but attempted to look calm and courageous while assiduously scrutinizing every other person in the room. We retired early. Outside, the waves beat against the sides of the ship and the rains splashed over the deck. I rolled into my damp pillows—which smelled of the ship, seawater, and tar—and tried to fall asleep. Towards morning, I finally succeeded.

I was awakened late by the breakfast bell and at first had no idea where I was. I looked out upon the grayish-green, endlessly surging water. Tall waves rolled toward the ship, broke, then topped themselves off with crowns of white foam. Wind howled and it was ever so cold and damp. I have always disliked the English Channel, and in the attempt to overcome my less than cheerful mood, I coerced myself into rising and contending with that subconscious power that forced me into such misery. Though I attempted to bring it under control, succeeding in this effort was another matter. I had a headache, I was frozen and dizzy, and my black mood had no wish to lighten. Sympathetically, my husband smiled and I lay down again. He comforted me and rang for breakfast. Soon the steward arrived with the food (the *Eduard* could not afford a stewardess!), but as soon as I smelled the muttonchops and the fried potatoes, my stomach broke loose with elemental force. That was the beginning of my seasickness. I had sailed many times before to England and back in all kinds of weather and had never even given seasickness a moment's thought. But now, for two full days it held me in its grip. I cared not whether I lived or died. The mere thought of food, let alone its smell, nauseated me beyond all measure.

Outside, the storm raged. Waves crashed against the portholes and washed across the deck. The engine pounded and boomed, and the entire ship seemed to tremble and groan. Inside our cabin each pounding caused the suitcases to tumble to one side of the cabin, then back again. Our crate of oranges broke and oranges rolled all over the floor, some whole, others bruised and smashed. The cognac bottle shattered and the cabin was engulfed by a disgusting stench. The tray bearing the untouched breakfast slid off the sofa, and in strangely coordinated movements, everything rolled around the floor in unison. Each new wave that pounded against the ship's stern raised the clamor anew.

When one is in the grip of seasickness, nothing else matters—no clinking, no deafening racket could have enticed me to rise up even for a moment. I clutched the protective board on the side of my bunk so that I would not fall out; I no longer cared about the tumult below me. Gradually, though, my apathy lifted and by the second day I felt a bit better and wished to go on deck to enjoy the fresh sea breeze. I crept upstairs, ashamed of my appearance, but instead of compassionately or derisively smiling faces, I encountered just a few persons, all of them, like me, looking more or less green in the face. Most of the passengers were not yet in a condition to leave their cabins.

According to the ship's navigational map, we were now close to Dover; indeed by afternoon the white cliffs of England appeared and our steamer changed direction. Heavy waves pounded our ship and the engines switched off. Poor *Eduard* took a turn for the worse, rolling so violently that had we not clung to the railing with all our might, we would have been jostled from one side of the deck to the other. Dinner was a most unpleasant event, for the raised lip-boards that had been fastened around the edges of each table to prevent dishes from falling to the floor proved useless. Even a boiled fish the steward carried from the kitchen to the dining hall jumped overboard!

The Channel appeared exactly as depicted in paintings: gray on gray. Light streaming from the lighthouse of Dover shone in the distance, and *Eduard* pitched and rolled into the morning hours. Our pilot returned to ship, bringing with him a Leichter—a covered barge of enormous size that transports freight from a steamship to a port located in waters so shallow that the ship itself cannot enter the port. The *Eduard* was to tow the barge to Port Nolloth on the southwestern coast of Africa. The barge was called *Agnes* and became a nightmare for the captain on this long voyage. The heavy rope that secured it to *Eduard* often broke, and several times *Agnes* was lost on the high seas. Each time we had to search half a day or longer to find the thing, and this exercise was repeated far too many times. During a storm in the Bay of Biscay, the freighter developed a leak. Two sailors from the *Eduard* had to stay on the freighter to man the pumps and ensure that it did not sink until it was properly repaired in Las Palmas. Because of that obstinate *Agnes*, we arrived four days later than scheduled.

After leaving the Bay of Biscay, the ocean became calmer and the air warmer. I very much enjoyed being at sea. For hours and days I rested on deck. It was difficult to read a novel, for the ocean was too grand, too marvelous. Nothing but water, water as far as the eye could see. Waves, coming and going, always crowned with bright foaming spray. Huge, threatening waves of aquamarine and lilac rolled against the ship. They rolled, groaned, and sent their froth spraying into the air or divided it amongst themselves. Often they created tiny, fawning waves that barely managed to crest before being reabsorbed by the vast watery desert.

Above all of this stood the vaulted sky, sometimes lead colored or an endless stretch of blue, but often spotted with graceful, white lambs. Sometimes they appear not to move at all, but then began their courses again like happy, exuberant children. The engine labors, moans, groans. This giant, breathing heavily, moves single-mindedly through the gleaming waters. All around us are waves: billowing and receding, always new, constantly changing, yet always the same. And thus it goes day after day.

As I lie on deck daydreaming, my childhood with its blessed fairy tales comes alive again. Does it not appear as if there, in the distance, an unhappy sea nymph sways amid the waves, who out of love for the prince took the fateful drink and became human? Immortal she could not become, for she did not receive a soul since her love was not returned, and therefore she dissolved into foam. I lean against the railing and watch. What I would not give to be able to penetrate the rolling waters with my own eyes and behold the shining depths? I would love to find the pale, shimmering pearls—and the rainbow colored oysters that, cool as the sweeping waves, gave birth to them. I would like to see all the creatures down there—how they live and work, the corrals, and the polyps with their innumerable tentacles. Perhaps a ship went down precisely at the spot we are crossing this very moment, and deep down there rest those who were once alive, combative, and happy. The corpses of drowned people are far more gruesome than those of the dead who did not suffer this fate. Once I saw the body of a little boy who drifted by so very quickly. His eyes were wide open and the expression on his little face was of a person needing to settle an important matter yet afraid of being late, for the little boy appeared so pale and so old.

The ocean encourages pondering. In the distance whales are exhaling and spraying wet breath into the air. Dolphins are dancing about the bow of the ship. They feed on the waste of the ship and accompany us for days on end. Rosy jellyfish, as large as a washbasin, are drifting alongside the ship as well as other creatures that look like paper boats with colored sails. The sailors call them Spanish men-of-war. What is that long blackish object that suddenly began following in our wake? It is nothing after all, just a wooden beam, followed by another and another. Just driftwood. The pieces come together and drift for a time until a wave separates them forever. I wonder how the wood finds its way to the ocean in the first place. Perhaps from a sunken ship, yet now the water has power over it. Thus is life.

> Life is like a piece of driftwood, ever
> Tossed in the restless main,—
> We meet and touch, and then we part
> And never meet again.

By the way, the British have a practical use for driftwood. Merchants travel along the coasts and buy it. They sell it to wealthy customers who burn it in their fireplaces to heat their homes. With a small, bluish flame, it burns slowly, crackling as it spreads its unique fragrance.

Now and then on the distant horizon, masts from other ships appear. If a ship comes close enough, it lowers its flag as a sign of greeting. When it is dark, the small Bengal lights sparkle and dance, trembling on the black water desert. These lights, ignited by the officer of the watch on the bridge, are burned in the color of the various steamship lines—the Woermann Line, for example, has a green-blue color—and the approaching ship then answers with its particular colors. Such an illuminated giant in the darkness of the night possesses an eerie mien, appearing, as it does, like a giant mythical creature with a hundred glowing eyes. Sometimes the ships came so close to each other that we could distinguish between single silhouettes and hear commands given. Then one ship greeted the other as dear friends do. Soon, though, the other ship disappeared into the dark horizon.

Life on board is both lively and monotonous. People are bored. One cannot eat or read all the time, so we play cards, roll the dice, and drink. Sometimes, if we are in high spirits, a wine glass is thrown into the sea. Faces are fresh and suntanned, and optimism and creative energy beam from many eyes. Some eyes, however, are dim and yearning, tired from the constant struggle of life in Europe. Now they anticipate making a fortune "over there," though they do not really believe it will happen. Still, they must move forward since looking back means little more than longing for a home they have lost. For that reason, forward to the south!

Despite the sun-sail that affords us a bit of shade, the heat is oppressive. Only in the evening does a cool, refreshing breeze begin to blow. Everyone lies back in his chair, tired and lazy. Fine cigarette smoke wafts through the air. By this point in our voyage we know one another very well. The black doctor over there, the one who always wears mousey-grey gloves, sold his practice and his villa because of his colonial dreams; he studies the Herero and Namaqua tribes diligently, and his wife and children are to come once he is settled.

The young veterinarian has yet to recover from his seasickness. We are trying to help by ensuring he has plenty to eat, but to no avail:

> He sacrifices to Neptune in the billowing sea
> The beef-steak, the cognac, and the jellied eel.

Then there is the heavy-set farmer. He will have a hard time in South West Africa, the land of thirst, for it is impossible for him to put down the bottle while aboard this ship. Two Boers, the younger of whom is unbelievably filthy, plan to return to their old homeland by way of South West Africa. There is a group of railroad commissioners as well, and we worry about the old chief construction officer—he suffers terribly from the heat. Also on board are several high spirited, elegantly dressed young men en route to business establishments in Swakopmund.

The most interesting passengers are the ones who have already been there: a business man from Swakopmund who is constantly shadowed by a silent helper, a man from an old Pomeranian family who has been and done everything imaginable—carpenter, railway worker, butcher, railroad engineer, merchant, and, finally, a successful farmer. Fortune smiles upon him now. He is taking home a kind and elegant wife, and they, like us, are on their honeymoon. We spend a good deal of time together. The corpulent lady over there is from the Transvaal. Her husband, who remained behind to look after the children while she traveled to Germany to raise a bit of money, lost everything in the Boer War. Now they will start again in South West. She appears to be more assertive than he.

I felt sorry for an aging and rather lonely woman who had worked hard as a teacher of young people. She decided to teach in the colony and was immigrating with a small, carefully tended myrtle plant. This plant was a special omen for her. "If I succeed in getting this to South West alive, I can make a crown of it for my wedding. If it dies, I will not marry." She was careless and innocent enough to mention this at the dinner table; several absolutely shameless scalleywags poured seawater on the plant. The buds fell off and the leaves turned yellow, then it died. The woman shed bitter tears over this, but worked on a lengthy entry in her journal and withdrew even more silently and miserably into her snail shell. She did, however, eventually find a husband and marry. Alas, she also became one of the first victims of the Herero uprising and was brutally murdered by natives.

One of the second-class passengers was a source of great amusement. He was a decidedly heavy-set gentleman with a shiny, bald head and a rosy, beardless face. He always walked about in his long, black, collarless coat, wearing either athletic shoes or tall riding boots on his feet. Besides his binoculars, he was well known for always carrying a green vasculum, from which he drew his cigarettes. He had been a Protestant minister, though later became an impresario for a ladies' band. Aboard our ship, he was the chaperone of two brides traveling to meet their grooms in South West, as well as the protector of a customs official and former soldier in the colonial forces. In company with the two brides, he waltzed up and down the deck whistling or singing a merry tune. In South West, the impresario became a well digger. Ironically, the poor man died of thirst in the Namib Desert.

Our ship's captain was a small man who held his liquor well, was certainly never a killjoy, and someone with whom we passed many a happy hour. As we lazily lounged on deck, a sudden breath of life fell upon the entire group when the captain announced we would arrive in Las Palmas the following day around noon. Heat and boredom were quickly forgotten as we devised plans for the landing, each of us giving advice to the others, all of us promising to meet up and tour the port together.

2. Las Palmas

When the great day arrived, we rose earlier than usual and dressed with far greater attention to detail—for amid the heat and boredom we had neglected our appearances. At the breakfast table that morning, ladies wore brightly colored dresses while the gentlemen had donned white tropical suits. They looked as neat as a pin, each one having attempted a more polished appearance than his fellows. The unsightly stubble that had started growing back home—to which no one had really objected—had been shaven. All of us were in high spirits and ready, after ten days at sea, to leave this old rust bucket and set foot on dry land.

On the distant horizon lay a bank of heavy mist. Ever so slowly, huge mountaintops arose from the mist, gradually taking shape until they appeared clear and distinct. Atop the mountains stood two gleaming lighthouses. Moments later the glimmering city stretched out below them. The cathedral towers stood silhouetted against the deep blue sky, and situated nearby was the sanatorium for tuberculosis patients. As our ship sailed slowly into the harbor, tiny white houses appeared before us—as if lovingly assembled by a child's hand. It was Sunday and everything was relatively quiet. Ships of all nations lay at anchor. Hardly had we docked when the harbor office boat carrying the medical doctor pulled alongside us. It was his responsibility to check each passenger and member of crew before allowing him or her to go ashore. After signaling that all was well, several dozen boats left the quay and rowed wildly in our direction, allowing us to carefully study the "poor brown sons of the south." Spaniards, Portuguese, Indians, and Negroes all screamed at the tops of their lungs hoping to attract our attention. With their few words of English, German, and French, they sought to make a profitable arrangement. The strangely woven baskets in their boats were filled with bananas and oranges, and they offered to sell us chirping canary birds in filthy wooden cages. (These birds never sing and are at least three times the price of the ones at home.)

Small, bronze-colored boys with arms and legs of steel stood naked in the boats begging for money. If one tossed a coin into the water, they jumped from the boats to the depths of the sea, swimming frantically to the spot where the piece of metal slowly disappeared. Below the water they fought, and smiling broadly, the lucky boy climbed back into his boat clutching the coin firmly in his brown fist. Like monkeys, the merchants climbed aboard, hanging on the stairs, the ropes, and the rope ladders. A virtual flood of these scoundrels washed over the passengers, offering everything from the inevitable postcards and photographs, to pieces of coral, filigree trappings, and embroidery. Those who wanted to be hired as guides yelled at the top of their lungs. Since they demanded four times as much as their services were worth, the bargaining began until the price dropped to an acceptable level: four shillings to be ferried by rowboat from steamer to quay, and from there by cart into town. The British influence was evident everywhere as everyone accepted English money and English was spoken all about town.

Once on the quay, we climbed into a typical two-wheeled cart, drawn by horses or mules, and quickly drove through the suburb. Our first impression was one of utter poverty: very small, one-storied row houses with green shutters but no actual windows. About every third house was an *estanza*—a third- or fourth-rate restaurant with greasy bottles haphazardly arranged at the front. Dark-eyed women, with distinctive black or white scarves wrapped about their heads, squatted beside the facades of their houses. We noted too that almost all the women of the poorer classes were, despite the heat, dressed in black. Spanish beauties lounged in the windows, their hair and bodices nicely adorned with flowers, flirting with the dark-skinned soldiers. They were a dirty lot in their sloppy yet colorful uniforms, courageously stroking their pitch-black beards.

For the first time in my life I saw palm trees. They were growing alongside the road and were as rough and full of dust as the begging children—of whom there were many. At once they recognized us as Germans and, cartwheeling beside to our wagon, began shrieking a phrase that was probably meant to be, "Hail to you, in your victor's wreath." Then rolling around in the dust, they concluded with, "Viva l'Allemania, viva l'Allemania!" Only once we had thrown coins and cigarettes in the road did we manage to get rid of them.

The town itself is located at the foot of a mountain. The mountains are desolate and bare, yet the valleys are, quite literally, a fruitful paradise: twice a year the fruit ripens and twice a year it can be harvested. We arrived at the market square, where between flowering bushes and trees stands the colossal statue of Columbus. At this place, life is similar to that of the Zoological Gardens in Berlin, with many benches for relaxing, sand boxes, children, nannies, and soldiers.

The streets are straight and narrow. The riverbed is completely dry, and adjacent to it stand the military barracks. Although the soldiers are rather small and far from physically robust, they nevertheless look handsome. Their eyes twinkle like well-polished buttons—made for loving I suppose. But whether they could hold their own in battle is open to debate. Situated next to the market is the cathedral, built in Moorish style. Inside, the sanctuary abounds with niches, side altars, and pictures of saints. The central altar is of pure silver, as are the numerous candelabrae, bowls, and chalices used for celebrating mass. In the midst of all this splendor stands a simple lacquered pedestal, made of blackened wood, that looks for all the world like an enormous sarcophagus. The whole scene gave me an impression of being on stage in a theater.

The beautiful old hymnals were bound in pigskin and adorned with jewels. The notes on the printed pages were the old-fashioned, square-headed ones. A well-fed padre sat in one of the confessional chairs, his tonsure shining like ivory in the midst of his black hair. A local farm girl covered with the white scarf knelt beside his chair. Distracted by our steps, she looked up and seemed taken by my husband. She smiled at him, nodded her head, then continued with her confession. As the organ was quietly played, clouds of incense wafted through the church like veils of blue. A priest clad in his colorful vestments celebrated mass on one of the altars with choirboys kneeling at his side. A painting of the Madonna, with sundry gems and waxen decorations hanging stiffly from it, smiled upon all the worshipers and curious visitors.

After our visit to the church, we drove to Quiney's English Hotel. It was properly named as everything looked very English: from the wide hall with Madeira chairs and colorful pillows, to the waiters, the owner, and the guests. The hall opened to a beautiful garden filled with flowers, a garden partaking of the oft-praised tropical splendor of the enchanting south! Such flowers—whose colors and shapes I have never before seen! And their fragrance! All of this mingled with exotic birds singing inside their cages and parrots on shining chains screeching. Beneath a tree, whose blossoms were the shape of large rosy bells that emitted a truly bracing and restorative fragrance, wine was served to us. Our goblets rang joyfully against one another: kling, klang—to a future that smiles upon us as does the blossoming south! Kling, klang—to life, to youth!

Later, we climbed in a small cart and were driven to the mountains. As barren and sad as they appear from a distance, how glorious and beautiful they are upon closer inspection. The road winds upward in serpentine curves. Magnificent banana groves alternate with slender palm trees and vineyards, and between these, the flat white roofs of farmhouses. Above it all stands the blue sky, and in the distance, a blue mist extends all the way to the marvelous blue ocean. Blue sky and blue ocean compete in their depth, dissolving into each other, only to separate themselves again.

We encounter curious people on the country roads: dark-brown women—and men who are even darker. Almost all of them ride donkeys or mules. I was constantly reminded of the old biblical image of Mary sitting on a donkey with the Savior as they fled to Egypt. I saw many such Madonnas. Odd sacks woven from straw, along with thick-bellied bottles of clay and baskets filled with vegetables and blooming flowers, hung from both sides of the saddle. The little half-naked children playing on the side of the road usually rode along on donkeys or were carried on people's shoulders. Every woman and girl—even the tiniest babies—had flowers in her hair and at her breast. Not a single woman was without them.

The mountain mail coach was a jolly sight, pulled by five donkeys, laden with priests, peasants, and beautiful women with smoldering eyes. The draught animals stepped lively. The coach seemed to stop in front of every *estanza*, and an earthen jug filled with wine made the rounds; then off they went, all the merrier! After an hour or so, we arrived at Monte, a favorite tourist spot. Cacti grow against the tall walls along the road. Their large, purple, starlike blossoms stand out radiantly from among the prickly spines. Lilac glycine and clematis creep along the walls of houses, and gardens are filled with fragrant blossoms. I stop. Over here mignonettes and stocks are growing; over there, palargonies, petunias, and carnations—alas, I do not know all the names of the flowers. Then the roses, so beautiful and so many varieties, as if King Laurin himself had climbed down from his Tyrolean mountains and planted them in his garden here. Tall acacia trees with white flower clusters and countless butterflies and beetles in bright, shiny colors! The air is filled with humming and chirping, and again, above it all, the blue sky, and far in the distance, the blue ocean. Deep below us in the valley lies the glittering city. Not a breath of air is stirring.

Here I stand, reverently admiring nature's beauty. How fortunate I am to witness this scene with my own eyes and to breathe it in with my own senses. There, far below us, we hear the blessed ringing. Festively, the bells sound through the still air. Down there resides the calm of Sunday. How magnificent, how festive, how wondrous and marvelously beautiful is this world! For quite a while I felt quite unable to face reality.

We stopped in at the Bella Vista Hotel to drink tea, after the English manner of course, and thereafter hurried down the hill to the port as the *Eduard* was scheduled to set sail around nineteen-hundred hours. We could not resist the call of the Las Palmas Souvenir Shop and purchased a few souvenirs that were both shoddily made and greatly overpriced. Boats stood ready at the quai, and ten minutes later we stood on the deck of our steamer. *Eduard* had taken on coal and was now most untidy, everything covered in fine coal dust. On the stern deck, Indian merchants had set up a large store with an accompanying a tumult of voices! Dealing, haggling, and bargaining with great vigor, these rascals swore up and down by every one of their saints that they would die on the spot if they even made as much as a single penny's profit on their merchandise. Nevertheless, with overly dramatice shrieking they set their precious merchandise into the waiting hands of their customers! I am firmly convinced that these merchants, on the whole, made rather handsome profits.

In the falling twilight the port came alive. Skylights had been lit on the masts of all the ships and their reflections quivered in the dark waters. The houses of Las Palmas now appeared as a swarm of tiny glowworms. The ship's siren sounded, and with further screams and shouts, the merchants scramble to clear the deck. *Eduard* turned, the dinner bell rang, and most of the passengers headed down to the dining room. Soon quiet was restored to the deck; only the constant, even pounding of the engine, combined with the sound of waves rushing against the hull, disturbed the quiet of the night. Las Palmas grew evermore distant from us.

3. Monrovia

Aboard ship, the holy day of Pentecost is hardly noticed. It is murderously hot. Everyone groans. The sky is deep blue; the ocean is deep blue. The air shimmers from the heat, yet it is impossible to stay inside the cabins. The air vents have been opened in the bull's-eyes, but still there is no airflow. The ship is like a large, floating oven. At six-thirty in the morning, the temperature was already thirty-five degrees centigrade, and the captain had a barrel of Zacherl-Bräu beer opened for our morning pints—but we were even too lazy to drink it!

We lie off the coast of Sierra Leone. Black, foreboding clouds gather on the horizon above the mainland. Lightning strikes and the entire sky seems to be aflame. Waves splash up high as fireballs shoot into the ocean. We are so close to land that we can discern the shapes of palm trees, yet I cannot take leave of the sun even as it slowly drops into the ocean. If the dark wall of cloud disappears, then the Southern Cross will shine in the night sky. The captain has made an exception in allowing me to stand on the bridge, where it is possible to see beyond the masts and ropes. Undisturbed by the others (the watchman does not count), I wish to enjoy the night sky.

Red lightning continues to strike all about us, though I hear no thunder, and not a drop of rain has fallen. When finally the clouds break up and the moon dares to come forth, it has a distinctly different appearance: it is larger than it is in Germany, but not quite as bright. As the clouds finally drift away, the first stars begin to twinkle, one here and then one there. Suddenly there is a myriad of stars shining like diamonds against a dark velvet pillow. I search for the Southern Cross—the compass for ships blown off course, for wanderers lost in the tropical jungle, for a single horseman on a never-ending plain. I cannot wait to discover it, for how many writers and poets have described its charm and the sheer power it exercises upon the viewer? I had imagined the constellation to be huge, fully dominating the entire firmament with its bright splendor. Yet all was in vain! I failed to uncover it anywhere in the night sky. When at last, after a long fruitless search, the officer on watch came forward and pointed out four dim little stars that were neither large nor bright, I felt bitter disappointment—a deep bitterness I have experienced but rarely in my life. My good friend Goethe is correct, "Be advised, do not love the sun and the stars too much."

Cape Mount, the northern border of Liberia, juts out far into the ocean. One sees clearly the fan-like tops of the palm trees and the surf crashing against the rocks. We spend the entire day sailing along the coast, when suddenly the mainland becomes wavy and indistinct, with layers of smoke hovering above it. A savanna fire! Who does not recall the ten-cent novels of Indian tales with their exciting images? How the flames blaze! How the prairie animals run for their lives trying to escape the inferno! How an Indian pulls a white maiden on his horse and vies with fire and storm! That same picture is with us the entire day—trembling, red-hot air, and in the distance, yellow sand, white surf, rocks, jungle. Towards evening we arrive at the peninsula on which Monrovia is located, the capital of Liberia.

Liberia was founded in 1821 by Negro slaves freed by Americans and recognized in 1847 as an independent republic by the United States of America. From that time to this, they have had their own state here, and an annoyance it is for the surrounding colonies! They have their black president, exactly like the American model, also a constitution, taxes, mail service, and last but not least, a navy, though their entire fleet consists of two old merchant ships they have purchased. Now and then they are used for long voyages to spite and intimidate their neighbors; however, they have never sailed beyond the Canary Islands. They barely reached one of their harbors before running out of coal. And because the government lacked funds and was thus unable to purchase coal for the fleet, the ships lay at anchor in Tenerife (or some other port) for at least six months, until a local government official supplied them with sufficient coal so these annoying guests could leave. One of these grand ships sank in the harbor of Monrovia. It had completely rusted through. When a top official was asked why the ship had not been towed into dry dock, he answered, "Dry dock—what a lot of nonsense! We painted the ship every year while it was in the water. A ship belongs in the water and it is stupid to draw it upon land. It is, after all, the fate of every ship to rust. That is its natural demise!" The other ship still lies peacefully at anchor in the waters off Monrovia. It is either incapable of moving, or Liberia simply has no coal; still, it makes a very decorative impression.

The Liberians do not have an especially good reputation on the west coast. Woe to the ship blown by a storm to this stretch of inhospitable coastline or that sails onto one of the countless reefs that act as barriers to the mainland. Instead of rendering help and saving a ship from distress at sea, the natives reportedly rob anything that moves. Indeed, passengers are fortunate if they survive at all after having been stripped of all their belongings. Apparently the Liberian authorities are helpless to curb these activities. Visitors to Monrovia are treated courteously and showered with kindness. The natives try to make them feel secure by acting like innocent children—so the visitors can more easily be duped. Like children who often imitate the customs and habits of grown-ups without having any idea why grown-ups do what they do, so these Liberians imitate our customs with no understanding of what they are doing. Since we arrived late in the evening and could not go on land, a mighty shot from a cannon informed the Liberians of our arrival. The night was peaceful. High on the mountaintop, amid the palm trees and jungle, the tiny glowing beam from the lighthouse shone through the night. Nobody stayed in his cabin. Everyone, both ladies and gentlemen, remained on deck, yet even on deck it was so unbearably hot that sleeping was simply out of the question.

The large, splendid sun rose at four in the morning. Soon thereafter we witnessed a rare but picturesque scene: dozens and dozens of canoes left the shore, swiftly rowing towards us. Watching them, I was reminded of Wilhelm Busch's verses:

> Pegasus, my noble steed,
> Carry me to Africa,
> There are many black men
> And many colorful birds.
> Clothes are not the custom here,
> At most one wears a hat
> And covers one's privates with a loin-cloth,
> Yet one is black and that's enough!

For the Woermann Line, Monrovia is a most important port. Not only are the offices of old Mr. Woermann's profitable business located here, but the Kruboys—natives whose help is indispensable in difficult landing maneuvers—come from here as well. As far as I know, the original Kru-Negros of Liberia were displaced and subjugated by the American Negroes. They have lived for generations along the coast and are therefore born sailors, so to speak. Their canoes are considered to be the very best along the entire west coast, their skill at fishing is second to none, and no one is their equal in guiding boats safely through the surf. With unbelievable skill they row up to the huge breakers and wait for the exact moment when the surf will quickly carry the boat forward. Here, a wrong move will prove fatal, for if a giant wave hits the boat sideways, the force crushes the boat, sinking it along with the men inside.

Since the surf in Swakopmund (South West Africa) is also very dangerous, the captain of every ship hires as many Kruboys as he needs in Monrovia and transports them to Swakopmund. Once the freight and passengers have been moved ashore in small boats, the steamer takes the hired men back to Monrovia on the return voyage to Germany. Kruboys are paid up to one German mark per day; the headman, a kind of overseer, receives two marks plus a percentage from each man in his crew. Because of the expenses involved in this, the company once hired natives from South West Africa to unload the ships—Herero, Hottentots, and the prefered Mountain Damara. However, they proved to be far less skillful and, unaccustomed as they were to the cool, damp coast, far more plagued by sickness than the tough Krus of Liberia (who are obviously accustomed to the fog and dampness of the coast). Our captain had need for about one hundred men. They came in their canoes to be hired, and their clothing was nothing short of remarkable. Most of them wore a kind of trouser made of large patterned silk, cotton, or wool (flowered cotton was very popular). These were actually sacks held together by a cord tied round the waist. Two openings had been cut for the legs. Their upper bodies were covered by flannel shirts riddled with holes; a terry towel or blanket, whose edges and borders were impressively finished and very beautiful, completed the ensemble. Their head coverings were random, often bordering on the outrageous, for any sort of hat would do: straw or felt, thick fluffy caps, sport caps, even the kind of red plush caps that grandfathers like to wear. The younger Kruboys wore much less, either a loincloth tied about the waist, or simply a cord—"One is black and that's enough."

I was appalled by the unsightliness of the children. They have extremely thin legs and very fat stomachs, precisely like frogs. They came along in the canoes and begged by clapping their hands, "Money, money, money." They jumped from steamship to water diving after the coins we threw into the sea. They dove for all kinds of rubbish our ship released into the ocean—mutton intestines, rotten lemons, old tin cans—and like fish, wriggled about in the clear water, tearing apart the slimy mutton bowels and pulling treasures from the water that eventually provoked some fighting.

The health authority boat carrying a black doctor, along with the black postmaster, pulled alongside our ship. The Liberian flag proudly waved, looking exactly like the American flag except that it had a single blue star. Both men wore uniforms, or rather liveries, abundantly decorated with gold fabric. The postmaster is a rather interesting character. He offers stamps for sale, but at a price far below their actual value. Reportedly, this is quite a lucrative business for him, though he inflicts considerable financial damage on the state—the state he supposedly serves—with these special sales. I am told his main activity consists of embezzling the letters deposited for mailing at the main post office. He simply removes the stamps, sells them again, and disposes of the letters. For this reason it is advisable to buy stamps in Monrovia, have them postmarked, then place the envelope in a new one, affix a German stamp to it, and pass it to the Woermann Office for delivery to Germany via the next sailing vessel.

There is a Dutch Trading Company in Monrovia, and their boat delivered mounds of small pineapples and a type of fruit unknown to me, alligator pears, to our ship. These "pears" are shaped like small pumpkins, whose centers contain a hard stone no larger than an apple seed. The flesh around the stone is edible—for people who have a taste for it. Actually, it tastes like nothing. One salts and peppers the thing, then gives it a healthy shot of Worcestershire sauce. After about fifteen tries, West African gourmets contend that nothing in the world tastes better; this fruit is simply marvelous! The Woermann Company boat also brought provisions to our ship. As the cooks on our steamer would need to feed the Kruboys, large quantities of rice were brought onboard to last through their return voyage as well. (Incidentally, several gentlemen who desired to see Monrovia joined the return voyage.)

The captain requested that the ladies remain on ship as Monrovia, with its many swamps, was a hotbed of fever, and he did not want us to become ill. But as it turned out, our day was quite interesting. A group of native women ventured out in the canoes. They were striking with their white makeup and unusual hairstyle. The hair on their heads is divided into sixteen equal parts, each of which is cultivated, with much love and care, into a single, longer, oily lock.

We anchored in front of Monrovia for an entire day. The first officer went ashore, but despite the many promises, our captain was unable to find the necessary number of Krus. Shortly before our arrival, two other steamers had hired most of the available men, so our captain had no other choice but to steam along the Liberian coast until he gathered the requisite number of men. The hirelings made themselves at home on the front deck, unpacking their soiled bundles, then eating their rice. They speak a sort of pidgin English with the whites that even Englishmen find difficult to understand. For example, instead of saying, "I am hungry," they say, "hungry catch me so much." When one of the Krus was sent to bring the ship's doctor, and was later asked where he had found him, he answered, "Him – fight – box – belong – cry," which meant something like, "he fought with the box that belongs to the screams," or in other words, he played the piano. Imagine the technique of our good doctor!

The Krus seemed to admire me with a sense of wonder. Perhaps they had never before seen a white woman with blond hair. With fingers extended, one after the other pointed me out to the rest of the Kruboys; eventually, though, it became quite tiresome.

When the ship began to move full steam ahead, some of the natives in their canoes tried to keep up with us, but finally gave up as the competition proved uneven. Like nutshells dancing on the waves, they became smaller and smaller until they completely disappeared. Their canoes had been skillfully made from a hollowed-out tree trunk shaped like a huge banana—a bit crooked and a bit curved and so narrow that the natives could only kneel in them with their legs pressed tightly together. The paddles are shaped like an angular shield. Now and again, a wave slaps water into the canoe and a narrow board is employed in bailing out the water. If a canoe tips on its side, which happens frequently, the occupant simply pushes it upright. As one watches the way these Kruboys operate in the water, one has the distinct impression that such chaps could live nowhere else but the sea and because of their skills and abilities they could not possibly drown.

With the arrival of the Krus much changed on board the ship. Our congenial peace was gone, for they chatter and yell, and lie about wherever they find a bit of room. They seem particularly fond of the Woermann Line, as well as of the ship's captain and officers. I was standing close to the First Officer when one of these chaps came up to him; he positively beamed, and handed him a soiled photograph of a corpulent black woman, presumably his beloved wife, Zissa. He asked if the officer would write a dedication on the back of the picture. He complied, and had the chap dictated the following line: To my dear Meyer, from his beloved little wife Zissa! Indeed, the chap was named Meyer. The Kruboys have the most amusing names. Since they have not yet been Christianized and still follow their native religion, they have given themselves names such as: Hamburg, Glasgow, Zweimarkfünfzig (two marks and fifty), Erbsensuppe (pea soup), Teepott (teapot), Halbergentleman (half gentleman), etc.

Aboard the steamship, they are most useful and productive. Every morning we are awakened by an indescribable racket as the Krus scrub the deck with large, square stones. They help lubricate the engines, work in the kitchen, wash the dishes, prepare vegetables, and polish all the brass on the entire ship until it shines and glows. There is even a black man who washes clothes. He does a fine job with the laundry, though when it comes to ironing he is hopeless.

Following the captain's plan, we continued scouring the coastline. The heat made it most unpleasant and many passengers suffered greatly from the high temperatures; had we sailed the high seas, it would have been much cooler.

4. Nifu

Around three o'clock in the afternoon of May 19th, anchors were weighed, and at six o'clock in the morning of May 20th, creaking and rattling, they were dropped again at Nifu, a vassal state of Liberia. We were obliged to remain three nautical miles off the coast because of the many dangerous rifts and shallows that made it difficult for a large steamship to enter the harbor. From the *Eduard*, we saw the white waves of the heavy surf splashing several meters high against the coast. The sky was lead colored; it rained cats and dogs. Unfortunately, the rain was not the least bit refreshing and instead made the atmosphere feel like a hothouse. Because of the high surf and rough seas, we passengers were not allowed to go ashore. The boats were lowered, but only the First Officer and several sailors were in them, along with various gifts, chief among which was a crate of common schnapps destined to serve as the pièce de résistance. The gifts were intended to soften the heart of the ruler of the Nifuanes so that he would send as many of his subjects as possible to be hired.

The rowers needed an hour and a half of arduous labor to bring the boats to the point where they could ride the waves to the shore of Nifu. Nifu is nothing more than a small village of clay and reed huts, ruled by His Majesty, the King of Nifu. As soon as the boats landed, they were encircled by some three hundred wildly screaming men, women, and children. No wonder three years had elapsed since a white man had set foot on Nifu soil!

Howling and screaming, the crowd led the officer and his interpreter to a somewhat larger hut at the center of the village. A blue cotton rag, the Standard of His Majesty, flew proudly on a pole in front of the hut. Immediately, the dignitaries of the Nifuanian Empire assembled there, wearing the shabbiest top hats imaginable. The Interior Minister appeared, dressed in a heavy Havelock coat, a symbol of his dignity. The temperature must have been fifty degrees centigrade. They seated themselves on the ground before the hut and stared at one another in total silence. Finally, after a very long wait, His Majesty appeared. He wore no trousers, but had donned a blue sweater and the obligatory top hat. He was extremely rotund, but also suffered from elephantitis (an illness that occurs frequently among the natives of tropical Africa, and attacks one part of the body, or rather the skin and the tissue underneath the skin of that part, which continues to grow, then hardens, and finally develops into an unsightly clump). Unfortunately, His Pitiful Black Majesty had an enormously thick elephant foot that made it impossible for him to walk.

The king was very gracious, and in exchange for the gifts, promised to do everything in his power to supply us with men. The exchange of gifts was then discussed: salted meat, tobacco, and spirits for pelts and amulets made of lions' teeth. The gifts offered by the natives had not been prepared ahead of time, but in front of the recipient were simply removed from one of the assembled persons. His Majesty very much regretted not being able to accompany his subjects to the steamship and sent his minister in his stead. Thus they took leave in intimate friendship, and the chosen workers, both voluntary and conscripted, prepared for departure.

We watched from our ship as the boats and canoes quickly left the shore and proceeded in the direction of the steamer. It was a colorful group that arrived, and we greeted one of the boats with roaring salvos of merriment. In the middle sat a tiny, dried-up man dressed in an oversized blue shirt and top hat, holding in one hand a genuine Prussian drum major's baton. A servant who had difficulty keeping his balance stood behind him and held a huge cotton umbrella above His Excellency, the Minister of the Interior. Because of the wind and the waves, this was no small accomplishment. To the right of His Excellency sat a young man who had a towel about his waist and a banana on the back of his head. To his left sat another youth with a large lion's tooth on his forehead. Both were members of His Majesty's entourage.

The captain had asked us to bring our levity under control and to show proper respect for the black envoys so that they might more readily agree to the wishes of the captain. Thus, we all stood most respectfully at the railing and watched. His Excellency was finding it difficult to board the ship with proper elegance and dignity in his blue shirt while balancing his top hat on his head and carrying the baton in his hand. Finally, though, he succeeded, and with his rather soiled hand he genially acknowledged our courtly bows and curtsies. At the side of the captain stood a number of sailors whom he majestically inspected before accepting the captain's invitation to join him in his cabin. There, His Excellency was straightway offered rum.

In the meantime, the necessary Krus were hired and, as a precaution in case they should regret their decision, were immediately taken below deck to prevent them from jumping overboard and swimming to the canoes that were slowly heading back to shore. It was a necessary measure. The Krus tend to follow, en masse, the lead of some of their peers; thus, had one of them elected to leave and jump overboard, the captain would have been in danger of losing his entire black crew. When order on the deck had been restored, the promised gifts for His Majesty were loaded into the canoes: spoiled salted meat, rice, and rum—the black man's death. Following this, the captain reappeared with his honored guest. For some minutes, His Excellency dwelt in that blessed state in which all men are brothers, for he had a devil of a time leaving us. Perhaps too the devil of arrogance had taken hold of him, as he could not understand why he would not be escorted back to land in the beautiful white boat, with the necessary complement of handsome white sailors, to which his dignity entitled him.

Finally, the captain had enough of this, and wishing to be on his way, signaled to the stout ship's carpenter, who spit into his hands and in his lowbrow German accent said, "It's time for you to get off the ship now, you black swine." Taking hold of His Excellency, the carpenter dropped him overboard. His aim was perfect, for though his shirt fluttered up as he fell through the air, and his top hat lost its dignified position on his head, like a ripe plum, and with an enormous bang, His Excellency landed squarely into the pitching canoe. The man with the banana on his head and the one with the lion's tooth on his forehead had quickly grasped the situation and had jumped overboard. They climbed into two other canoes. We cheered mightily.

His Excellency did not hold his hasty departure against us, but waved graciously. I imagine he will praise us lavishly and report all the honors bestowed upon him by the whites—after all, not everyone can be the Interior Minister of the vassal state of Nifu (which must pay tribute to the free State of Liberia)!

Not a breath of wind ruffles the surface of the leaden sea; not a cloud is seen in the azure sky. Circling the masts, a few seagulls proclaim the closeness of land with their gravely cries. Schools of flying fish flee before the approaching ship. How easily they glide across the silent plain of water.

A few days later we crossed the equator, called in sailor's language "the line," and participated in the frivolity of that event. All the first-timers crossing the line had to be "baptized," and buckets full of baptismal water drawn from the ocean were poured upon them. We informed some of the truly naive first-timers that the "line" actually existed. By attaching a single strand of hair across the wide lenses of the binoculars, these simple persons stared through them in wonder, later swearing up and down that they had, in fact, seen the equator! The day was celebrated with drinks from a huge pineapple bowl. We ladies retired early while the gentlemen stayed up until the wee hours of the morning chatting and jesting with one another.

Gradually, the sultry, oppressive heat began to subside and rain showers became fewer and fewer. With the advent of the cooler temperatures, we slowly became human again—as well as curious as to which costal town we should next visit.

5. Mossamedes

Mossamedes is the capital of the Portuguese colony of Angola, service to which has only recently been included in the route of the Woermann Line. On May 29th, we arrived. The sea was marvelously blue, and schools of fish surrounded the ship. Soon the rope ladder had been let down and fastened in place, the doctor and the agent of the Woermann Line had come on board, and the health certificate issued. Cheering loudly, we rushed to the boats longing for land under our feet again.

Not much was to be seen in the harbor. The black landing bridge protruded into the deep blue sea, and directly in front of us lay the town. The southern part of Mossamedes is built on cliffs and strongly fortified. The mighty fortress reaches into the sea itself, and below the fortress lies the wreckage of a Portuguese naval vessel. A throng of curious people had assembled on the beach to watch the arrival of the steamship. Growing along the beach is a row of beautiful coconut palms. These, in contrast to the ones in Las Palmas, appear fresh and green and are a magnificent accent to an otherwise monotonous stretch of coastline. All of the simple white houses have their fronts to the sea, and here and there, someone has made a modest attempt to cultivate a bit of a garden in the sand—a few blooming rhododendron bushes, climbing clematis, and flowering pomegranates, all thickly covered with dust. That is the layout of Mossamedes, one main street with a few small lanes and alleys branching off of it. Behind this strip of buildings stretches an endless desert, without a single bush or tree, just fine yellow sand and a deep blue sky as far as the eye can see. Small wonder that water is in short supply in Mossamedes and that what they have of it is pumped by motors from deep below the sand.

As typical Germans, we strolled along the main street hoping to find an establishment that could be identified as an inn, but our search was for naught. At the very end of the street we chanced upon a small shop that offered bananas for sale. We rummaged through the merchandise searching for something unusual, but were again frustrated. Communication was also difficult as the owner only knew Portuguese (something that proved frustrating for both sides), and in the end we were unable to find the name of a restaurant. On we strolled and came to a pharmacy where I was able to ask, in my perfect French, for directions to a restaurant. The owner, a tiny, yellow, dried-up man, was most gallant and had his black servant escort us, at long last, to the desired restaurant. There we ordered dinner, including coffee and wine, for 2800 Reis. To the uninitiated this price sounds horrendous, but since in Angola, as everywhere in Portugal, a rather humorous grandeur pervades the monetary system, we soon calmed down. In all, it was the equivalent of about seven marks in German currency; and since the drinks were included in the price, I have no idea how these people generate a profit.

While waiting for our food to be prepared, we decided to see the other side of town. It was a weekday, yet the streets were curiously deserted, but the people who were about looked as if they were dressed in their Sunday best. The Angolan Negroes made an especially fine impression. We eventually discovered that it was the Feast of Corpus Christi and were able to observe the procession as it departed the church. Leading the procession was a rather degenerate looking Portuguese man who shot a rocket into the air every five steps. Without any color, the thing exploded in the air with no visible effect whatsoever. I had no idea of its purpose, but following him came the short, yellow priests and then the officers. Their uniforms bristled with gold, and on each man's chest hung countless decorations and honors; their hands were enclosed in cotton gloves and they carried their helmets on the handles of their swords. The native soldiers came next, dressed in khaki uniforms with a red fez on their sidearms, followed by the high society of Mossamedes, and finally the black inhabitants.

At the botanical gardens we admired the beautiful flowers, shrubs, cedars, magnolias, cacti, orchids, and different varieties of palm trees, as well as other species of tree completely unknown to us. For all that, however, the gardens were dusty and forlorn. The water in the fountains was green and muddy, and the pathways were covered in weeds. Indeed, as with the whole of the town, the gardens were sadly neglected. Beyond the coastal regions, Angola is said to be paradisiacally beautiful. There is a railroad with inland connections, but on that day no trains were scheduled. We returned to the hotel where an inviting table had been laid in the manner customary of the Portuguese. At each of our places stood a decanter of country wine and on each of our plates steamed a freshly baked roll as large as a child's head. In place of napkins there were terry washcloths. The silverware was awful. Made of dark black steel, the forks had well-used, bent prongs, while the knives were calamitously notched.

A corpulent black lady with a red sash around her capacious waist served the meal. She waddled into the room followed by her three children. The food was good—provided one was accustomed to the oily taste. Ten different kinds of food were served, eight of them fish, and one of the others, a huge bowl filled with spinach heaped into a tower. Something gray, looking curiously like ill-prepared cutlets, was placed all around the table. It turned out to be the sixth fish dish. The chicken was excellent, and we attempted to ask for a second helping, gesticulating our desire in various ways, but to no avail. Finally, we pointed to our empty plates and crowed, "cock-a-doodle-doo," but the answer was a kind of baring of the teeth. The table was quickly cleared away. Drinking the dry red wine put us in good spirits. As there was a piano in the establishment, before long we were dancing a fine German waltz, enjoying our merriment to such a degree that we failed to hear the shots fired by the *Eduard* summoning us back to the ship. The captain, tired of waiting, finally dispatched his First Officer to fetch us. He had a most solemn face, but it made no impression on us, only to remind us of the traditional German children's verses:

> Master sent Jochen to cut the grain,
> But Jochen did not cut the grain
> Neither returned he home again . . .

We locked the door so that the officer had to stay and drink and dance with us while the *Eduard* continued blowing her whistle. At ten o'clock, the captain himself appeared. At first he pretended to be annoyed, but immediately gave up when he realized nothing was to be done! By eleven o'clock, we reluctantly obliged ourselves to return to the ship. 'Twas a pity we could not have stayed longer!

The entire sea was shining as though it were spraying masses of sparks. With each dip of the oar, silvery drops splashed about. I kept my hand in the water and it looked as though I were spilling hundreds of thousands of sparkling diamonds. The boat's wake was a brilliantly white stripe, and the bodies of individual fish were clearly distinguishable. Thus we rowed through the clear, starry, tropical night. I felt nostalgic but could not determine why. Was it the buoyant mood that had enveloped everyone, or the singing of familiar songs from our faraway home? I should like to have been alone at this moment, far away from the others whose gaiety now annoyed me.

Soon the colossal steamship with its luminous red-green lanterns on the masts and flickering bull's eyes floated before us. On board, the good times continued as the silence of the night was broken by music, singing, and glasses clinking together. I somehow gathered the strength to shake my doleful, nostalgic mood and joined the others.

A group of Portuguese had arrived on board to visit the captain. Champagne corks popped at every table. Although we were all complete strangers, toasts were offered on every side in very poor French. As these guests felt so very much at home with us, we had difficulty in saying good night. Only after the captain switched off the electric lights did our guests begin to depart, all the while swearing eternal friendship and hugging several of our gentlemen who stood along their path. For a long time, they called and shouted from their boats until it sounded like nothing more than a weak echo. "Vive l'Allemagne!" they called, and we replied until we burst out laughing, "Vive le Portugal!"

The anchors were lifted, the flag was lowered, and finally peace returned. I lay on deck and could not sleep for watching the stars that shone far brighter than I had ever seen them at our latitude. An endless peace lay across the dark, silent, watery plain, and on this quiet tropical night I felt the ocean speaking to the depths of my soul more strongly than ever before.

6. Port Alexander

Two days later, very early in the morning, we saw a gray strip of land on the horizon above the blue ocean. It was Port Negro. The weather had changed and the waves were very high. Port Negro does not have a harbor, and thus the few bits of freight the *Eduard* carried for Port Negro could not be unloaded. As we moved closer, we were able to make out tiny rows of wretched houses. Atop the larger ones, probably official buildings, flew the Portuguese flag.

Suddenly, several colorful little flags appeared on a flagpole signaling that landing was impossible on account of the surf and that a lighter would be sent to the next closest harbor to fetch the freight items. This was fine as we were already scheduled to stop at that harbor. Onward steamed the *Eduard* and the outline of the coast gradually disappeared. The rocks grew smaller and smaller until everything disappeared in a bluish fog. Only a few birds followed in our wake.

The next day we arrived at Port Alexander, the southern border port of Angola. Recently, especially in British colonial newspapers, the name has appeared repeatedly—though, undoubtedly, half the readers had no idea of Port

Alexander's location or of its future significance. Port Alexander is one of the most beautiful harbors on the southwestern coast of Africa. Even the largest ships can anchor there, and it is not at all difficult for passengers to disembark. For this reason, the port is being prepared for inland trade. A segment of the Cape to Cairo Railroad, branching from the main line to Port Alexander, is in the planning stages. This way freight could be unloaded more quickly and, conversely, the many products from the inland regions could be shipped quickly and easily to the sea. At this point in time, however, nothing has materialized primarily due to the indolence of the Portugese—and their lack of both capital and a proper sense of enterprise. If only Angola were British! Then Port Alexander would be a hive of activity within a short period of time. But we must be patient, for Portugal is reported to be so heavily indebted to England (according to a myriad voices) that it will only be a matter of time before Portugal has to cede Angola to Britain in order to settle its debts.

South Africa is a British possession, as is the Walfisch Bay (Bay of Whales), thus the British possess the key to southwestern Africa and could make good use of Angola. It is unfortunate that we Germans have wedged ourselves between the two. If only we had our own good harbor in South West Africa! If only Port Alexander were situated in South West! Were this the case, we would have a splendid natural harbor instead of the sandy mole of Swakopmund. We could look down on Walfisch Bay with a smile, instead of envy, as we now do, bothered by its proximity. Unfortunately, it is not so, and we can only hope that Portugal will find the ways and means to resolve its financial difficulties with England. May Port Alexander remain as it is for a long time: a small harbor where every three months a Portuguese ship calls, for it is a grand occasion when a steamship arrives in port.

The sea was wondrously blue in the crescent-shaped harbor. On the narrow promontories situated at each end of the harbor, thousands of flamingos found their rest. At first we were uncertain as to what this endless, and oft times moving, black stripe could be. But after studying it through our binoculars, we discovered it was a huge gathering of birds, appearing to us as if they were standing in ranks on parade. Port Alexander has no more than ten houses, a few native huts, and several huge thatched-roofed sheds and barns, erected on stilts. It is said that the harbor is one of the richest fishing grounds on the western coast of Africa; indeed, a strong breeze coming from land convinced us of this fact, carrying, as it did, an intense odor of fish and oil. The entire white and native populations live from fishing. It was early in the morning, yet thirty or so fishing boats were already in process of pulling in their nets. A Belgian fishing company has leased the entire harbor in order to take advantage of its rich marine life. With the help of a large iron jetty, the *Eduard* was able to berth with ease. The company owns all the barns and sheds, and makes living and breathing—indeed, the very existence of Port Alexander—possible.

We were greeted by a handsome government vessel, rowed by native sailors wearing clean, white suits. Due to the habitual leisurely pace of the Portuguese, we were to remain in port for thirty-six hours. When we asked the captain to allow us to go ashore, he consented, and immediately the Kru-Boys rowed us to the beach, with my special friend, the corpulent "Glasgow," at the rudder. He was dressed in his festive, red-silken, checkered, and baggy breeches. The beach was so flat that the boat had to be landed in the water quite some distance from the actual shore. The Kru-Boys jumped in, took us on their shoulders, and trotted away, carrying us to land. Of course we got wet in the process, and if the boys dislike the persons on their shoulders, they will drop them into the water. Smiling broadly, my friend Glasgow grabbed me and carried me safely to *terra firma.*

Once on shore, we nearly fell ill from the smell of fish. Everywhere on the ground lay fish skeletons, bones, whale flippers and vertebrae, rotting fish shimmering like mother of pearl, large and small fish scales—everywhere it was just fish, fish, and more fish! On top of this, the landscape appears godforsaken, hopelessly dreary. Nothing at all on which the eye can concentrate—not a single tree, not even the tiniest blade of green grass on the endless plain. Nothing but blue sky and dreary sand. During the whole of their lives the local population eats fish. Their entire existence depends on fish. And there sat natives, in their sheds, preparing the morning's catch. They worked with unbelievable speed and simplicity. Squatting in the sand in front of a board, they took freshly caught fish from the piles next to them, cut them open and threw them onto the next pile. Women spread the prepared fish on the low thatched roofs of the sheds where the sun dries them and where they are allowed to stink to high heaven. (The odor changes according to the variety of fish and the weather.) We visited one of the sheds in which the dried fish are stored. In this huge building, the fish were set out in countless rows, then piled on top of one another until they reached the roof—just as they were in all the other sheds. It was most interesting, though. With all this work accomplished, were it not for the dreadful stench, I suppose one of the native men might even have enjoyed smoking a fine, new cigar!

The settlement also boasted a small general store where the men of the place gather to drink their rose liqueur amid the stench of fish and sand. I cannot even comprehend how anyone could possibly live there. To this day my heart still shudders when I think of that hopeless isolation. We also met the physician of that settlement. He was a grand old man who presented me a rosy flamingo wing and several dozen rare and valuable muscles that had "eyes." I reacted happily and thanked him profusely, and have kept these items, though I have no idea of their worth.

Just then, a camel driver appeared, coming out of the desert with his pitifully emaciated animal. The beast seemed to have journeyed from afar because its master collapsed from exhaustion. The load on the camel's back was interesting and consisted of a waterskin, a curious saddle with massive pockets, and many strange little sacks and packages. The driver was dressed in colorful rags and carried a long thin stick in his hand. His skin was shiny, as though lacquered black; his eyes, also black and deeply set in his forehead, were flaming, melancholy, and symbolic of all the misery typical of this region.

The natives of this area look quite strange to me. They dress in knee-length frocks made of a curious cloth resembling soft, flaky sackcloth, interwoven with red stripes. I was able to converse with the physician in French, and he told me the material was imported from Belgium by the fishing company and is specially manufactured for the Port Alexander Negroes. The natives' hair looks like a ball of wool turned completely grey from the sand. They are taciturn and almost consider it a punishment when someone asks them a question and are thus compelled to answer. Strangely, there are many albinos among them. One such child startled me. God Almighty, what a sight! The entire body is white, almost pink, and covered with bright red lichen, while the hair is snowy-white and woolly, the eyes are red. The children have angular heads, bloated stomachs, and spindly legs.

Most of the natives of Port Alexander are not Christians and pray to a wooden figure placed in the corner of their huts. The figure is as large as a four-year-old child, and as the figure ages, it becomes more potent as to its ability to perform wonders. If a member of the family is ill, a nail is hammered into the figure, the size of the nail depending on the kind of illness. Through this exercise, the figure frees the afflicted person from his or her illness and receives it unto itself. Apparently, these figures have become extremely rare and museums are willing to pay extraordinarily high prices for them. Most of the ones seen in the museums are said to be imitations. The Ethnological Museum in Berlin has several fine (and genuine) figures on display.

We had very much hoped to see one such figure—and even more, to get our hands on one. We rummaged through every hut, and the doctor even led us inland a way to search for one. Finally, we arrived at a lonely hut situated between two huge sand dunes. Bending low, we entered through a small doorway. Leaning against the wall in a corner of the house stood our much-searched-for figure, a nearly shapeless wooden doll with a roughly carved face so filled with nails that it looked like a rusty porcupine. A while later the owners of the hut appeared. With the doctor acting as interpreter, we told them that we very much wished to have the figure. They began howling in a terrible rage, shaking their fists, rolling their eyes, and standing in front of the figure to protect it. This performance, however, did not last long; for the moment one of our men took a shining ten mark gold piece out of his pocket and offered it to them, the shrieking stopped. There was greed in their eyes and the shining gold performed its age-old magic.

The native men were the first to begin bargaining, and when a second gold piece appeared their will power weakened entirely. In this region as well, gold—the most powerful of all idols—was victorious. The oldest man seized the twenty marks from us and rushed to the corner where the figure stood, but this time his intention was not to defend it against the foreign invaders, but to hand it over. At this, the women became hysterical. They fell upon the man, wrested the figure from him, and attempted to hold it fast in their arms. They tore their arms bloody on the innumerable nails, screaming and howling as they did so. Pleading with the figure for help, they implored it to demonstrate its miraculous powers by sending all the illnesses contained within itself to slay the sinners on the spot. Fortunately for us, the figure remained stiff and silent. Its rusty nails became wet from tears and red from blood, and at this point we would happily have rescinded our offer, but the old chap insisted. Once the desire for gold awakened within him, there was no changing his mind. Pushing back the women, he handed us the figure. I shuddered with horror and was deeply touched by the grief of the women, and would gladly have abandoned the figure they loved so dearly, but the head of the family was adamant. In a rather depressed mood, we took our leave, carrying off the wooden figure.

We decided to inspect the local watering place—a disgusting, slimy, muddy hole. While some of the locals washed their clothes (or what they took for clothing) in it, others drew their drinking water from it. After two hours of exploring, we had seen enough and signaled for the boat to come fetch us. Waiting for it to arrive, we decided to play some children's games. The tide was coming in, so the waves were the Boers and we were the British courageously running towards them across the wet, deep sand. But as soon as the waves threatened to catch us, we turned round and swiftly retreated, calling at the top of our lungs, "The Boers are coming! The Boers are coming!" while attempting to reach dry land before they captured us.

We were happy to be aboard ship again, and placing the wooden figure at the center of the dining hall, we celebrated its arrival with champagne. Yet at the same time, somewhere out in that desert, several bereaved women were probably sitting in their hut, staring blankly into the corner where it had stood, toward the now vanished wood figure that had patiently accepted all the pains and illnesses they had nailed into it. For quite some time I was haunted by these thoughts.

All night long the dreadful fish odor caused us discomfort. Some of our sailors had been fishing during their free time and had caught enough for everyone on board to have a meal; but many of us, myself included, were unable to take even a little bite. During the evening, the upper crust of Port Alexander came aboard playing their mandolins and drinking to excess. Soon our captain had had his fill and dismissed them, an action they did not appear to resent in the least. To their credit, we must assume that in their godforsaken desert they have gradually lost all sense of propriety, even when they have had too much to drink!

Finally, we put Port Alexander behind us and with impatience looked forward to the next few days. At last we would arrive at our final destination! Life on board became dull, the food had grown worse, and we longed for solid ground under our feet. Strangely, upon leaving Port Alexander the weather turned bad. The world was gray on gray: the sky was gray, the ocean a greenish-gray, and the festering smoke rising from the chimneys was a darkened gray. Waves as high as a house, bearing white frothy crowns, beat angrily against the ship's sides. Often, they washed across the deck, sweeping away everything that was not securely fastened down. They even tore away several air vents. It was difficult to walk or even to stand, and the laws of reasonable equilibrium seemed only a hypothesis.

Despite the supports, our tables were swept empty several times and sitting comfortably at the table in chairs bolted to the floor was not always possible. At times it was so bad that we were not even allowed to be on deck and the portholes were kept tightly closed to prevent the raging seawater from entering. *Eduard* rolled and the ship's propellers spun noisily in the air. A goodly number of the passengers became seasick again, and Neptune enjoyed many a quick sacrifice. You can imagine how we longed to arrive at Swakopmund.

It was three days before the storm subsided, but a fog so thick and impenetrable that we could not even see our hands held before our eyes settled upon us. The siren howled incessantly, and the captain feared we might sail past Swakopmund. Then, suddenly, the fog lifted, and before us lay the town of Swakopmund, bathed in the loveliest sunshine. We were out of our minds with delight. We embraced, danced, screamed, and laughed. Of course, we all wished to disembark immediately, but Swakopmund signaled that landing was impossible due to the rough surf, whose roaring and thundering we could hear from the ship.

At that time, the mole (a breakwater or jetty made of stone) had not yet been completed, and we still landed the old-fashioned way—with the help of the Kru-Boys and small landing craft. But even before the mole was completed, many people doubted its eventual usefulness and the contractors were blamed for numerous oversights and mistakes.

7. Swakopmund

Under brilliant sunshine, the settlement of Swakopmund appeared friendly, full of promise. Our steamer was welcomed by brightly colored flags that fluttered on many a house and contrasted with the darkness of the sand dunes. It was with happy anticipation that we awaited our removal to the shore, though the wait proved to be a long one. Standing at the railing, we looked at our land of promise with warm, evocative feelings: Over there, your new home beckons you! Will you find happiness there? What lies in store for you?

The day passed and the seas grew increasingly rough. That evening we celebrated our impending departure from the ship late into the night. Morning came. A heavy fog settled along the coast and taxed our patience. By noontime, the heavens took pity on us by driving away the fog and calming the waves. Swakopmund sparkled under the warm, glittering sunlight, and we prepared to disembark.

Our belongings were quickly packed, and we readied ourselves for landing. Dressed in our oldest clothes, as we fully expected to be soaked by the spray of the breakers, we took a cordial leave from the captain and the officers, and waited our turn to be offloaded in the most primitive manner into the landing craft. These boats rocked mightily in the waves below. One after another, we were set in a basket that dangled from a chain as a crane lifted and then lowered us into the boat. A Kru-Boy stood ready to pull us from the basket, after which it was drawn up for the next passenger.

I must admit that I have experienced far more pleasant moments than hovering between heaven and ocean in a creaking basket that landed with a jolt inside a small boat tipping headily fore and aft beside the *Eduard*. Hardly had we taken our seats when a warning shot was fired from the coast signaling that no further boats would be allowed to bring passengers ashore. We were hoisted back on ship and had the pleasure of waiting another two hours. It was a necessary caution, however, as almost immediately the dreaded fog settled again, making all movements impossible. The boat that preceded ours nearly capsized when the crew wrongly assessed the situation of the breakers because of the dense fog. Only men who are entirely familiar with the surge can make it through it unharmed. In nearly every landing several boats will capsize; lives are lost, and many pieces of luggage, as well as bales, crates, and railroad ties are immediately buried under the silt.

Since the customs house closed at three o'clock in the afternoon, we were most anxious to get ashore. Many things had to be settled before journeying inland on the passenger train that scheduled to leave the following day. If we missed that train, we should have to wait three more days for the freight train, as well as paying twice the fare, or wait another eight days for the next passenger train. Wondrous prospects! At last the signal was received that landing could resume. Again, we were lowered in the basket, and with all our luggage, auspiciously landed on target in the boat. After the long ride across the calm sea we approached the giant breaking waves. I admire the skill of the Kru-Boys who plunged their oars with determination and cool heads into the waves and quickly moved forward. But the moment I saw the colossal breakers at close range, towering far above us and then plunging into the depths, I could not help but feel concerned. So, I thought, we were to land over there on the shore, but would this tiny craft survive a battle with the elements? I cannot say that I was exactly scared, but I somehow felt that peculiar tingling I always experience in moments of great danger, when all my nerves must function correctly in order that my powers of reason can remain clear. I was to have that same peculiar feeling many times in my life.

The tension and excitement were so piqued that I had no time to be scared. Like an arrow, we shot along the breakers until, with a mighty thud, the front end of the boat slid onto the beach just as the boat's stern was lifted up high into the air. Several natives came running from the shore, and before I knew what was happening, one of them loaded me on his back and swiftly carried me to dry land.

Now that I was on land, I looked about for my husband. He was far too tall and heavy to be carried by one man alone. Thus, several natives fought over this "sweet load" (almost one hundred kilograms), and since not one of them was willing to forfeit the handsome reward paid once they arrived on the land, they pushed and shoved one another until my husband was finally dropped into the surf and disappeared under the water. Soon, however, he emerged, shook himself like a dog, and eventually dried out in the sun.

There was simply no time to change clothes since we had to declare all our belongings at the customs office. In those days, this was quite a cumbersome procedure in South West. During the German-Herero War, for example, all officers arriving on shore had their personal weapons impounded. They were only returned after days of negotiations and, as far as I can remember, only after Colonel Dürer had arranged free entry for them. Hence, all our belongings were given a most thorough inspection upon arrival, though I must admit that we passed through customs quickly enough and that the customs inspector treated us most politely. Indeed, his kindness knew no bounds after I offered to help "lick" the customs declarations to expedite this official procedure. Before long, all the proper stickers were resplendently displayed on each piece of luggage. Even so, the entire procedure took nearly two hours to complete. Afterwards, we waded through the deep sand to our hotel, the Fürst Bismarck, which does not in any manner resemble the splendid edifice found there today.

In those days the hotel was nothing more than long, black, wooden barracks. To access the guest rooms, we were required to traipse across a filthy courtyard where several grunting, squealing animals ran wild, continue past a manure pile, and dodge the clotheslines on which the wet, dripping laundry was hung. It was unsightly, but it was practical. Located inside the wooden barracks were the tiny hovels—our rooms—laid out in a row. Each hovel contained two camp beds, an inverted crate with a tiny bowl on top, serving as a wash basin, and a towel. The floor was genuine, bottomless dune sand. The bedding, the crate, the basin, in fact everything in the hovel, was covered with a thin layer of sand. And the tariff per night: twenty German marks.

Since we had a bit of time before the meal was served, we took a walk through Swakopmund. The diligence and endurance of the people who lived there had accomplished an unbelievable feat. Wind motors are constantly running, pumping water. In the bigger streets, the walkways are surfaced with stones. Many houses have a little garden in front with carefully tended flowers. The town also has a few large buildings in which various companies have located their offices: the Damara and Namaqua Trading Company, the Colonial Company for South West Africa, Erhardt & Schultz, Rascher & Thielecke, Wecke & Voigts, and many others. In some buildings, one can find second-, third-, and fourth-rate restaurants, along with the apartments of private individuals such as the harbormaster, as well as government officials serving in the colony. And last, but certainly not least, there stands a rather pretentious-looking railroad station from which the narrow-gage train departs for the interior.

Over the past few years Swakopmund has grown considerably, and since the beginning of the war (the German-Herero War) it has truly blossomed. This blossoming is due in part to the increased demands of the people living in the interior, in part to the railroad line that links Swakopmund with the mines in Otavi, and also from the heavy traffic in the harbor on account of the war. But this rapid growth has, on the other hand, brought with it some negative undercurrents. At present there are elements in Swakopmund that, to say the least, cannot be considered to belong to the upper crust of society. All sorts of shady characters and general riff raff carry on their dubious business. If one of these rogues needs to leave town in haste, within twenty minutes he can easily pass into the British territory of Walfisch Bay—with no police guard to prevent him from crossing the border. Many a crook has spent a night in the little guardhouse on the border, but recently the shack has completely disappeared.

We had a great deal to do in Swakopmund: draw money from the bank, check our luggage, purchase all the necessities, and visit a number of people.

Soon it was time for dinner. My first meal on South West African soil did not at all agree with me. The meat from an old draught ox was too tough, I never cared for prunes, and at that time I was still accustomed to proper cleanliness. From a single glance of the soiled tablecloth one could read the entire menu from the previous week!

A special evening edition of the local newspaper informed us of the peace treaty made between the Boers and the British, ending the Anglo-Boer War. Like all of the new arrivals, we were still infected by the idealization of the Boers then so common in Germany; but once we became better acquainted with the Boers ourselves, we quickly changed our minds. After dinner, a party began in the hotel with the typical heavy drinking and vociferous carousing, so we left and retired to our beds. As the evening progressed it grew so cold we simply could not sleep. Some will ask how it is possible for it to be so bitterly cold here on the coast of Africa, one of the hottest continents of all, but it is caused by a cold ocean current that comes from the Antarctic and flows north along the southwestern coast of Africa. I heard the thunder and rage of waves breaking on the shore, and the bitter cold penetrated the openings in the thin wooden walls. I missed the up and down swinging of the ship to which I had grown accustomed and restlessly tossed and turned on the damp pillows.

The waves finally awakened the six-legged tormentors, the plague of Swakopmund! They pounced on me, the bloodthirsty creatures, an entire legion of them! The next morning, as I dressed in the reddish flame of a smoking candle, I wondered if I had been infected by the measles. It was five-thirty in the morning, and without breakfast (in Swakopmund people rise quite late, even the owner of the hotel), our limbs stiff and our bodies shuddering from the cold, we tramped through the sand to the railway station. There we bought some wonderfully hot coffee and baked goods. It was still dark, but the moon shone brightly and stars flickered high up in the cold sky. Suddenly, an icy wind arose and blanketed everything in a deep, dark mist. When the signal for departure sounded, the train pulled away from the station and entered the thick gray morning.

8. The Journey to Karibib

While we stood on the front platform of the railroad car, the fog gradually began to lift, and we were able to differentiate more clearly between various objects of the landscape. To be sure, there was still little to be seen: sand and dunes, dunes and sand. Now and again rocks of varying shapes appeared, as did single bushes almost fully weighed down by a coating of sand. With difficulty, the train climbed a small rise, and to the left we discovered green bushes and trees. Our eyes saw them with great pleasure. It was the Swakop Riverbed, which, during the dry season, has no surface water at all, yet when the rainy season comes, it carries such an immense amount of water that the river becomes dangerous. At this moment it was completely dry, and we surmised that the green bushes were fed by underground water. If one knows where to dig near the bushes, one soon finds water.

South West Africa is not as dry as typically depicted. Of course, on the surface of the land very little water is to be found, but beneath the riverbeds there is plenty of water. Natives follow birds that fly to water sources morning and evening; however, if they are unable to locate open water, or if they are too thirsty to search for it, the natives carefully observe the vegetation of the area. They know very well the particular bushes and plants that allow them to assume the presence of underground water. If they find a deep riverbed they simply do not begin digging anywhere in the sandy bottom, rather, they walk along the bed until they come to a bend or rocky stretch. For they know by experience that water is dammed and stored more abundantly at such points than in places where the flow is unhindered by natural impediments. Accordingly, water is often found in these spots.

Lately, one reads accounts of some person who has died of thirst over there (South West Africa), or is lost or missing. These words imply a tragic message, though they are far more meaningful to those who truly understand them. The agony of dying from thirst is supposed to be the most horrible thing one can imagine. Nature is gracious if she shortens the time of dying. But this dreadful form of death is preceded by madness. I have known several persons who were close to dying of thirst and who gave gruesome accounts of their experience. Even my husband was once close to dying in this manner.

In the year 1897, Themis had to travel to Franzfontein in the northwestern portion of the Protectorate. At that time he owned neither a horse nor a team of oxen and cart, and had waited in vain for days and days to join a wagon train trekking there. As he had little time left, there was no other choice but to go it alone. A Hottentot accompanied him—carrying his own three-year-old child as well as his supplies. My husband carried blankets, his gun, and his water bag. They left Okombahe and went first to Ais where, at that time, many Herero had their homesteads. To give the Hottentot a rest, my husband left him there with the supplies and went on to Sorris-Sorries by himself where he intended to visit the Englishman, Mr. Sabatta, who was in the process of inspecting his cattle stations. Themis wished to borrow at least a riding ox from him, but Sabatta had no suitable animals. However, the latter assured Themis there would be plenty of water on the way to Franzfontein as that region had received abundant rains. So my husband returned, choosing the Ugab riverbed as a trail, and collected his Hottentot companion in

Ais. The Herero, to be sure, shook their heads as they had heard no news of heavy rainfall, yet my husband had little choice but to leave immediately if he wanted to reach Franzfontein in time.

They spent a dull Christmas Eve and began walking at three o'clock in the morning on Christmas Day, marching cross-country so that they could reach Dawib-Urib by four in the afternoon. Since they fully expected to find water there, the inadequate supplies they had brought with them were completely exhausted by late morning. The closer they got to Dawib-Urib, the more dubious the claim of heavy rain appeared. They searched fruitlessly for green grass and leaves—the telltale signs of recently fallen rain. When they reached the water hole it was completely dried out, and only after they had dug for an hour did they reach a bit of wet sand. There was nothing to be done. They had no bread, no water for boiling, and none for drinking. Following a short rest they continued, hoping to find water in the mountain flats; however, upon their arrival shortly after midnight, they found that the flats were completely dried out. They soothed the child as best they could. His lips were cracked and his tongue had turned brown and was also cracking. After a twenty-hour march, all three of them were not only thirsty but deathly tired as well, and quickly fell asleep.

Early the next morning, well before sunrise, they resumed their journey in the hope of reaching the next watering hole at Khorixas by evening. As long as their strength and the cool morning air sustained them, they made fairly good progress; but once the sun rose, thirst began to overtake them. Their bodies burned so that they could hardly make progress. Very slowly, they walked on. By now, the Hottentot was so exhausted that he no longer wanted to carry the child, and in desperation thought to leave him behind. My husband took pity on the child and tied him to the blankets on his back. There the child hung on him like lead, which made walking all the more difficult.

All day they walked the plain of Khorixas, but from noon to four they had to rest, for in the afternoon heat they were simply incapable of taking another step. Later that evening they reached the foothills of the mountains where Khorixas is situated, but they still had hours of walking before they would reach the waters. Desperately, the Hottentot threw himself on the ground and refused to take another step unless they rested for a while. The child hung on Themis's back, exhausted and whining. Themis stopped to mull over what he should do. The night was dark and he did not know the way. His Christian sense of duty, of which he was reminded at this crucial moment, did not permit him to leave the Hottentot behind. Still, thirst was becoming unbearable, hunger agonizing, weariness overwhelming. Themis reasoned with his companion, "If I had a balance and placed thirst on the one scale and weariness on the other, tell me which one would be the heavier?" The Hottentot groaned,

"Sir, both scales are equally heavy. But remember, our bodies require rest, and we can rest here and now. Our body screams for water, but we must walk many more hours before we get it, so let us rest!" They spent the night at the foot of the mountains, and what reality had denied them, a merciful dream offered them while they slept. Themis had a beautiful dream. He dreamt that he was bedded on fresh grass in a forest next to a fountain and that he had only to open his mouth to drink the clear, precious water, only to move himself toward the water to cool his burning body. Next, he saw himself sitting at a richly set table in the house of his beloved father. The glasses clinked, the pineapple bowl was inexhaustibly filled. . . .

By the time they awoke, the sun stood high in the hot summer sky. Their bones rattled with fever, their feet were bleeding, and their words were but a quiet, hoarse croaking. The child was whimpering. The Hottentot had to be pushed forward with the stock of a rifle. But he threw himself on the ground, clutched Themis's knees, and pleaded with him to have mercy and shoot him, to put an end to this misery. My husband moved on with the child who from time to time was unconscious but now and again began to struggle. Themis could no longer think clearly. He saw red in front of his eyes and continued his mechanical steps, instinctively holding onto but one thought: Carry on until we reach water or all will be lost.

Finally, when the sun had reached its highest point in the sky, they arrived at the water hole. Shouting with joy—as if they had been given the rarest, most precious gift—they rushed toward the disgusting hole with its cloudy, muddy water. They did not take time to scoop it up; they simply lay down and greedily slurped the shadowy liquid. But their stomachs revolted. Themis took three drinks; three times he regurgitated. Only the fourth drink stayed with him. As soon as they had they drunk their fill they collapsed and fell into a deep, dreamless sleep. It was in this state that a patrol from Franzfontein found them. The soldiers had orders to look for enemy Swartboois (this was during the Swartbooi Wars) at the water hole. They loaded my husband on one of their reserve horses, and he reached Franzfontein safe and sound. The Hottentot arrived shortly after him. The child was very sick for a long time.

To this day, that Christmas remains unforgettable. I hope no one else ever will have such an experience; but do our brave soldiers down there have it any better? Here in Germany, we enjoy drinking beautiful, clear water, and often, when we see how this most precious of all drinks is ill regarded at home, we wish from the bottom of our hearts: If only we could take a glass of it to the brave men out there. On their knees they would thank us for it!

In this vein, our dear friend, Herr Ertmann, was once hunting with his servant when in the distance they spotted a creature running speedily around a thorn bush. At first they could not tell if it was a human being or an animal, but as they came closer they saw it was a man. He was frothing at the mouth and screaming blue murder as he continued diving headlong into the thorn bush. It was an Englishman who had lost his senses and thought the thorn bush to be a fountain. Again and again, he dove into the bush, cheering and beating himself angrily, screaming as he ignored the thorns and his own warm blood that streamed down his face, neck, and chest. Ertmann tried to help by offering him water, but the Englishman thought Ertmann was an enemy determined to prevent him from getting to the fountain. Like an animal provoked, he fought off Ertmann with superhuman strength. Finally, when the Englishman had completely exhausted himself and lay lifeless in the bush, Ertmann succeeded in giving him a few drops of water. Alas, it was too late. He fought desperately against an assumed enemy until cruel fate ended his suffering. They buried him in the nearest settlement beside other persons who had likewise died of thirst.

Many of our own brave warriors face the same fate, dying miserably because they do not know where they are. Many of them are found dead within a quarter hour's walk of water, yet they knew nothing of the fountain's existence. Comparatively few natives die of thirst. Of course, I am talking about normal times when they do not have to flee from an enemy and hide in the most desolate, inhospitable areas. If a native cannot find water, he digs up pulpy roots and plants that will quench his thirst. It is extremely difficult for a white man to learn which plants these are and where they can be found, for the native does not divulge his secrets. He disappears suddenly and then reappears with the plant in question. We Germans have to be on our guard, keep our eyes open, watch and observe. The soldier in South West must learn a good many things that no instruction manual and no noncommissioned officer can teach him.

Our journey continued, and as the sky became lighter and lighter, we recognized the characteristic bushes of the Namib Desert. Joyfully we greeted the first hackidorns. We saw great quantities of small, golden-yellow pumpkins growing in long stretches beside the track. They were strange looking things, resembling yellow balls with which children might have played before thoughtlessly tossing them here or there. The vines quickly wither away, leaving the fruits lying on the ground for a time. They have a bitter taste yet contain much liquid. These so-called pampusen are the salvation of the teamster and ox driver, for when no water, or even a tiny blade of grass, is to be found, the thirsty animals can eat these as a last resort. Besides being the favorite food of ostriches, they provide both nourishment and water, and thus sustain draught animals in difficult times.

Alongside the rails, we suddenly noticed a black, ghostlike, immovable object made of iron. It was an old Troost steam engine that had got stuck there years ago and now can neither be moved forward nor backward. Apparently, it has been here for a long time fully resisting any attempt to move it—and will likely remain in this position until rust destroys it or until it has been buried in sand.

Our little train (the track is only sixty centimeters wide) strained as we moved on, passing the huge, black cliff that marks the border between German territory and the British Walfisch Bay. Then all of a sudden, a steep, dark, and massive range of mountains lay in front of us: the Hanoas Mountains. Groaning, the engine wound its way through the range, and just as we reached the top of the pass, the sun rose magnificently. Immediately, the landscape was filled with light and warmth.

At noon we passed the Khan Mountains, devoid of any vegetation, but picturesque on account of their fantastically shaped rock formations. Here the tracks are laid along the bed of the Khan River. During the rainy season, it is said, when the river is swollen, floodwaters tear the rails from their sleepers and wash them downstream. When that happens, traffic to the interior grinds to a halt for days and even weeks. When the Otavi Railroad Line was constructed, the engineers debated whether or not they should place both tracks next to each other. Finally it was decided, as far as I know, to lay the Otavi Line tracks around the Khan Mountains. Although it was far more expensive to do so, the danger of having the tracks washed away by the raging river was averted. About two o'clock in the afternoon we reached Jakalswater where the train stopped for the long-awaited rest. We were served a fairly good meal that had been ordered beforehand via telegraph. Gradually, the traveling became less and less enjoyable. It was extremely hot, our eyes burned, we were thirsty and unable to find relief, the engine noise was extreme, and we were shaken back and forth to such a degree that we were in constant danger of landing on someone else's lap. Thus it continued for hours on end.

The whole of the area we traversed reminded us of the kind of high mountains—with their canyons, rocks, and ravines—that sane men dare not enter. We passed a flock of ostriches, hordes of springbok, and a family of baboons sitting together on a rock. How quickly the African night pressed upon us with its purple glow. Stars glimmered brightly and seemed much closer and much larger than they were at home. Shortly after nine o'clock in the evening, we arrived in Karibib, the last stop of the day. We were glad to have this exhausting journey behind us. Hungry and thirsty, covered with dust, feeling absolutely shattered, and my ears ringing inside my head, how happy I was to rest my weary body on a narrow bed in the Hotel Khan.

9. Karibib

Karibib is the single most important town along the Swakopmund-Windhuk Line. It was settled, then quickly expanded. Until 1899, the area was a complete wilderness. Neatly tucked away beneath tall camelthorn trees stood a lone farmhouse belonging to Herr Hälbich, former craftsman of the local mission. His son currently resides in the house, and all the land roundabout Karibib, stretching for kilometers in every direction, belongs to this farm. Years ago the old Herr Hälbich purchased this huge tract from the Herero for almost nothing. The natives had been heavily indebted to Hälbich, and instead of paying with money, the Herero captain, Zacharias of Otjimbingue, paid his own debt, and those of his black brethren, with this enormous piece of land.

Towards the end of 1899, my husband had joined a large Boer trek consisting of fourteen wagons that had come from Grootfontein in the north and had been on the road for four hard weeks. Because of the numerous herds of game in the area around Karibib, they camped next to the marshland for ten days and hunted. Natives had also brought the news that the steamships that were supposed to deliver freight had not yet arrived in Swakopmund. Such hunting camps are very interesting. I have experienced them several times. But in the camp of 1899, a group of twenty white men, mostly Boers, and a native company of about fifty load bearers from different tribes (Hottentots, Mountain Damaras, Hereros), hunted together. They had five horses, six dogs, and something in the order of two hundred seventy-five trek oxen. Well before sunrise the hunters were up. Several natives and two or three Boers stayed behind to guard the camp and the animals. The hunters quickly made their way into the bush while Mother Nature slumbered. Only now and again could the sleepy but malicious call of a mocking bird, its Narr! Narr! be heard from the bushes growing along the narrow rivers.

Quickly, but quietly, the hunters hurried through the thicket, crossing the cliffs and ravines of the mountain range before reaching the area where the game was grazing. The various herds were clearly visible on the horizon. The hunters then surrounded a very large area and dispersed in several different directions in an attempt to encircle at least two of the herds. The springboks, which seemed to have gotten wind of them from one side, galloped away to the other side and were immediately shot. Several animals fell to the earth, while the rest, obviously frightened, fled in a different direction. Yet from a distance, the animals' every movement was carefully monitored by the hunters, and soon their path of escape was blocked. Shots were fired and several more animals were hit. At this point, as is typically the case, the herd dispersed, and in a state of panic, the antelope broke through the hunters' line.

Obviously this kind of hunt can only take place on a savanna sufficiently large to create a tight circle of riflemen. Both my husband and I strongly oppose such mass slaughter. Hunts undertaken in accordance with civilized hunting practices are far more exemplary, humane, and even beautiful. However, if one has to provide food for many people whose provisions are completely exhausted, and for whom other kinds of food are simply unavailable, "one must act according to need, not one's own desires," and assent to this type of hunting. Much of the meat from hunted animals is cut into strips, salted, and hung to dry in the branches of trees around the camp. Meat prepared in this manner can be kept for months and is called biltong.

Biltong played an important role in our farm life as well. Firstly, it is needed for the hirelings, for whom a handful of dried meat is a valuable edition to their daily rations of flour or rice. Besides this, biltong is prepared and eaten by us. The strips of meat are ground between two stones, mixed with rice, and flavored with cayenne pepper or curry for a meat pudding. It can also be added to vegetables and pumpkin cubes, in place of potatoes, to make soup. Meat dumplings prepared from this ground meat powder are fairly good—especially when nothing better is available.

When it was decided that the railroad from Swakopmund to Windhuk should be constructed via Karibib, a few speculators purchased construction sites from the older Hälbich. Hotels, bars, stores, and even a brewery shot up like mushrooms from the ground; an office building, houses for the officials, a railroad station with electric lights, and a modern hospital—all able to satisfy current demands—were also built. Where once antelopes and ostriches roamed, now the whistle of the locomotive is heard and thriving businesses and craft enterprises have sprung up. The game has moved to surrounding areas not yet affected by civilization (interestingly, the poetry of that era has likewise shifted to the wilderness).

Karibib is important for other reasons as well. The main road to the north leads from there via Omaruru to Outjo, to the Otavi Mines, and finally to Grootfontein. Because of this, Karibib rapidly blossomed as all trade and freight traffic to and from the north of the colony begins here. Now, when I describe South West African roads one must not think of a German country lane. Most of the roads here are only simple paths created by game animals, mainly zebra, but are now used by human beings. To ride or drive on them is dangerous, and for years the teamsters of ox carts have tried to follow in the ruts of their predecessors, thinking, perhaps optimistically, that where they made their way through, we will do likewise!

The next morning I had a thorough look around Karibib. The little houses are closely jammed together so that everybody can easily observe the comings and goings of his neighbor. Gossip abounds! People also seem to drink heavily, and the racket on the square at night would probably not be permitted in Berlin. The inexperienced newcomer often falls prey to the competition between hotels and stores, and I noticed more than a few dubious-looking characters. One meets the so-called prospectors, many of whom convey an impression of depravity with their shabby clothes, red shiny noses, and bloated faces. Some kind of a worthless stone has been knotted into their filthy handkerchiefs as they cast about looking for a victim. When they have found one—and that happens frequently as there are many "who haven't their wits about them"—they take him into their confidence. In broken English these hucksters say, "Look at this stone, an excellent stone. And nobody but me knows the place where they are found! The only thing I want is to form a partnership with someone who has just a bit of money. I assure you, a hugely profitable business is at hand, etc., etc."

If the contact is naive or overly ambitious, he is quickly persuaded, accompanies the rogue, and then is thoroughly fleeced. We met one such dull-witted devil who, in addition to losing his money, had been given a piece of cynical advice by the prospector: "Now you've seen how easily one can make money in this place, go thou and do likewise."

These prospectors also visit farms and proceed in much the same way. They introduce themselves by asking if they might be allowed to prospect on the farmer's land—without pay, of course. If the farmer falls for it and gives permission, they often lie around for months, enjoying the farmer's food and hospitality. Occasionally they disappear with an air of importance, then return with even greater pomposity. If one asks where they have been, they mysteriously answer, "Why, I have been out prospecting"—when the simple truth is they were simply sleeping under a bush.

There is another curious sort of town charlatan who summons an unsuspecting person—always in the evening—with a rather mysterious invitation. If that person appears at the appointed time and place, he says: "I have heard you wish to buy provisions. I can get them for you at a much lower cost, though you must pay in advance to be able to do so. Early tomorrow morning I'll meet you at such and such place." If he makes the payment, he loses his money. He either waits in vain the next morning or is given stolen provisions or ones that "were no longer needed" by some kind of general store. Involuntarily, such a person becomes a criminal since he dares not to denounce the rogue for fear of the village gossips and the possibility that the crook might seek revenge. They count on this, and once they have a person in their power, they squeeze him for all he is worth.

One storeowner wanted to charge us 6 percent interest per month if we did not pay cash immediately. He thought we were rich and ought to be fleeced. Another one wished to sell us an old restored ox wagon for two thousand four hundred German marks, and a third wanted to get rid of his old, broken-down oxen. Since we were very cautious with our purchases, we were spared falling into too many traps.

From the wagon maker, Redecker, we bought a cart with all the necessary equipment, trek goods, and fourteen oxen for eleven hundred German marks. The equipment consisted of two water barrels, a front crate, tools including a hatchet, pails, a "kattel"—a kind of primitive mattress with a supporting wooden frame covered crosswise with ox hide straps on which one could rest quite comfortably. The trek goods we purchased were a chain for the oxen to pull the cart, the yokes, along with yoke wood, the leather straps needed to hitch the oxen, and a whip called a swip. We also bought food to last us several months, kitchen items, things to use in the bush, and finally selected goods for use in bartering with the natives.

Though our business transactions were complete, we were forced to stay on a while longer because our suitcases had mistakenly been shipped to Cape Nousseu. It would be difficult to take possession of them if we pressed on to our farm, but just a few days later they were delivered to us in Karibib and, surprisingly, had suffered no damage at all. Finally we were ready to depart. Our cart was packed to the hilt and beyond. Cooking pots and their tripods, along with the water kettle and the teapot, were suspended by straps attatched to the sides of the cart. The water barrels were filled to the brim and stood at the back. On both sides of the cart hung waterbags made of watertight canvas. The rifle was placed close to the front crate, which contained enamel cups, bowls and plates, tin spoons, and dreadful knives and forks made of steel. Our tightly rolled blankets lay atop it all.

We had one final need: Natives to drive the cart and to guard the oxen, and one to be our personal servant. This was rather difficult in Karibib as most of the natives living there worked for the railroad, the government, the hotels, the stores, or tradesmen. After lengthy negotiations, my husband did succeed in "borrowing" three such chaps from an acquaintance. However, he had to promise to send them back as soon as possible.

At long last, our trek began! My husband and I sat on top of the front crate, while on a large bag of rice tied to the shaft, our ox guard (whose task it was to watch the oxen at night) comfortably positioned himself. Mattheus, our driver, together with a Swartbooi-Hottentot, walked to the side of the cart. Fourteen oxen pulled our cart, and each of them had a name and recognized that name when it was called.

It is no easy task to get an ox to know its name. While still a young calf, it is tethered to a tree and struck by the whip while its name is called out. This lasts for about half an hour, and the procedure is repeated the next day, indeed, for several weeks thereafter. Then the ox is hitched to a wagon with oxen that already know their names, and is again struck while being called by its name. After this manner the beast gradually learns that as soon as its name is called it is struck. Instinctively, the creature tries to avoid the strike by pulling forward. Soon it learns that it will not be struck if it pulls; thus a well-trained ox has only to be called by name. Of course, the driver training the oxen must be patient, and above all, the animals must never be beaten lest they become insensitive to the point that the whip no longer elicits a reaction.

Later in our stay, we had an excellent and obedient team of oxen. The whip was hardly ever used. The names of the oxen are typically borrowed from the Dutch language. Some of ours were called Dunker, Bloemveldt (flowery field), Swater, Whitsand (white sand), Twelfuer (twelve o'clock), etc.

Michael, our guide and leader of the rope (reins), walked in front of the oxen to show them the way. Once the oxen are accustomed to one another they will stay on course without a guide, who thereafter only takes hold of the rope when traveling over dangerous terrain. By rope I mean the two leather straps that have been tied to the horns of the two lead oxen. They are knotted together and thus become a type of rein. The rope leader takes these reins and walks in front of the oxen with the straps in his hands or on his shoulders.

So looked our cart when on Thursday, June 12, 1902, we departed Karibib. Our acquaintances waved good-bye, yet we were only too happy to leave the town behind us. Constantly having to be on guard, always having to be careful and cautious about everything had finally gotten on our nerves. Besides that, we plagued ourselves with worries—had we done everything right, had we done everything in the most practical and economical way. Now, though, nothing could be undone. We simply had to grin and bear it and try to meet the future with confidence. To be sure, the road before us was fraught with difficulties, cold nights and periods of thirst, but these toils were only physical. We were in good spirits and looked longingly at the distant peaks of the Erongo Mountains, covered as they were in a bluish haze. Our journey would take us along the foothills of these mountains.

10. By Ox Cart from Karibib to Okombahe

Our trek would eventually take us to Okombahe, on the banks of the Omaruru River. It was one of my husband's favorite places from his previous visits, and he had both a bit of land and a small cottage there. It was our intention to settle permanently in Okombahe and to begin a farm.

At first the path wound itself along the plane. On either side of us grew various bushes with long green, gray, or white thorns, and as our cart moved along we often had to lean far back to avoid being caught by them. The clear air was blue and trembling. The Erongo Mountains remained forever in the distant haze.

If we were able to find good though somewhat dry grass around three o'clock in the afternoon, we unhitched the oxen and allowed them to feed. Because it has been completely grazed over by the local animals, there is no grass in the vicinity of Karibib. Animals simply go hungry or the owner is forced to buy poor but expensive fodder. I found it incredible to see all of this desiccated grass growing in the soil. Having only recently arrived from Germany, I quite naturally thought of our lush, verdant pastures and could hardly imagine the existence of grass growing in scanty bundles on sandy soil, resembling the absolute driest of hay—all yellow in appearance and entirely wanting for moisture—yet which serves as precious fodder for draught animals in Africa, both oxen and horses. Native horses, not the recently imported ones, get very little in the way of oats and survive primarily on dry grass.

Soon enough we hitched the oxen and began the journey anew since we wished to reach Ongoati, the next watering hole, by evening. The landscape hardly changed. It was an almost unending though thoroughly charming picture of genuine African savannah. On we trekked and finally reached Ongoati shortly after sundown. An old Englishman lives there in a tiny stone house. One frequently encounters houses of this kind. The walls consist of fieldstones that have been placed one on top of the other without any kind of mortar between them. The roof is most often constructed of small tree trunks and larger branches that are woven together by thin bushes, and then covered with clay. The roof of this particular house was nothing more than a piece of canvas from a wagon-top or a tent.

Half a dozen naked children romped and played about the front door. The old gentleman, a bit of an odd duck to say the least, had a tiny, dried-up Hottentot woman welcome us with a cup of goat's milk, but did not allow us to visit.

In the meantime, night had fallen and a wonderful moon spilled its light across the dawib trees, with their long, fine, quaking leaves. We pitched our camp under one of these trees. My husband and our driver took the oxen to water. The guard was given his orders for the night and off he went with his blankets. Michael, the rope leader, had gathered a pile of dry wood and started a crackling fire. I cannot imagine anything more picturesque: the magnificent night sky with its shining moon and myriads of stars, and the tall ghost-like trees with their long, low-hanging branches rustling back and forth in the evening breeze. The fire glowed red and the flames danced up and down. For me, it was magical—and so very, very beautiful. In the years to come, I spent innumerable nights around campfires, and they always pleased me with that same beguiling charm. I now long for the days when I might once again be privileged to see the moon shining through the hackithorns or to ride across the savannah when the camelthorn trees are in bloom.

Two three-legged cauldrons stood over the fire, one for the natives, who were cooking themselves a salty flour mush, and the other for us, also a flour mush but ours was mixed with American pork fat. Bread was also served, as well as tea in tin cups—something to which I gradually became accustomed. The evening turned bitter cold and felt doubly unpleasant after the heat of the day. We prepared our camp for the night. Two large bags were hung on one side of the camp to help keep the cold wind from blowing on us. A canvas sheet on the ground protected us against ants, scorpions, and thorns. On top of this came our countless blankets and pillows, all of which we shared. We climbed into our bed so that our feet were near the fire. On its other side, our two chaps slept so close to the flames that the blanket of the nearer one was scorched during the night.

I lay awake and allowed all the magic of the African night to work upon me. The moonlight shone through the branches of the trees. The fire crackled, and when it had nearly burned itself out, a figure rose, picturesquely clothed in a light-colored animal skin, and revived the flames. The night was so completely silent, yet it seemed as if it were alive with a thousand voices. The night wind rustled the leaves in the trees, and the small shafts and pods and grasses around us nodded and waved. A mocking bird cried, "Narr! Narr!" and owls and other birds of the night flew past us. In the far distance, jackals barked. Everything was so new and so strange I simply could not fall asleep. Finally, the moon took pity on me and slipped away; I forgot Africa and everything around me, and slept softly and sweetly till morning. My husband awakened me, and though golden sunlight now flooded over the rocky peaks, so cold was the night that he was blue from the deep chill. The chaps at fireside were trembling as well.

With the exception of our blankets, the cart was already packed, the oxen had been given their morning water, and waiting for me was a large cup of warm goat's milk—again brought to us by that same little woman. There was no time to refresh oneself properly, so on we went into the Erongo Mountains. The path was steep and sandy. The cart's wheels ground deeply into that sand, and with heads bowed, the oxen pulled hard and the cart moved forward, encouraged again and again by the driver's voice. We walked beside the cart until the heat became unbearable, for it is not strenuous to walk during the cool morning hours. I soon forgot that I had not washed myself that morning and no longer looked at my hands whose lines and creases had gradually become black. The air was unbelievably dry, so dry, in fact, that our lips and our faces and our hands cracked as they often did at home in the icy winter winds should one fail to properly dry oneself.

The Hottentot had eagle eyes and pointed out springbok spoors, then the spoors of a steenbok or a kudu. Suddenly, before I realized what was happening, I found a long whip had been laid in my hands. The Hottentot had quickly and without bother slipped it toward me as he pulled his rifle from the cart and dashed across the veld as fast as his legs could carry him. Immediately a shot rang out, and triumphantly he carried his two guinea fowl back to the cart. We were delighted to have meat for our midday meal.

The guinea hens are as large as pheasants and have exotic feathers. We hung them on our cart and drove on until our noontime rest. A steenbok was sighted among the cliffs, and again the driver ran after him, but the animal was faster and got away. After the team was unhitched, Michael plucked the hens by simply pulling off their skins, and did it very quickly. Next, he removed their insides and then put them to roasting. Though we roasted the hens for a long time, they were still as tough as shoe leather and remained so most likely because they were freshly shot. Eating them was torture, but both my husband and I commented on how delicious they were. But finally, enough was enough, and we started laughing about the situation and tossed the remainder to the driver who picked them up grinning, and enjoyed them. After eating his share, he gave the rest to Michael, who also appeared to enjoy them. We had cooked some rice in a new, unwashed iron pot. It was golden yellow in color and had a strong iron taste.

Unfortunately, we had neglected to buy a coffee grinder in Karibib, but our servants were able to grind coffee beans between two flat stones, though not without difficulty and not particularly well, so we decided to prepare tea. The water we used was stagnant and the tea left a few slimy strings between our teeth, but we drank it just the same. On we continued and soon found ourselves in the middle of the Erongo Mountains. Chipper little rock-dassies played among the cliffs. They are related to marmots and some regard their meat as a delicacy.

While passing through a kloof (canyon)—one so beautiful it might have inspired Böcklin—the oxen suddenly came to a dead stop. Nothing we did successfully urged them forward; neither shouts nor the whip had any effect on them. Standing near a thicket, they just as suddenly came back to life and began running in an absolute panic. I had great difficulty holding onto my seat. Racing across rocks and stones and through heavy thorn bushes, the oxen pulled the cart at top pace, and for a moment or two I feared it would overturn. Considering whether or not to jump off, I concluded it was better for me to sit tight, come what may. It was nearly impossible for my husband

and the driver to keep up with the oxen. When finally the oxen stopped, it was only because the cart had wedged itself between two trees. Thorns had shredded my clothing and thoroughly scratched my face and hands, yet I was still sitting tight on the box! Unfortunately, the shaft was broken in half and had to be repaired with screws; a yoke was also smashed and the trek goods were heavily damaged. One of the oxen was trapped in its harness, and since it had been pulled along by the others, it had nearly lost its life. The poor animal was so exhausted that it and its companion ox had to be unhitched at once. The following day we allowed them to walk freely behind the cart so they might recover from the tumult.

After everything was set back in order, we were, of course, more than a little curious as to the reason for the sudden flight of the oxen. Themis and the driver returned to the thicket the animals had been so reluctant to pass. There, they discovered fresh leopard tracks that ended in the thicket. Since twilight was rapidly falling and there was too little light to hunt for the animal, they decided to return to the cart and were still discussing what they should do when the Hottentot screamed as a huge leopard vaulted into the open and dashed between the two men, carrying some kind of small animal in its teeth. Apparently, the leopard had been disturbed while enjoying its meal. Instinctively, my husband pulled up his rifle, shot, but it failed to meet the leopard. It was now too dark and the shot's only effect was a loud echo in the mountains. After that, silence returned, and this great expanse of nature fell under the magical spell of the wild.

The men made their way back to the cart. In the meantime, I had had the boys unhitch the team and instructed one of them to keep and guard the animals close to the camp. The other one started a fire, and so the two were welcomed with steaming coffee. They were still quite excited from their adventure. The Hottentot stretched himself, and with sparkling eyes finally exclaimed, "The leopard is close by—I can feel it because icy cold shivers run down my spine. I always feel like that when a beast of prey is in the vicinity."

The night was cold and our simple meal consisted of coffee, cheese, and bread. We waited for the moon to rise before hitching the team to the cart. It was essential that we journey onto Ameib this night, a Herero village in the mountains, because the oxen were thirsty and had to be watered.

Peacefully before us lay Ameib, meaning the green face. For the first time I saw Herero pontoks, or houses, which looked like large gray molehills. Beautiful cattle stood inside kraals made of thorn bushes, and close to the water they had cultivated gardens with melons, pumpkins, and maize. A mission station, falling into ruins, was a witness of the transitory nature of all earthly things. Some Herero were still awake when we arrived, and I was especially regarded with particular curiosity. It had been a long time since a white woman had come to their village, and their leader, a man called Theophilus, sent me a low chair so that I could sit down, and his wife, Mary, brought plenty of soured milk, called omaere, for us to drink. In return, we gave him tobacco and sugar. Others came along as well, bearing milk cows, sheep, and goats as gifts for us. We accepted them, of course, although there were too many, and we gave them presents in return.

I was exhausted from the toil and stress of the day and was much pleased when my husband prepared our bed beneath the cart. It was far better protected there than under a tree, and the night proved to be damp, foggy, and very cold. I was frozen. But it was the month of June, and it is common knowledge that June, July, and August are the coldest months in South West Africa. For a long time my husband sat at the campfire chatting with the village chief and some other Herero. Since they knew him from his previous stay in South West, they enquired about his visit to Germany and Scotland. He told them many things, and they often expressed their astonishment loudly and appeared unwilling to believe everything he said. Exasperated, he finally said to them, "It's really no use telling you of the life and activities of the people on the other side of the salt water because your only response is that such things cannot be true." They answered, "Ii [yes, it is so], when you were back home you must have been drunk and seen everything double and threefold, and now you believe it is truly so. It is impossible that the German Emperor has as many soldiers as you claim. But keep on telling your stories, it doesn't hurt a thing if they're not true. We also like to sit together like this and swap stories; it's both pleasant and interesting to have such conversations. Does it do any harm if we lie? A Herero ought to lie to such a degree that blood spills from his tongue!"

When the icy cold wind started blowing from the East, a sign that daybreak was nearing, Themis finally took his leave of the Hereros and lay down. His rest was short-lived, for at sunrise I awoke to voices uttering the sustained, "Ii, ii, ii," the sound of truest agreement. A dozen or more Herero were already sitting round the fire, and their conversation must have been most interesting, for again and again I heard the, "Ii, ii, ii." The leopard had come to the water during the night. Its tracks led closely past the cart, and several times we were asked if we had not heard it hissing. The creature had taken a sheep from a nearby kraal and disappeared with it.

Neglecting our usual morning toilette, we quickly dressed ourselves, oversaw the packing of our cart, and stood waiting for the oxen to be brought forward. Because once we left Ameib the oxen would have to trek for two days before coming to water, we were anxious to leave early. The oxen would draw the cart during the cool of the morning, then rest through the noon hours, only pulling again once the heat of the noonday sun had abated to some degree. It was seven, eight, and finally nine o'clock, and not a single ox was to be seen. My husband, the teamster, and some of the Herero fanned out to search for them. At last they were found. Around half past eleven, the driver brought them to the cart. Sometime during the night the guard had secretly disappeared and left the oxen to themselves. Unfortunately, the animals had grazed on very poor grass, and not enough of it, and had roamed so far from camp that they were already in a state of exhaustion before they were hitched to the cart. Yet tired or not, we had to go on. During the heat of the day with these tired animals, we doubted we would reach Okombahe.

The path that led across cliffs and bushlands, through deep sand and dried-out riverbeds, grew worse and worse. There were times when the track was so steep our cart was in danger of tipping over, other times when the wheels cracked so loudly we expected the worst; strangely, everything went better than we had dared to hope. During the first part of the night we trekked in the bright moonlight; it was cold but marvelously beautiful. It was strange how the moonlight made all the objects about us appear larger. I could have sworn I saw ghostlike figures racing from bush to bush, but as we moved closer, it was the bushes themselves that eerily stirred in the night wind, or rocks whose shadows were penetrated by moonlight that cast pale images onto the sand. It was so light that my husband suddenly discovered a heavily used game path that led towards the massif of the Erongo Mountains. As he was quite familiar with the countryside, he concluded that there would be water among the rocks. We stopped, and together with the driver, he followed the path and his instincts proved correct. The water was located deep within the cliffs and had to be removed by pail, and even then, only a partially filled one—but at least we had water. Three hours and a good deal of work later, the oxen had been watered.

A Herero village was situated near the water, and the natives later told us they had heard us and figured we were in urgent need of water but chose not to contact us because they wished not to have to rouse themselves and show us to the water in the dead of night. The water hole was called the Aub Puddle. Though the natives had been most unhelpful, it did not prevent them from begging for presents and accepting them. We were so delighted to find the unexpected water that my husband pulled out his concertina (known in South West as a trek-organ). Music playing in the quiet night attracted natives who were scattered about the surrounding pasturelands, and soon a dozen or so had gathered round the fire. We finally managed to get rid of them by giving each a handful of coffee, which they bound into the ends of their leather clothing. As silently as they had come, they disappeared into the darkness. The rest of our night was disturbed only by hyenas and baboons.

In the morning, we trekked onward and I noticed we had traveled all the way around the Erongo Mountains. The peaks and hills and rocks looked quite different from this side, far grander than they do from the side from which we had come. About halfway up the slopes we saw smaller stone blocks placed like walls on top of the massive rocks. It was as if in prehistoric times a gigantic fortress had been erected by huge hands. This natural wall surrounded the entire range, always at the same height. The higher the sun rose, the closer we got to our new home. Now the double pyramid peaks of the Spitzkoppies rose mightily from the flat, dusty land that surrounded them, and soon we saw the Okombahe Mountain before us, starkly jutting upwards, shimmering and all blue—a sign that we were still many miles away from it. In the shadowy distance, a broad swath of a shiny ribbon of green wound itself through the otherwise featureless plain. It was the Omaruru River.

Thirst came to plague us during this strenuous part of the journey, but water was so scarce we were unable to even clean the harnesses that evening. As our water bags were almost empty, we were forced to ration the servants' water. We anticipated arriving in Okombahe the next day, and in the effort not to overtax our oxen, we started very early in the morning. My husband offered me a cup of tea as refreshment, but it was prepared with the last water from the barrel, water we had taken from a slimy, smelly puddle. I tried to be brave and put the strange brew to my lips, but it was greenish-brown, and no matter how hard I tried, I simply could not drink it. Three times I made the attempt, and three times my body rejected it. I finally gave up and decided to be thirsty until we arrived in Okombahe.

The road was dreadful beyond any description. Cliffs were so high our cart jumped with a bang, then groaned as it raced down the other side with such virulent thorn bushes lining the path that my dress was shredded from the unexpected turns of the cart. We proceeded but slowly; the poor animals had become so footsore they frequently stopped to give themselves a bit of relief. My face was cracked and unbelievably filthy, and besides being thirsty and exhausted, I felt myself entirely unkempt and unpresentable, so it was with much happiness that we spied the station house of Captain Cornelius, located on an incline.

At last we reached our destination and stopped in front of the station house. The captain welcomed us cordially and straightway offered me a liter of cold coffee. Never had a drink tasted as refreshing and delicious as that coffee! I regained my strength, but remained at the station while Themis rode into the settlement to check if his house was still in a liveable condition. After a while he returned. Termites had badly damaged the house, the rains had undermined it, the natives had done their part as well, and we were faced with the prospect of having to spend the next few weeks in the open. Just at that moment an old settler, Herr Ertmann, appeared and most kindly offered us one of his two rooms. As he was an old friend of my husband's, we most gratefully accepted his kindness. In the course of the next few years, Herr Ertmann became one of my most loyal and devoted friends, and it was a great personal tragedy for me when he was killed in that miserable war (the German-Herero War of 1903–1906).

OKOMBAHE

1. Something about the Natives of the Colony

Before I describe our life in Okombahe (meaning: giraffe) in more detail, I would like to briefly summarize the characteristics and qualities of the various native peoples with whom we had contact. During the years of the last war (the German-Herero War), the native tribes in the Protectorate have become quite well known, but here I report only from my own personal experiences and those of my acquaintances. The individual tribes in South West are distributed as follows: The Ovambo are settled in the north, Herero dwell in the northwest and in the center of the colony, while in the south live the Hottentots. Mountain Damara are scattered across the entire country but do not possess contiguous areas of land. Small groups of Bushmen lead a nomadic existence in the eastern and western sections of the Kalahari Desert. The Bastars, a mixed race of Hottentot and Boer (Dutch South Africans), live in the south close to Rehoboth.

The Ovambo and Herero belong to the larger collection of Bantu-speaking peoples, and their two languages are related. Though the Ovambo live in the Protectorate, the government has not yet been able to subject them to colonial rule, largely because the government has had its hands full dealing with other tribes. Little is known about them. Now and again, Ovambo break the law, but the government has issued warnings to white settlers not to move into Vamboland since the government is unable to protect their lives and property. A few traders have dared to venture there, yet the only whites living in Vamboland are a few missionaries from the Finnish and the Rhenish Missions, and they are in constant danger of being slain.

A military campaign against the Ovambo would be very difficult, for there are no roads leading to the area and soldiers would have to march through very rough terrain. The climate and the numerous swamps and marshes make it dangerous for man and beast alike. Mounted troops would encounter special problems since horses, which are not indigenous to the area, would be subjected to common, life-threatening diseases. Thus a campaign against the

Ovambo could only be successful if mounted troops used the so-called salted horses that survived that dreaded disease and are now immune to it. Still, it is highly questionable whether the Ovambo would be inclined to oppose German guns in larger groups.

Because the border between German South West Africa and Portuguese Angola was determined by simply connecting the Kunene River with the Okavango River by means of a straight line drawn on a map, actual tribal lands were not taken into consideration. Thus it happens that the Portuguese Ovambo move into our protectorate when they are threatened by Portuguese troops, and it can be expected that German Ovambo will cross the border into Angola when they learn that German soldiers are approaching. Only a combined military action by both nations would stand any chance of success. In 1904, the Portuguese staged an attack on their own but were soundly defeated in the Battle of Humbe. A second military expedition planned by the Portuguese has been postponed.

Of the estimated one hundred forty to one hundred fifty thousand Ovambos, about sixty thousand live in South West, and eighty thousand dwell in Angola. In contrast to Herero and Hottentots, the Ovambo are excellent workers, gardeners, and even craftsmen. In their own way, they are masters of the art of wicker and iron working. Vamboland is a huge plane with few hills and is thus well suited for agriculture. Cattle raising takes second place, and the cattle of the Ovambo are characteristically small because they have been consistently inbred. Cultivating the fields by hoe is a woman's task. Indeed, all the heaviest tasks such as cultivating millet, harvesting and milling the grain, brewing the beer, etc., is done by women. Men's work is limited to herding the cattle, milking the cows, building huts, and so on.

The richer an Ovambo man is and the more land he owns, the more wives he must keep so that all the work can be performed. For this reason, missionaries will have little success preaching against polygyny. The social realities of Ovambo life make polygyny a necessity, and a very long time will be required before Christianity changes these realities. It is almost expected that a man will replace a weak, old wife with a healthy, young worker. It is typical that a young man purchases his future wife by presenting his mother-in-law with four hoes. If she accepts them, he presents his chosen bride with four hoes of her own—signifying that they are now married. From that point onward, a woman faces a sorry lot. Only during the Efundala festivities are women venerated, but then only the unmarried ones. Efundala festivities are a kind of officially sanctioned marriage market, and every young girl must attend so that the young men can select the ones they deem most desirable.

The staple food of the Ovambo is kafferkorn grain prepared in various ways with a fermented milk called omaere. The most popular delicacy is dog meat, especially when the dogs are purchased from white traders. Thirst is quenched with Ovambo beer, as well as a drink made from the eengongo fruit that has a strong, intoxicating effect. Portuguese merchants do an excellent business selling sparkling southern wines to the tribal chiefs. To preserve their grain and save it from vermin it is stored in large grain baskets elevated above the ground by stilts. Their clothing consists of a loincloth, sandals, and glass beads. Women wear a loincloth woven from narrow straps.

All land is owned by the family of the chief and is distributed by the chief to his subjects. He has the right to demand its return anytime he pleases. In fact, the chief has nearly unlimited power and is only subject to the will of his mother. When a chief dies it is not one of his male descendants who reigns in his stead, as is customary with us; rather, it is the son of the deceased chief's eldest sister who inherits the chieftainship. For this reason, the grand old Matriarch of the chiefly family is able to exercise a powerful influence not only over the chief but over the entire tribe. The main chief of the tribe has the right to appoint and remove minor chiefs at his own will and pleasure, to demand the payment of tribute, and to order the execution of persons he dislikes.

We cannot ascertain whether or not the Ovambo possess a genuine religion. They have only a confused concept of a higher being by the name of Kalunga. They imagine this god to be tall, handsome, and as pliant as a palm tree. He wears a beautifully woven belt around his waist, with a large basket hanging from either side. One of these baskets contains good; the other contains evil. If human beings please him, from the basket of goodness he scatters rich, abundant harvests, beautiful cattle, an array of beverages to drink, and a happy life free of illness. If their actions anger him, he takes his other basket and sends starvation, failing harvests, gruesome disease and pestilence, death, and the spirits of the dead who will torment and afflict human beings. Superstition and magic play an important role in the life of the Ovambo; for this reason, the magician, the man of medicines, the mixer of poisons, and the rainmaker earn a very good living.

The Herero, like the Ovambo, belong to the great tribe of the Bantu. Years and years ago they migrated from northeastern Africa to the Protectorate. They are beautiful people, tall and slender, often with European features. Their skin color is coffee brown. They are intellectually very active and much superior to other natives when it comes to their ability to learn. Perhaps this fact has contributed to their corruption as contact with European civilization has stimulated their demand for clothing and other goods. Educated Herero love luxury goods such as starched shirts, white suits, and handkerchiefs; and the Herero lady owns more toiletries than the wife of a white settler. Of course, the entire household is patterned after a corrupted caricature of

European society. Many a Hennessy cognac bottle, many a tin of fine food is found in their pontoks—their circular huts. For the sake of alcohol, a Herero man is willing to sacrifice his honor and anything else. In the city squares, one often sees a Herero man forcing his wife into prostitution so that he may buy a wretched bottle of liquor. It is my distinct impression that the Herero who settled near our towns looked rather miserable. I was unable to detect much in the way of Christian influence.

With the heathen Herero, on the other hand, who lives on the land, the story is quite different. As clothing, he wears animal skins, and his wives are arrayed in traditional dress. Women wear a leather headpiece fitted with three leather horns, a loincloth made of animal skins, and arm and leg ornaments of iron pearl beads (which weigh up to fifty pounds) of their own manufacture. All the household items and all the clothing of these people are smeared with red-colored grease. In addition to a pulverized tree root called bochu, scent of moschus is added as an ointment. The red color of the grease derives from the bark of Gerber acacia trees. With some aging, grease prepared in this manner emits a pungent odor nauseating to the newcomer but not particularly unpleasant to us old Africans as we have become quite accustomed to it. The odor is, however, so strong that one can smell a Herero settlement from a fair distance away. Woe to the Don Juan who embraces one of these heathen beauties while dressed in a white suit. He will undoubtedly find her complete body contour impressed on his suit in red grease! Herero women who cover themselves with this red oinment are generally known in the Protectorate as "sardines in oil."

Before the Herero War of 1903–1906, the number of Herero was estimated to be one hundred thousand. We do not know how many are still living today. The main characteristics of the Herero are arrogance, cruelty, and laziness. He disdains anyone not of his tribe. Only with reluctance did he accept the yoke of the Germans, whom he calls Otjirumbu, meaning, "yellow thing." For centuries, he has been involved in tribal warfare, notably with the Nama. Long ago, the Herero completely subjugated the Nama who lived in the mountainous regions, and today one only finds members of this tribal group toiling as slaves among the Herero.

Only when it comes to love for his cattle does the Herero overcome his laziness. The one thing that truly interests him is farming cattle. Though the Herero man possesses very few admirable qualities, one must admit that as a cattle herder and husbandman he has no equal. Only cattle matter to him. At sunset, when the cattle return to the settlement, they are received with song and dance for the purpose of enticing them to give milk. The singing consists of a monotonous song (they are quite unmusical) in which the refrain *Ozon-gombe* (cattle) is repeated over and over again. Only the men dance while the women form a semicircle and clap their hands, as their heavy clothing makes quick movements impossible.

The staple food of the Herero is omaere (soured milk). Only children and the oldest men enjoy sweet milk. The Herero also gather wild foods: wild berries, roots, and grass seeds. Rarely do they hunt, but they do catch game in traps. Cattle are only slaughtered on ceremonial occasions or during severe famine. The Herero man leads the comfortable life of a lord of the manor. Most of the time he lies inside or in front of his hut and has his women serve him. His work consists of little more than carving wooden vessels for a variety of purposes. Occasionally, he fits together the iron beads his women use as ornaments in their clothing, but most of the time he obtains them from the Ovambo who are expert in the art of metal working.

The Herero woman has a difficult life. She must build the pontok, tan the animal skins, milk the cows, look for food, prepare the meals, etc. Herero scarcely ever tend gardens. Only on rare occasions does one find a garden, planted along the banks of a river, in which maize, pumpkins, and melons are grown. As the huge cattle herds prevent the Herero from settling down in any one place, they wander about the entire country as nomads.

Out on the countryside, Herero live in pontoks, while those who have settled near the towns often dwell in houses constructed of clay bricks. The pontok is conicle, and the roomier family-sized pontoks have a large central pole set into the ground that supports the roof of the hut. Long, narrow boughes are placed in a circle and set in the ground at a distance of approximately one and a half meters from the pole. Their ends are tied to the top of the pole with strings of bast or straps of ox leather. The individual branches are tightly interwoven with twigs and shrubbery. At the centermost part of the roof the materials are less tightly woven, allowing smoke to escape. When the framework is completed, the women prepare a thick plaster of clay and cattle dung and smear it on both the inside and the outside of the framework. When it is fairly well dried, they spread a second layer of plaster to cover any remaining cracks and openings.

A good pontok is entirely smooth and looks like a grayish-brown molehill. An opening on one side near the ground serves as the entryway, and is sealed with thorn bushes, animal skins, cloth bags, or even a wooden door. Though windowless, the inside of a pontok is very cozy. The interior floor is made of smoothed clay, dried until it is very hard. The fireplace is located at the center of the floor. Sheepskins and blankets belonging to the dwellers lie along the side, and the few bowls, pots, and utensils they possess hang from the walls. Wooden boxes and other belongings are set on large, heavy stones. They take special care of their milk pails that are either carved of wood or made of hardened skins or gourds. The latter, the so-called calabashes, are held in great reverence and must never be cleaned as they believe that tampering with the calabashes will cause the cows to cease giving milk. Calabashes belonging to the master of the homestead are for his use only and are strictly separated from those of the servants—even the visitor is not offered a drink from the master's gourd. When an important council is held, the chief makes an exception to this rule and permits others to drink from his large calabash. At sundown, all gourds are placed in a circle around the fire, and the calabash that serves as the container for the milk of an unusually valuable cow, one that is about to give birth, is hung upside down on a tree.

Herero pontoks provide shelter from the cold, yet during the hot season they are pleasantly cool. Though they are to be kept clean, this is not the case with most of them, which makes it rather unpleasant to enter them when one pays a visit to nurse the sick. If a pontok is vermin infested, or if the dweller has had an infectious disease or died inside, the family abandons the pontok and burns it to the ground since it is easy enough to construct a new one.

As with the Ovambo, all the land is owned by the chief. He distributes tracts of land to his subjects according to his whim and pleasure. It is his prerogative to sell parcels of land to white settlers, though he needs the consent of the government to do this. In matters of inheritance, the Herero also follow the matriarchal order. The Herero concept of God is not clear, and because they are suspicious of strangers, we have yet to gain any real insight into their religious practices. They venerate the sun, Ejuva, as the giver of light and life, but no special cult is attached to this worship. Magical spells, conjurings, magic sticks, amulets, phantoms, and ghosts of deceased persons all play a role in their beliefs. They hold that in the spine of every person exists an immortal worm, and when a person dies, it creeps out of the grave and plays mischievous tricks on the living. To prevent this, the worm must be killed by smashing the spine of the deceased, after which the body is tightly wrapped in a cattle hide and interred. The sacrifice of a black sheep is performed to cast out evil spirits.

The Herero have their own physicians who mix poisons and ointments. A widely used remedy for the healing of wounds is cow dung. Like many other African tribesmen, the Herero do not value life too highly. The suicide rate is high. Poisons, especially euphorbian poisons, are often ingested by old and enfeebled persons who are no longer able to follow the nomadic lifestyle, and they seem quite content to be gathered to their forefathers in this rather violent manner.

In Eharui lived an old, weak Herero man. The pastures were bare and the water holes had dried up, so arrangements were made for a move to the north where they might find better grazing for their cattle. The old man noticed the preparations and called all his children together to ready them for his death.

"I have seen many cattle grow up," he said, "and have watched the regular return of the rains, but now my legs are weak and my sight is poor. I know that you stay here merely because of me, and I can see that the cattle are emaciated because there is not enough grass to feed them. You cannot take me along. I will be but a burden to you, hence, I have decided to poison myself. I believe you will have no objections, will you?" When the children voiced no objections, he took leave of them by gulping down a large quantity of sap tapped from the milkbush. He became quite ill, yet the wanted death did not come. His children counseled together and sent a servant into the mountains to gather the euphorbian poison that brings a quick death. The old man was given such a massive dose that he died immediately. His family started with the loud lamentations as they ate the ox that had been slaughtered in his memory. The lamentations of the women did not cease.

Scholars still debate the question of the origin of the Hottentots (Nama). Because of their yellowish skin, their delicate physiques, and their high, protruding cheekbones, some scholars maintain that they are of Mongolian origin. They lack, however, the most common characteristic of Mongolians: the shape of the eye. Their language differs markedly from that of the neighboring Bantu tribes and is characterized by click sounds. Frequently words that are spelled the same way receive a completely different meaning by different clicking sounds.

Hottentots originally lived in the area now known as the Cape Colony, and they were displaced in the seventeenth century by migrating Boers. The Hottentots retreated northward and in the process repeatedly fought with Bantu tribes advancing southward from Central Africa. By their action against Hendrik Witbooi, in 1894, the German government attempted to end the continuous fighting between the Herero and the Nama in South West Africa. The main tribes of the southwestern Nama settled in Namaland and in the Kalahari Desert. The northernmost regions of the protectorate were settled by the Topnaar and Swartbooi Hottentots. They lived in Kaokoland and in the region of Franzfontein.

In general, Hottentots are small, lean, and wrinkled. Their hair is tightly matted into small but thick knots and are consequently called *peperkop* (peppercorn heads) by the Boers. They do not relish work and are weak and lazy, abhorring all activity that requires physical exertion. Hence, they do but little farming and prefer to sit rather than stand when carving pipes, tanning animal skins, making shoes, and fabricating objects of horn, wood, and beads. This much having been said, it is surprising how Hottentots immediately abandon their natural laziness when they go off to hunt or to fight and are hence known as excellent hunters, marksmen, and scouts. Their diet consists mostly of game meat. They prefer to smoke dagga (cannabis) rather than tobacco, and are passionate consumers of alcoholic beverages. Their skillfulness in battle against the Herero—and there have been many longstanding conflicts between them—has yielded a good deal of booty that allows them to live with a certain measure of luxury. They wear mainly European clothing, and as long as they own at least one cow they maintain a decent standard of living.

When the German government finally put an end to this continuous fighting, the primary source of Hottentot prosperity—booty—dried up, and they have progressively become poorer and poorer. Pampered, as they were, in their previous manner of living, they are now unable (because of their natural indolence) to rise to greater heights. Instead, they sink lower and lower, and I believe they will eventually die out as a people. The original Hottentot religion consisted of ancestor worship and supplication to the moon. Now most of them are Christian, but those who have yet to convert have nonetheless lost all memory of their former religion and have no beliefs at all.

For many decades the Hottentots have caused problems for our government and have no appreciation whatsoever for the efforts made on their behalf. Following the rebellion of 1895, the Swartbooi tribe was taken to Windhuk where the people lived as prisoners under the supervision of the government. The entire group very nearly perished. Only a small minority of the Swartbooi Hottentots had not taken part in fighting, and they were permitted to remain in Franzfontein under the rule of Lazarus Swartbooi, arch crook, liar, and criminal. Despite the warnings of the white settlers living there, he masterfully gained the trust of the government. During the last war (the German-Herero/Hottentot War), he naturally committed high treason and murdered white settlers. When he was captured, he escaped punishment by taking poison that had secretly been given him by his tribesmen.

We still remember the greatest of all Hottentots, Hendrik Witbooi, the "faithful friend" of Governor Leutwein. His son, Samuel Isaak, wisely surrendered his arms because of the counsel his dying father gave him: "Go to the white men and tell them, 'I come from Hendrik Witbooi, who is dead. He sends you a message. My children shall enjoy peace.'" Then Samuel Isaak left.

Because of the protection of our government, the Mountain Damara now live as a united tribe, protected from the Herero, along the banks of the Omaruru River. Previously, many of them lived as slaves among the Herero. Mountain Damara are the hardiest and most industrious natives in our colony—albeit because of external pressure. Due to their cooperation, the colonial government was able to make a treaty with the now deceased Chief Manasse of Omaruru, and as a result of this, the Mountain Damara were given the region of Okombahe as a reserve, an area that had previously fallen under the control of the Herero.

Mountain Damara are predominantly broad-nosed and thick-lipped, with dark, coffee-brown skin, and ape-like long arms. Over the course of time, they have lost their own language and adopted the Namaqua tongue. The Damara who live on the reserve are healthier than the Cliff Damara who roam the countryside, so the women can gather grass seeds, roots, and onions—a laborious task, to say the least—for them to eat. While Cliff Damara wear animal skins as clothing, the more distinguished Mountain Damara try to imitate the whites by dressing elegantly. In Okombahe, a single black coat existed as common property. At baptisms, weddings, or funerals, the closest male relative of the person for whom the ceremony was being performed was permitted to wear it. On these occasions one observed some of the strangest people.

Like the Herero, Mountain Damara live in pontoks. The more distinguished among them construct small cottages normally consisting of two rooms. Captain Cornelius, the chief, owns two cottages—an old, dilapidated one in which his wife and mother live, and a beautiful, new, and airy one consisting of two rooms in which he lives by himself. He owns a genuine iron field bed with beautiful blankets, a washbasin placed on a wooden beer crate, and a wobbly wicker chair. In his living room he has a sofa, a small wardrobe in one corner, and a roughly hewn table with a tablecloth over it. The old gentleman does not have a happy marriage and his wife is not permitted to enter his house. Instead, the young women of the village are required to clean and maintain it according to his moods and wishes. Josua, the second in command, likewise owns a small cottage, but in this household it is the wife, the "black" Lydia, who rules her husband, child, and grandchildren with an iron fist.

The life of the Mountain Damara is dominated by a system of class and prestige. They compete among themselves to determine who is the most distinguished, whose family is the oldest, and which families have tapped more deeply into Herero bloodlines. This latter distinction, Herero bloodlines, is crucial in determining high rank and high standing in Damara society. The most distinguished persons do almost no work at all. They keep servants since it is taboo to work with one's hands—to plait baskets, to carve wood, or anything of that nature. Agriculture and cattle farming are necessary to keep one from starving, but once the fields and the cattle have been cared for, they lounge lazily about in front of their pontoks, smoking homegrown tobacco or drinking sugar beer.

Drinking beer brewed from sugar and honey has ruined the people. It is prepared from some kind of a cider whose ingredients remain secret—though green peas and chicken dung supposedly form part of the infusions. Sugar water or honey is poured on top and the mixture is allowed to ferment for about twelve hours. When finished, the brew is a cloudy water that tastes like sweet, rotten white beer, yet possesses a stronger intoxicating effect. In the past, before traders began doing business with the Mountain Damara, they had to undertake the laborious task of searching for honey. Now they can purchase a pound of sugar for forty pence, making life that much more pleasant. Many a tribesman has traded entire herds, or his very last goat, just to get his hands on the ingredients to make sugar beer. Men and women alike are addicted to it.

The inhabitants of the Okombahe region, comprising about eight hundred souls, raise cattle and tend fields on the banks of the Omaruru River. Usually in July, once the rainy season is over and the flow of the river has significantly diminished, they cultivate the riverbed soil and grow maize, wheat, pumpkins, and melons. When the grain has reached its full height, the riverbed offers a lovely view: as far as the eye can see, fresh, green grain billows in the wind. Individual fields are marked by thorn hedges, and the entire river bank is fenced in this way so that cattle cannot enter and graze there. Rabbits are a major pest as they eat both stems and roots and cause further damage with their underground passages. The Damara catch them with traps and attempt to frighten them away with loud noises. Each night, the tribesmen take turns standing guard and beating, for hours on end, an old pot or a cement barrel with sticks and stones, pausing only for brief periods. This dreadful racket does in fact scare off the animals, but has robbed me of many a good night's rest!

As harvest time draws near, Mountain Damara become progressively merrier. Most of them seem not to know how to manage their harvested goods. They simply eat, drink, and make merry until their harvest is exhausted. Their grain they trade for sugar in order to brew more beer, as well as for suits for the men and clothing for the women, and then live happily ever after. But once all their provisions are gone (and these are often quite extensive), they begin to starve and beg for food. Some of the more commonsensical Damara work for settlers or at the military outposts. The others eat grass seeds and wild onions or resort to begging, sponging off others, and stealing. Many people, mostly the old, the weak, and the infirm, die during this time. The harvest is always preceded by a period of terrible starvation—something we witnessed while living there.

The government doled out flour and rice, and on alternate days the mission provided a meal for the men, then a meal for the women, children, and old men. In my own backyard, I cooked flour into mush in my largest pot. But to what end? Certainly, their plight was relieved to some extent, but next year it was the same story again. When the natives were hungry, they came around by the dozen and begged to work for food, but as soon as the crops were harvested they disappeared. No amount of money could persuade them to return to work. "What do you mean? I don't need to work, 'my pen is full' (i.e., my stomach is full), just feel it. When I'm hungry again I'll come back." Because they have no concept at all of duty, harvest time was always problematic for us settlers because our workers abandoned us. "I don't feel like working. At home with my mama we have lots to eat."

It seemed not to bother them that we had gone out of our way to help them when they were starving. Perhaps it is fitting that their language lacks even a single word for "gratitude" and "thank you." They have no understanding of this concept at all. One had to overlook and tolerate so many things just to retain them; we also had to provide them more and better food than they could get at home. Yet since the law did not permit it, we were unable to give them a good thrashing when that was what was wanted.

When harvest time comes, everything is readied for the new grain. Each settlement of reasonable size has a threshing floor: a round area covered with a layer of smooth, hard-packed mud. The edge has been raised with bricks and stones so that the mud-plastered floor declines slightly to the center point. The sheaves of grain are set around this center point with the stalks facing out, the heads of grain gathered to the center point—and by this time many are so hungry they do not even allow the grain to ripen completely. Next, the girls and young women of the settlement gather and dance in circles on the heads of grain until every last kernel has been separated, a process that can take a week or more to complete.

A fire burns close to the threshing floor where a black man roasts grain in a pan. Another man pounds this into small bits inside a hollowed-out tree stump, then boils the mixture in fresh sweetened milk, while a third cooks coffee in a huge kettle. Whoever wishes a cool drink walks over to the chief of the settlement to fetch omaere (soured milk) or sugar beer from a calabash.

When the energy of the young dancers declines, the schoolmaster—the most gifted musician of the settlement—comes and plays his harmonica concertina, or the girls liven their dance themselves by singing a monotonous, often melancholy sounding song. The young men sit around the threshing floor smoking, drinking, and joking with each other. It often happens that one of them jumps into the circle of the dancers and is rewarded by them with screams of delight; then the party really warms up! The threshing lasts until evening. But when the full moon has risen above the Okombahe Mountain, when the cattle have been cared for and the lambs penned for the night, the settlement rises to an entirely new life. Everyone congregates at the center again—where the day's work has taken place—and clay beakers filled with sugar beer are passed around. Then commences the wild, furious, fanatical dancing that only the uninhibited black races are able to do.

The dances of the Mountain Damara are no longer unique to them. As with their language, these must have been lost when they were forced into servitude by the Hottentots and Herero. All their dances have been borrowed from the Hottentots. Typically, they gather in a circle and while men and women clap their hands, sing in beat, and stamp their feet, the dancers begin to turn round and round, slowly at first, then faster and faster until the speed reaches such a fevered pitch that some of them lose consciousness and fall to the ground.

The rooster dance is especially popular. A girl dances in front of a young man with her arms firmly pressed to her bosom; in her movements and gestures, she imitates a frightened chicken. The man, for his part, imitates a rooster with his walk, by his gestures, and in his sounds. This dance almost always rages out of control, ending in the most savage gyrations and excesses. It is probably for this reason that the missionaries have a profound dislike of dancing. Among the natives, converts to Christianity are strictly forbidden to participate in dances; hence, dancing in Okombahe has nearly come to an end. Yet when the light of the full moon is too tempting, when the harvest is abundant, when the flames of the fires burning in front of the pontoks grow more intense, and the sugar beer numbs their senses, missionaries, law, and Christianity are forgotten, and everyone dances until early morning.

Mountain Damara do not have a musical instrument peculiar to themselves. When dancing, they sing, clap their hands, or use a six-sided concertina and a harmonica introduced to them by white traders. They are a musically inclined people and even incorporate melodies learned from our own soldiers. To these melodies they add their own texts. One day I was quite amazed when my black nanny sang my son to sleep using the melodies from "The Singers from the Dark Woods" and "Castles on the Moon." They also play very simple reed flutes obtained from the Hottentots. These instruments are capable of playing two or three high- or low-pitched notes. When ten or so men play these flutes together, the sound is shrill but distinctly melancholy—and loud enough to be heard all the way to the Kalahari Desert!

Mountain Damara do not possess special laws. If they live together in larger groups they will choose a leader. The government has appointed the chief of the entire Cornelius tribe and pays him a small annual stipend.

Their concept of God is mysterious and unclear. They worship the good god Heisab who, in times past, freed them from a monster. In the whole of South West, so-called Heisab graves are to be found. These are nothing more than piles of stones, and when passing a Heisab grave, a Damara will toss another stone on the pile or, lacking a stone, will pitch a rag upon the pile. This is a kind of sacrifice, and the person unwise enough not to do it must go without God's protection. Evil ghosts and terrible sickness befall him, and his amulets of teeth from wild dogs and leopards can do nothing to protect him. Superstitious heathen mothers bite off the first joint of their babies' right little finger.

Damara also tell many fables and fairy stories, most of which are set in the animal world. They love to tell of a golden past and make every effort to preserve it in memory. In remembrance of those times, they call themselves *hou khoin*, meaning "man of men" or "the right men." Namas, however, call the Mountain Damara *xoudamab*, meaning filthy, "soulless" creature, or *ozondjima*, black baboon. These are the Nama's favorite insults.

Unfortunately, I was not able to become personally acquainted with the mysterious Bushmen. Though many experts maintain that the Bushmen were the original inhabitants of South West Africa, most of them are uncertain as to their origins. They seem to be related to the Hottentots in appearance, body size, and skin color. Every now and then I chanced to get a look at one or the other of these groups. Shyly, they retreated into the Kalahari Desert. They mainly hunt game, have very few wants, seldom or never own cattle, and wander throughout the Kalahari. They are greatly feared because of their violence and their raids.

In the early morning hours, well before sunrise, the Bushman women move into the grasslands with the small, hard pumpkin shells they use as water containers. From the blades of grass they suck the droplets of dew, as well as from other bushes, and collect them as drinking water for the day. They are skilled in this, and in no time at all the container is filled with water.

If one is hunting in the Kalahari, believing that one is all alone in the endless desert, heads suddenly appear here and there, and follow the hunter for hours or days on end. They stand in front of him as if they were growing out of the ground. As soon as the shot is fired, they rush forward from all sides, gather up the dead animal, and carry it off to camp. They must, however, watch the hunter for a long time before they are convinced that he has come to them with friendly intentions. Bushmen can quickly fire their arrows with deadly precision. An arrow consists of a thin reed shaft with a sharp, detachable head of bone. A sticky, gummy paste containing the poison is smeared on the head. The head is made so thin that the shaft fits over it like a sheath, and in this manner they safeguard themselves against being poisoned by the arrowhead. With the arrowhead thus safely secured, the hunter cannot be harmed by it and the poison is not lost. It is not until the very moment when they spy an animal that they quickly pull an arrow from a leather quiver, remove the head from inside the shaft, and attach it to the arrow.

Often an animal wounded by their arrows runs for quite a distance. The poison, however, causes the wound to expand so that the animal gradually loses its ability to move. When the hunters close in on it, the beast is quickly dispatched and the body cut into pieces. The poisoned flesh is carefully removed and discarded.

The language of the Bushmen is unlike any other. It consists of hisses, lisps, and clicks. It cannot be described. When my husband was posted to the lonely border stations of Otjituo and Okatjeru, he became a good friend of the Bushmen and often witnessed their dances around huge fires at the time of the new moon. On those occasions they adorned themselves with the horns of a kudu, or tied the large, curved beak of the pepperbird to their foreheads, and these strange men jumped about the fire with grotesque movements—like devils performing a satanic dance of hell.

The Bushmen have no laws. If several families live together, they elect a headman, though he has precious little influence over his people. Rarely do they build huts of grass but are most often satisfied with bending the branches of a bush to create a shelter.

All Bushmen remain heathens, and magicians and medicine men hold considerable sway over them. Once an officer fortuitously happened upon a larger village of Bushmen in the Kalahari. After the customary exchange of gifts, he retired to his nearby tent. A little while later, the medicine man—the omnipotent one of the village—appeared at his tent. Amulets and magical objects hung on his body, and he bowed repeatedly to the officer.

"I've come to receive my gift from you, Sir!" he said.

"Get out of here you yellow monster; nothing is what you get!"

"I'm telling you that I've come to receive a present."

"I'll give you nothing. Now, get going."

"You don't want to give me anything?"

"No. As I've told you, you're not the chief."

"That's true, but I'm far more powerful than the chief is."

"Well, here's some tobacco for you, now get out."

"That's not enough for me."

"Get out, I tell you. I want to go to sleep."

The chap grinned, "Good," he said, "go to sleep. I'm leaving. It won't be long before you'll call for me. Then you'll welcome me back, heap presents on me, and ask me to stay!"

"Only the devil could make me do that."

"Yes," he stated, "and he will, tee, hee, hee. I'll set rocks in your stomach and they'll keep you down, tee, hee. You'll writhe in pain and get very, very sick. I'll leave now, but I'll be back shortly, tee, hee, hee!"

The monster left the officer's tent, his thin, spindly arms curiously stretched and his fingers widely fanned out. Feeling tired, the officer lay down and went to sleep. A short while later, he awakened. He felt uneasy and decided to get up. His limbs were as heavy as lead; his stomach ached beyond endurance. At first he was angry; then he started to laugh. "Is this yellow monster supposed to have caused these pains?" He thought to rid himself of the pains with a mighty swig of cognac, but it did not help; indeed, the pain increased until it became unbearable. Running up and down gave him no relief either. He was about to give in to despair and decided to empty the cognac bottle, but it gave him no relief. Finally, he called for his orderly—not wanting the medicine man to know of his troubles. The orderly greased his master's body and massaged him so vigorously that sweat ran from his brow, but it was all in vain.

After a while, the servant said, "Baas (boss), I can't help you. If you want to get rid of the sickness, let the medicine man come." But he resisted.

Then, with painted skull, the monster entered the tent.

"Well, how are things now? Do you need me?"

"Get out, you scoundrel," came the reply.

"Certainly, but I'll be close by—and you may call me."

He disappeared, but eventually the pains became unbearable, and the lieutenant ordered his servant to fetch the magician. It was hardly necessary, for he slithered into the tent like a serpent, grinning all the while.

"Now it's my turn. You are calling me and here I am. I'll take those pains away from you—but only if you give me what you previously denied me."

Groaning, the suffering officer answered, "Make your choice, take what you want and however much of it you will, only help me get rid of these pains!"

"Yes, just wait a moment, I'll help you, tee, hee, hee. I know you're feeling very ill, but wait just a bit longer."

Taking his time, he snatched up exactly what he wanted. Rushing around in the tent, he heaped little sticks into a small pile on the floor, and by means of muttering and conjuring, he lit the bundle on fire. Next, he set all sorts of essences and herbs on it; soon the tent was completely filled with smoke and the officer feared he would suffocate. The medicine man danced around the fire, screaming and raging until he was bathed in sweat. Suddenly, he stopped. "Are you feeling a bit better?" he asked. "Yes," came the reply.

The whole ceremony began again, but this time more wildly than before. The jumps and contortions grew more frenzied. "You're still feeling the illness, but now it's far away?"

"Yes, I do feel better."

New magical sticks burned and fresh smoke rose.

> "You're still a bit ill, correct?"
>
> "Yes, but just a bit."
>
> "Give me now what you have given the chiefs!"
>
> "Take anything you want."

Again, the rabid incantations began but suddenly came to an end.

> "Now," he said, "tell me that you are well."
>
> "Yes," said the officer, "I am well."
>
> "Now do you understand how powerful I am? My power reaches far, much farther than you can imagine. It even reaches the white man who must submit to it."

Utterly exhausted, the medicine man gathered up his presents and left the tent.

The lieutenant breathed a long sigh of relief, for the pains had left him and he felt like a newborn babe. To this day he has not forgotten the Bushman magician.

Old Ertmann was once asked by a Bushman medicine man to give him one of his fattest sheep. Ertmann laughed at him and said, "Sa danas ge! eisenha" (You're out of your mind)."

The medicine man replied, "If you don't do it, the ram will give you no pleasure."

Ertmann responded, "Leave that up to me; now, away with you!"

The magician left, but Ertmann's servants were afraid and urged him to give in to the magician's demand. By now the old man had become adamant.

"You pretend to be Christians yet believe in the arts of this rascal? If only the missionary knew about you!"

They were ashamed of their fear. Ertmann, gun over his shoulder, was helping them herd his livestock to a different pasture. All of a sudden, in broad daylight, a large hyena jumped in the middle of the goats and sheep. The animals panicked and scattered in all directions before Ertmann was even able to get off a single shot. The hyena disappeared with one of his rams. The servants were terrified, and amid their screams and cries they shouted, "Arikana, arikana (please, please), spare us great magician! We acknowledge your power, but please do not harm us!" Frightened, they ran about aimlessly. Finally when they had calmed down, they ran to Ertmann—who had been watching the general chaos with typical African calmness. "Oh Master, if only you had given in to the wish of the great magician. Now he has taken the form of a hyena and pursues us so that he may destroy us."

Ertmann tried to convince them that it had been pure coincidence, but was unsuccessful. When they had gathered together all the scattered animals, only to notice that the fattest ram was missing, there was no end to their dismay. "Can't you see that it was the magician? Let's hurry and escape his ferocity." Running as fast as they could, they drove the animals on for hours, not even pausing to eat or to rest until they reached the Hottentot settlement. They thought the power of the medicine man could not extend quite that far.

Stories such as these frequently told and retold in South West have not particularly added to the popularity of the Bushmen. One fears them, their insidious poisons, their whirring arrows that suddenly hit the unsuspecting hunter in the Kalahari Desert.

2. The Activities of the Missionaries in the Protectorate

In the following pages I wish to recount a few things I have observed about missionary work in South West. My account is based on my own perspective and has no significance beyond that.

For more than fifty years ambassadors of various mission societies have worked hard to bring the light of the gospel, along with the blessings of our culture, to our poor black brethren. The Barmer Mission has been here the longest and works among the Herero, Mountain Damara, and Hottentots. The Finnish Mission serves in Ovamboland in the north. The Catholic Mission has its office in Windhuk and is preparing to spread their activities over the entire Protectorate. The hard work of these missionaries is undeniable, neither is the seriousness with which they devote their lives to this task, yet to this point in time they have enjoyed little success. In no way do I want to belittle the missionaries' efforts, but I strongly disagree with the manner in which they work to convert people. I believe I may safely assume that others will agree with me, yet I acknowledge that my judgments are based on my own personal observations.

The Rhenish Mission Society serves as a good example of what I mean. Before their missionaries are sent out to their fields of labor, they receive only the most basic seminary training. Many of these missionaries have not been educated beyond the elementary or lower secondary level, and have pursued employment as tradesmen or small-time merchants before deciding to take up the cross. It is quite impossible for them to master, in so brief a time as their training allows, the vast amount of material contained in modern theology. And this lack of adequate schooling and theological preparation makes it impossible for them to teach the natives convincingly. They are also prone to very poor decisions. The missionary who has not himself mastered the teachings he is to convey to others is certainly in an impossible position to educate entire tribes, especially when he must do so in a completely foreign language, against the backdrop of foreign customs and entirely different religious traditions.

Often, though not always, the power of personal example is absent in these missionaries. I know several missionaries who devote all their time to caring for their own family. They do the best they can to teach their own children. They work in their own gardens and with their own livestock. Their Sunday sermons must be prepared, but time passes and unintentionally the congregation takes second place. The missionary wife has much to do in her own house and in the care of her own little ones. The day of a good German housewife is entirely spent in domestic activities. She starts a school for native women to learn needlework. One missionary wife taught the native girls to crochet fine lace items, yet they were unable to mend clothing or to cut out patterns and sew. Another one carefully unraveled woollen socks that had been knitted by her friends and then had the Nama girls knit them anew. Gifts are most often presented to the wrong people. And never is the native encouraged to hypocrisy as much as when he steps in front of the "emissary."

The task of a missionary is difficult, extremely difficult. For no matter how determined he is to succeed, he will only be able to influence the men. Native women in all the Bantu tribes of South West represent, in the eyes of their men, not much more than a somewhat more intelligent work animal. Women must be won over by women, not men. Now and then a missionary wife attempts to fulfill this high commission. She tries once, twice, but ultimately fails; it is nearly impossible to avoid sinking into despair. Bound by her daily routine, as she is, it is difficult for her to rise above herself.

The climate encourages laziness and fatigue. Servants are cheap and the missionaries employ many. A lack of initiative and work ethic on the part of the missionaries likely makes a deep impression on the native. All that remains is the Sunday sermon. And to what end are they directed! The salvation of the soul, the promises of an agreeable future existence if one is but good in this life. But that does not suffice the native? He is much too grounded in the material things of this world and expects material advantages from Christianity. Is the Rhenish Mission able to provide these? Occasionally, woollen socks or shirts sewn at a kaffee-klatch are given to the best-behaved natives, but this is never enough; no, the native always wants more. His idols were handed down to him by his forefathers, they are objects he can plainly see; thus, according to his way of thinking, he should honor them, and he does honor them. He clings to his customs and superstitions. There is no apparent material advantage in Christianity that would entice him to convert. The natives are taught not to drink, not to take multiple wives, not to kill or commit adultery at the same time as they are taught to share with their neighbor and to help him and love him. To what end? For a place in heaven? What meaning does heaven have for the native? It is merely an ethereal concept for which he must give up tangible things.

The Mission Society provides native helpers for the missionaries, so-called evangelists. They are cleverer than the average native, and for this reason they are able to fool the naive and trusting missionary. Because they have learned more than the average native it is assumed they are able to teach others. But when it comes right down to it, they can only spell a few words and know their numbers from one to ten.

I had to pass by the mission school a number of times for several weeks, and every morning, day in and day out, I heard all pupils, ranging from the very young to those who were nearly adults, spell the very same word in a monotonous cadence. When, after weeks on end, I requested to have a look at the slates of my servants (they had left their work so that they could enjoy the lessons), they were either on vacation or the same word was still listed. The natives have better success with their singing. The choral pieces sung in Namaqua do not at all sound bad, though they are much too loud (the natives sing with the intensity of a yell or a shout). But is not South West a German colony? Why is the German language totally neglected in these schools?

Money is collected in Germany to enable "suitable" natives to be trained as evangelists at the Augustinium academy. For years, their every need is attended to—yet what is the result? I have only known of a very few of them who had a good influence on their fellows. Gottlieb from Omandomba was excommunicated several times from the church for adultery and other similar egregious transgressions. Timotheus from Karibib was imprisoned for theft; Emmanuel from Okombahe was able to explain away any involvement in the theft of wine from a military post until the stolen bottles, some half empty, were discovered hidden under the altar of the church to which he alone had the key. It was proven that for several days prior to the discovery he had preferred to be inside the church, and each time he left the church he was heavily inebriated. Now, what is the moral of this story? Among his schoolchildren he was like the goat that had been made the gardener! Besides this, he suffered from a disgraceful and disgusting illness, and came to me rather often seeking my help. Such exemplary evangelists do the cause far more harm than good.

I have neglected to even mention the Auderling. *Auderling* is a Dutch word more or less equivalent to "elder." One recognizes them immediately because even in the greatest heat they wear either a long, threadbare, black jacket, or a long greatcoat. It is legitimately questionable whether or not they always wear a shirt and underpants beneath the long jacket or greatcoat. In any case, however, they appear very dignified and have an uncommon talent for acting—an almost innate ability peculiar to the black race. They imitate the missionary in their walk, in their bearing, and in their unctuous manner of speech. Yet someone who knows this situation intimately maintains that these Pharisee-like Auderlings have a devastating influence on the morals of the local beauties.

To me, it is a brazen mistake when the missionary befriends these hypocrites and discusses politics with them, even the comings and goings of the white settlers, singling out the whites as poor examples of Christian living. Even if the missionary does this in a childlike manner, it makes no difference—it is still inappropriate.

One day an Auderling came to us and said, "I'm coming from the missionary, and behold, the Lord delights in him, for the Lord causes the flowers to bloom all around the missionary's house, though no one has planted them there and no one tends them. The path to the missionary is always very soft, as if it has been carpeted with the softest animal skins. And I asked him why the Lord did not plant such flowers in front of the house belonging to Herr von Eckenbrecher. The teacher answered me, 'Herr von Eckenbrecher is a godless man. He reviles, curses, and does not attend church, and because of this Heaven denies him these flowers.'"

As a matter of fact, the flowers in question (dübbelcherdorns) have ugly, thorny seeds that we call morning stars, and in order to keep the area around our house free of these vexatious things, my husband had all these flowers removed from our property. It was a difficult task to complete, and I wonder how these people arrive at such ridiculous conclusions.

Another time, one of these missionaries came to my husband, Themis, and asked him to make good on a voucher he had signed for a native girl quite some time before but which had mysteriously "gotten lost" in the shuffle. Themis replied, "Of course, I will gladly pay you, but you must first show me the voucher. Otherwise, as has happened many times before, if I pay you now, in a few days you'll find the 'lost' piece of paper and expect me to redeem it as well."

"But really, Herr von Eckenbrecher, this poor girl is crying and swears that she had it in her hut, but that when her child suddenly burst in, the child began playing with the voucher, and before she could prevent it, the voucher was thrown into the fire. Of course, she quickly ran to the fire to save it, but it was too late. The flames turned it to ashes before her very eyes. Now, think what you're doing—surely you don't want to cheat the poor girl out of what rightfully belongs to her?"

But my husband stood his ground. "I must see the piece of paper before I redeem it. I'm sorry, Herr Missionary." Saddened by the hardness of my husband's heart, the missionary left our house.

Two hours later, the evangelist Emmanuel arrived. "The missionary sent me," he said. "Trevina has found the lost voucher, here it is."

"Well," replied my husband, "where was it?"

"All day long she read from the Bible," came the answer, "and suddenly she found it inside."

"I thought it burned in the fire—she swore that she had seen it burn with her own eyes."

"The will of the Lord caused her to find the piece of paper in her Bible, to His honor, and to the poor girl's delight!"

It is also instructive to note how the natives make use of Christianity for their own advantage. A Herero man in Otjimbingwe was accused of having stolen schnapps and was made to appear before the presiding officer in the court of law.

> "It is said that you stole a bottle of schnapps?"
>
> "No Sir, not I, but Satan—who took possession of me and tormented me until I did it."
>
> "How did he manage to do that?"
>
> "Well, at first I felt a twitching inside of me, then a strange person forced me to look at the schnapps—though I didn't want to do that."
>
> "Then what happened?"
>
> "Then Satan stretched out my hand and closed it so forcefully around the bottle that I had no choice but to hold on to it as tightly as I could. I struggled and fought against him, but he wouldn't let me go, so I was forced to follow his command to put the bottle in my pocket, to run away, and then to drink it. I was given no choice. You simply cannot punish me, Sir! Instead, you must punish Satan—who made me do these evil things."
>
> "Why did you not resist?"
>
> "Look Sir, there is nothing one can do against Satan!"
>
> "Well, yes, it is a terrible thing to be under the spell of Satan. But what do you expect me to do?"

"Why, punish Satan, Sir. Drive him out of me so I won't be tempted in the future."

"Yes, but that's not very easy, you know. Where precisely do you feel Satan?"

"In my back. When I see in someone's possession a thing I'd like to have for myself, it begins to twitch inside me and my arms stretch out to take it."

"Calm yourself down and I'll think of something that might just help you."

The officer then commanded, "Take this rascal down and drive the devil out of him by administering twenty-five lashes to his back." It was done. And smiling, the officer said to the old rogue, "Take note. Do not think that *you* are receiving the punishment. We only want to lash at the devil until he no longer feels like pestering you and ruining your good reputation."

After that, I think, the devil no longer felt like tempting the old man.

Only a few people are such hypocrites; many take their Christianity very seriously. Old Zachäus Goerze of Franzfontein, for example, was once asked what he had planned for the following day. Without any hesitation, he answered, "As de Herre will und die Osse kum zall ons meskien morrow Mist rei." (If it pleases God and the oxen come, we shall cart manure tomorrow.)

The Catholic Mission has some definite advantages over the Protestant Mission. The Padres support the lay brethren with skill and circumspection. The garden of the Padres is one of the most beautiful in the entire colony. They cultivate fruit trees and vines, and the natives who have learned a craft or trade from them are highly sought after. They do not preach Christianity while at the same time forbidding every errant behavior; rather, they say, "Come to us and we will show you how you can also learn to work and cultivate a garden. And if you do this, you will not go hungry." They force no one to attend prayer meetings or worship services, but gradually and steadily encourage them in this way so that the natives are not even aware of it. Yet the natives do notice the blessings that come from this. Thus the Padres lead them to Christ through work: "See, you are able to do this thanks to the Saviour who wishes to help all men. Come to him, believe in him, and he will give you the strength to accomplish even more."

I am certain the Protestant Mission in Omaruru could likewise have influenced the natives in this manner. The missionary in charge has been living there for twenty-five years. He also has a beautiful garden and sells his produce to the whites, thus generating a good second income. There were many Herero living in Omaruru at the time, but why did he not use his influence by telling them, "It's good to have cattle, and I am delighted that you have them. But create gardens, cultivate dates, figs, vines, and vegetables, and sell them. I'll show you how to do that; I'll help you. I'll order fruit trees from the Cape and you can pay me with cattle." Gradually some would have become interested during those twenty-five years, some would have followed his example, and at last the people of Omaruru would have been persuaded to give up their nomadic lifestyle and settle down and tend their gardens. It is likely that this could have been repeated in Okahandja and on the Waterberg.

Sufficient money was available to cover the costs of watering systems, dams, and wind-driven pumps. The Herero were certainly more affluent than the whites, but the right man was missing—a man who had earned the trust and respect of the people such that they would be willing to carry out such projects. When once I made such a suggestion to an old missionary, he became indignant and said, "But Frau von Eckenbrecher, I am a missionary after all! Such impudence! How is it possible to establish the Kingdom of Heaven with wind-driven pumps and dams?" I, however, still maintain that much could have been accomplished and much could have been done differently. When peace is finally restored in South West, a new era based on new ideas will replace the tired, old misconceptions. The mission is presently in dire circumstances. It spent a great deal of money and invested much toil and labor in its previous efforts, and it will tackle a new and difficult task with renewed vigor. Although the mission did not succeed in saving the native tribes from the effects of that terrible war, it will have to exert all its strength in the future. Wise leadership coupled with carefully considered methods, as well as a close, working relationship with the government, will aid the mission in helping lift the remnants of a once-proud people and edify them both culturally and morally. This is, after all, the duty of human beings.

3. Our First Abode

In Okombahe, we moved into the room that was kindly made available for us and arranged our belongings as best as possible. On the same evening, the natives of the area came by to greet us. A group of them said to my husband, "We thought you'd come back, Herr von Eckenbrecher. You've become so accustomed to living here in Damaraland that you won't feel yourself permanently at home anymore in verdant lands of Germany. However, not one of us anticipated that you'd return with a white woman." To celebrate the occasion, my husband ordered a distribution of rice, coffee, and sugar among the people. The natives of Okombahe danced and sang late into the night.

The next day the local missionary and his young wife called on us, inviting us to use their guest room until we were able to make more comfortable arrangements. We gratefully accepted this offer and stayed with them for twelve days. In the meantime, we succeeded in renovating my husband's little house and making it a bit more habitable. Then Old Ertmann moved into our house and we into his—Ertmann's two-roomed house was somewhat larger than our one-roomed home.

We moved our boxes, provisions, the harness for the oxen, and the rest of the household effects into the larger room. The other room became our living room and bedroom. My husband built a table from wooden planks, and in the corner of the room he built me a bed from planks and branches. The mattress was a sack stuffed with long, dry grasses. When the whole thing was finished and covered with a blanket it looked somewhat like a European bed—provided one had a vivid enough imagination. Weather permitting, my husband slept outside in front of our hut. Crates functioned as chairs, and our rifles, my husband's ammunition belt, other weapons, and various antelope horns decorated our walls. Against the lower part of the walls we built shelves on which we stored milk containers, bread, and other food supplies. The floor was carpeted with the skins of antelopes and sheep.

Despite the rather spartan decor, the room had a degree of quaintness. Many a night we sat at the crooked table in the light of a flickering candle, and while drinking a cup of tea, listened to Old Ertmann's mesmerizing tales of his life in South West. He loved living as a hunter in the wilderness, far away from civilization. But when we heard the sounds of the night, when the leaves of the old trees rustled in the night wind and the howling of hyenas and jackals in the nearby mountains penetrated the blackness of the sky, we appreciated having a home in which to live. Under these circumstances, one becomes aware of how little a person needs to be content.

4. Worries of a Housewife

It was far from easy for me to become accustomed to my new circumstances, as the expectations of my capacities as a housewife were considerable. I had problems with the servants. They did not understand me, neither could I understand them. My husband thought it would be easier if I had a girl as my own personal servant, so he hired Pauline, the daughter of the headman, Paul Goseb. From the very beginning she wished to impress me. Her father shared that intention and repeatedly told me, "You know, Madam, we come from a good family. We are royalty, and just as noble as your German emperor and old Navilion (he meant Napoleon)."

Though of royal descent, I was horrified by Pauline's lack of manners. Without asking permission, she entered my room flanked by her many sisters and girlfriends, squatted on the floor as the natives do, and cackled ceaselessly. With nonchalance, they chewed their tobacco, smoked their short pipes, and with even greater *sangfroid*, spit on the floor. When I dared to forbid this behavior, they vehemently objected to my request and carried on as before. Pauline herself objected vociferously when I uttered the humble wish for her to clean the room a bit on Sunday. But the missionary had said that Sunday was a day of rest and she was not permitted to work. Instead, one was obligated to dress finely and go to church.

The young girls of the village beautified themselves by placing a white linen band around their forehead. This band was the idea of the local missionary and symbolized moral purity. If one of these young ladies fell into moral transgression and it became public knowledge, the Council of Auderlings, presided over by the missionary himself, assembled. On the following Sunday, before the devoutly assembled and morally dismayed parishioners, the girl stood at the altar as her white band was removed. The missionary pronounced the curse of the church upon her, and she was not permitted to attend school or church until she repented for a period of six weeks. Even after repentance, she was forbidden to wear the white headband ever again.

I did not tolerate Pauline for long. It was very expensive for us to feed her relatives (from hither and yon) their daily meals. Her performance in the house was without any merit whatsoever. Very early each morning her royal mother disturbed me by begging for *treckselki*, a handful of coffee. I made up my mind and gave her the sack, and since that time I have only employed boys as domestic workers.

Cooking outdoors over an open fire was no easy task. The constant bending, together with long periods of standing by the fire during the heat of the day, was most uncomfortable. Old Ertmann finally built me a sort of sunscreen of rushes and reeds in front of the house so that I at least had some shade. Under these conditions I really had little idea how to prepare a meal, and in a very kind, benevolent manner, Old Ertmann helped me with this as well. I tried hard to be an eager student, and he taught me how to bake bread and zwieback, how to prepare many different African meat dishes, and even how to make a pudding.

It is not easy to bake bread. Since yeast is unavailable one has to use a sourdough starter to make the bread rise. We made sourdough starter by stirring flour into lukewarm water and setting the mixture in sunlight. As soon as it fermented it was ready for baking. This I mixed with flour, added water, kneaded it thoroughly, and then let it rise in baking tins made from old cement barrels. We started a big fire of wood mixed with dried cow dung (the latter material creates a far more intensive heat), then the servants formed a cavity in the middle of the fire where the tins would be placed. Hot ashes were set over the bottom of the cavity. The baking tin, which also had a tin lid, was set on top of ashes. More hot ashes were heaped around and on top of the tin itself, and in this manner the bread baked for about two hours. Sometimes the bread is not completely baked through because the ashes had not been hot enough; other times, the ashes prove too hot and the bread comes out as black as coal. I had a good many failures baking in this manner, but gradually I learned and later became so experienced that I could bake the most delicious bread over a campfire at night when we traveled. If I had no mold, I simply placed the loaf in the hot ashes. When it was done, we removed the ashes and burned parts as best we could and we ate the rest.

Most of our meat came from sheep and goats, though I wearied of that before too long. For more delicious tasting meat we hunted antelope and wild fowl. Early in the morning we waited at the river for the quail to come and drink. Since ammunition was very expensive, we tried to maximize the value of each shot. In order to do so, we placed thorny twigs on the water so that the birds, which came flying in large flocks to drink, had to land in a rather small area. We recognized their flight pattern in the distance, and when they came closer, we heard their distinctive screeching high above us in the air. A few birds, probably the leaders, flew ahead of the flock. They circled back and forth scouting the area. If we made a careless movement, they noticed us and immediately sounded a curious warning call, and in an instant the entire flock would be gone. Soon we became aware of their patterns and remained under cover, not moving in the least, and allowed the birds to drink for a while. However, as soon as a large group had gathered on the edge of the water, we fired a salvo of buckshot. Immediately, our servants caught the wounded birds and killed them. The entire kill was placed in a sack. Sometimes we had as many as forty or more birds; most often, though, we caught far fewer than that. Plucking them was not difficult; the servants simply skinned and cleaned them—but only when I expressly asked them to do so. We hung them on a long string and cooked them the next day. Soup made from these fowl was nourishing, and we added dried cabbage or some other vegetable, along with cubes of pumpkin in place of potatoes.

In the evenings we hunted guinea fowl or we went into the wilderness for several days to hunt kudu and gemsbok. We usually returned home with a great load of fresh meat. To find the larger animals we had to go deep into the veld, as in the immediate environs of Okombahe game had all but disappeared. The natives had scared it away. Every month, each native was given five rifle cartridges, which were often cut in half. Match heads were used as fuses to detonate the cartridges. These were all short range, but since the natives did not shoot from long distances anyway, the cartridges served them well enough.

Fresh meat was roasted on a spit. Unfortunately, though, roasted meat did not keep very long. After six or eight hours of the daytime heat it was spoiled, thus for economic reasons we preferred biltong (dried meat). As an African housewife, I prepared a great variety of meat dishes: pieces of zebra meat fried in zebra fat, ground squirrel, turtle, kudu roast, springbok, wild dog, iguana, ostrich eggs, guinea fowl, locusts, etc.

During our first year we had no vegetables other than the pressed vegetable pills imported from Germany or what was given to us. Since we had no potatoes to go along with the meat, we prepared rice or pumpkin boiled in salt water. I also made excellent mashed pumpkin. After the pumpkin had been boiled, it was put through a strainer, to which I added sugar and vinegar. It tasted very nice. We did not have much butter or any other kind of fat. My husband had bought large cans of American pork fat in Karibib, but it was rancid and quite inedible. The natives make a kind of butter by shaking thickened milk in a calabash, but it always tastes like the calabash—sourish and peculiar. Hence, I had no choice but to make butter myself. I placed cream in a mixing glass and patiently shook it until it became butter. Later on—once I had the Hünersdorf butter machine and a churn—it was much easier, though it was always inconvenient since I could never let a servant do this kind of work. Servants always licked and lapped up the cream until it was gone.

Because of the difficulties of procuring butter, I soon learned to appreciate the uses of sheep fat. Over here, the sheep have beautiful fat tails, and as often as a sheep was slaughtered, the tail was brought to me. I skinned it immediately, cut it into pieces, and fried them. In this manner I was able to procure a good amount of grease, and besides its use in cooking, the rendered fat tasted very nice on a slice of bread. I used it for baking cakes and pastries as well, and I always regretted very much that sheep have but one tail!

Now concerning the beverages available to us in Okombahe, our water was simply the ground water of the Omaruru River. Servants carried it from the river in pails on their heads, or drew it with a pail and poured it into a small barrel. When the barrel was filled, the bunghole was closed with grass or cloth, and the thing was rolled home with the help of leather straps that had been nailed to the top and bottom of the barrel. Our main drinks were weak tea, coffee, and milk—the latter coming to us mainly in the form of omaere. As mentioned earlier, omaere is a thick, soured milk and the national drink of the Herero. Well-established settlers also drink it with their meals, but I had to overcome an initial dislike for it. Omaere has an intoxicating effect and causes a feeling of fullness. But once one is accustomed to it, one does not want to miss it. One must be careful not to drink omaere before or during a horse ride, for it lies like solid lead in the stomach. It was a very rare occasion when we had beer or wine, but we did on occasion have an excellent Hennessy.

The law that prohibits the sale of alcoholic drinks to natives, unless one has a government permit to do so, is an excellent one. Still, they annoy one with their constant begging, though one can successfully refuse them by referring to the governmental decree.

Directly after our arrival, my husband purchased a flock of goats and sheep, and a herd of cattle, and continued adding to their numbers in order to create, as quickly as possible, the basis for our future farm. He primarily traded with the Mountain Damara for sheep and goats, while our cattle came from the Herero, with whom he often spent more than a month at a time in this effort.

While he was away, I had difficulty with the servants. I had no language guides and was thus slow in learning the Hottentot and the Herero languages. They often misunderstood me or became unruly; sometimes I had to resort to my riding crop to arbitrate the difficulty! Typically, the young people were completely out of control. Karl, our head cook, had fastened a nail to the back of his shoe to act as a spur. Whenever he passed by the bags of sugar in the storage room he "inadvertently" tore them open. By "unintentionally" moving his foot back and forth, he caused the sugar to flow abundantly. Then he bent over—as though he were going to pick something off the floor—and filled his pockets.

Kombani, the Ovambo gardener, stole into the kraal at night and suckled the lactating goats dry. I wondered and worried why it was that within two days twenty or more of the little kids died. I suppose it was his way of showing gratitude for the care I had given him, day and night, when he was sick with pneumonia. Isaak most diligently plucked the birds and skinned the rabbits. He was economizing and efficient. When serving us, in order to save the clean dishcloth from getting soiled while wiping off the plates, he pulled his shirt out of his trousers and used it to clean every plate during a dinner party to which we had invited a large group of neighbors. When I asked him what on earth he was doing, he calmly answered, "Don't worry, my good Lady, the shirt's not quite clean, but that's all right."

Ruben, my husband's personal servant, smoked my husband's best cigars and drank all my Cape wine. Theodor slaughtered the best animals of our herd and swore up and down that they had gotten lost. I got along best with Kasupi, an Ovambo, who could perform any kind of work. Although he was thirty years old (or perhaps even older than that), he had never worn any clothes in his life. I gave him a blue linen suit, and trustingly he came to me to learn how to put on the trousers and suspenders. He was always looking for a wife, and to make himself more attractive, he tied a thick kitchen towel round his neck as his cravat.

All the servants carried vermin; none was truly clean. And the differences in sanitation drove me to wondering about my European ideas. I gave the servants soap and a washbasin; they gave them away. I gave them towels; they used them as decoration around their waist or neck. Sleeping blankets soon had holes in them or were burned. Only the common metal comb was popular with all of them. Before going to church on Sundays they fought for it, and it made the rounds with all the men and local beauties who had gathered in our yard for the purpose of combing their hair. With eagle eyes, my servants watched to make certain nobody stole it "by mistake."

We enjoyed our life in Okombahe and planned to build a larger, more comfortable home, as our current "luxurious" abode was not entirely satisfactory. Because we did not have glass windows, we had to nail the tiny window opening shut with crate covers. Inside, it was almost completely dark when the door was shut. Once a week the floor was polished African style—with very wet and well-mixed cow dung. I had no objection to this, though until the polish dried it was a bit awkward. Between the beams in the ceiling one could see the corrugated sheets of tin that comprised the roof, which was, unfortunately, in a state of disrepair. Sunlight streamed through the openings; at night, it was the moon's turn. Rain fell on me and my bed, and formed wide streams in the plastered walls.

Along the tops of the walls, just below the roof, large holes had been bored to provide cross ventilation, but the wind blew through them incessantly and carried large clouds of yellow dust into the house. There were times, whether day or night, that everything, including the roof, was in danger of being blown away. To prevent our roof from being blown away, we piled heavy stones on top of it. Birds and hideous thick-bellied lizards entered through the openings. Entire families of these creatures sat in the corners—but in a way I was glad that they were there, since they were quite useful in catching the innumerable mosquitoes and flies. Owls and bats circled our heads at night.

The corrugated metal roof was constantly crackling, and made a terrible racket whenever a bird hopped around on it. Often, these noises startled me at night and I wondered what was going on up there. Rats and mice rustled about on the ground and termites constructed their mounds. To protect ourselves against them, everything we owned was set atop stones, though this hardly afforded much protection.

5. Building Our Home

Given these circumstances, it is easy to understand why we were desirous of having a different house. We had already selected a place on rocky ground (in front of my husband's old house) that faced the wooded riverbank. Together, we prepared the house plans, and with the help of some natives, marked out the foundation. Once we decided on the size of the lot, we sent the plans to the business office in Omaruru and sought permission to buy it. Permission arrived quickly, and we became African landowners.

We now had to hire laborers who could help us build the house and had little trouble in finding them. One group of them was to make the bricks. Corporal Scholz, a skilled carpenter, made the brick molds for us. As clay was practically everywhere to be found, my husband showed the workers how to do the work. One of them dug the clay out of the earth; another fetched water. A third mixed clay and water, and thoroughly kneaded it with his feet. Another pressed the mixture into the molds; the rest carried the full molds to the drying field and placed them carefully one behind the other. They then pulled away the board that supported the wet bricks and left them in pairs, and then rows, upon the ground. The bricks were dried by sun and air. When the tops were dry, they were turned over, and when the backs were dry, they were placed in an upright position to complete the drying—we did not fire them. The laborers were paid twelve German marks per one thousand bricks.

Every Saturday it was my job to count the bricks that had been made during the week and to pay the workers. I had to be careful not to count bricks made in the previous week as the workers liked to set old bricks among the new. Every thousand bricks had to be specially stacked and covered with sheets of corrugated iron to protect them from the rain.

Another group of workers, supervised by subcaptain Joshua, gathered rocks suitable for the foundation. These were fetched by cart from the distant mountains, and we needed about twenty cartloads in all. It was unbelievably difficult to prepare the foundation in a competent, proper manner. The workers had no concept at all of a straight line, accustomed, as they were, to their round huts, and for weeks I had to stand ready with level and square until we were able to get the foundation lines straight. In one corner of the foundation we placed a sealed bottle with all the documents relating to the house. When the stone foundation was completed, we poured a thick layer of concrete over the whole of it to prevent termite problems in the future.

Termites are a kind of ant frequently found in tropical America and Africa. They live together in huge colonies. The "soldiers" are responsible for defense; the "workers" for the construction and upkeep of the common abode. The "queen" has an extremely large abdomen. Termites build pyramid-shaped community nests or dwellings out of clay and sand mixed with their saliva. Such termite hills are often five meters high and so hard that they withstand the strongest tropical rains and can only be destroyed with a pickaxe or crowbar. Termites are both a nuisance and a real danger for human beings. At night they march into homes and destroy wooden beams, furniture, crates, and other things made of wood. But they only hollow out the inside of the wood so that destroyed items look fully intact yet immediately crumble to dust when just barely touched. Termites also eat clothes, curtains, and towels hanging against the wall, and when one takes them from the wall in the morning, they simply fall to pieces. I once saw how these creatures had hollowed out an entire bookshelf together with all the books on it. There the books stood, beautifully set in a row, but when I touched one of them they all crumbled to nothing.

At times we lifted a heavy crate only to find it surprisingly light—the bottom and all of the contents had been devoured. To protect a picture against these pests, we nailed a sheet of tin on the wall behind it, though this did not make much of a difference. The termites made tunnels all around the picture and marched along inside them until they arrived exactly where they wanted to be. It was in this manner that they were able to reach things set away, sometimes well away, from the wall. The safest measure against this plague is concrete, for the termites cannot penetrate it. But if they are inside the house, all items on the wall must be inspected daily. If we find a termite path, we douse it with gasoline, which keeps them at bay—until they reappear somewhere else!

While the foundation of our house was being built, Themis drove two wagons to Karibib and purchased everything needed for the construction of the house itself: many barrels of concrete, wooden beams and corrugated iron sheets for the roof, various colors of paint, turpentine, oil, nails, window frames, windows, glass, doors, and door locks. Some of these items came from Germany; others were locally available in Swakopmund and Karibib.

Once all the supplies were on site, we began building the walls of our home. Themis hired twenty or thirty women of various ages to carry the bricks to the construction site. Each morning before work began, they assembled outside our door and waited for my husband, and though not always properly dressed, he stepped out and gave a ticket—a simple piece of cardboard with his seal—to each woman present. In the evening, they returned their tickets and were paid. We followed this practice because sometimes women who had not worked came to collect wages. After the noon break we counted the women once again to determine whether or not any had left. Every woman carried five bricks on her head, and in single file, the group moved from the brickyard to the construction site and back again. The female overseer, Old Lenas, led the procession, clothed in an old sack that had an opening for her head and neck. The others were similarly attired. Rhythmically, they moved about and sang, for hours on end, the same monotonous tune with the same Nama words: “Eh, we are carrying bricks for Herr von Eckenbrecher; eh, carrying stones is hard work; eh, but we now have enough to eat.”

Vice-Chief Joshua and his son-in-law, Manuel Timbo, built the walls, though my husband did the corners. Clay mixed with water was used as mortar and ferried in buckets to the "masons" by several women. Now and again, I checked the work with my level and plumbline; I also handed bricks to the workers, and generally supervised the entire project. Constructing the window openings was unexpectedly difficult. Because of termites, using wooden beams as lintels was unwise. We therefore decided to build brick arches above the windows even though this presented some major challenges. I drew an arc on the wooden lid of a crate and my husband carefully cut it out to use as our pattern. Next, we used a chisel to shape the bricks correctly and set them in place against the top of the wooden arc. Once we were satisfied with the arrangement on the ground, we set the bricks in place at the top of the window frame. It worked splendidly. After a while, we became quite proficient in building arches. After all the window arches were finished, we began the more difficult task of constructing arches for the doorways. In this task we succeeded as well. Not only were they sturdy but fully symmetrical and pleasing to the eye.

Our house had four large rooms plus a kitchen, a pantry, and a large storage area. To keep the interior nice and cool, we built the walls quite high. The ceiling was made by laying narrow beams across the tops of the walls. In some of the rooms, we set reeds across the top of these beams; in other rooms, we nailed cotton fabric from beam to beam, which made an excellent ceiling as well. The walls extended another meter or so beyond the ceiling, and it was at this height that we set in place the beams that carried the weight of the corrugated iron roofing sheets. By constructing the house in this manner, we created a large pocket of air between the ceiling and the roof that kept the house cool and pleasant for us to live in. If the corrugated roofing is placed directly above one's head, it conducts the heat to such an extent that a house becomes unbearably hot.

The corrugated sheets we placed on the rafters were of varying sizes, and we used screws to fasten them to the beams and to one another. To prevent water from dripping on the threads and rusting them, we set a small piece of tin under the head of each screw to act as a washer. The entire roof sloped about half a meter from front to back to facilitate the easy flow of rainwater. At the rear of the house, because we had no rain gutter, we extended the roof well beyond the back wall so that rainwater could drain without ever coming into contact with that wall. We also extended the walls a bit above the roofline along the front and sides of the house, and on the sloping fourth side we set large stones on the metal sheets—all of this done to prevent the wind from blowing the roof away.

Now that the roof was in place, we decided to test it for watertightness. Using ropes, we pulled up several pails of water, and while I remained inside the house to check whether or not the roof was watertight, my husband doused the corrugated sheets. In one of the rooms, the ceiling became wet, but only after a good deal of searching did we find the leak—a crack in one of the sheets—and fastened a second sheet on top of it.

We installed the interior windows and doors, plastered the walls, and painted them. After this was done, we turned our attention to the floor. For budgetary reasons, we decided to do the common clay floor in some of the rooms. Local native women carried the clay mixture inside where we spread it over the cement. The mixture had to dry for fourteen days, and during that time no one was allowed to step on it or even to touch it. After this period of drying, it had to be compressed to make it hard. Again, the women were engaged. My husband sat down in the middle of the room and played a lively dance tune on his accordion. At once, the women began stamping in circles with rhythmic movements, and the wilder the music, the more vigorous the stamping. These exercises continued for two mornings, after which we had a first-class solid floor. Once the clay was completely dry, we poured unbleached flax oil over the top of it and let it saturate the clay. When at last the oily clay was fully dry, our floor was finished.

In the other rooms of our house we mixed one part concrete with four parts sand and spread the mixture about ten centimeters thick on the ground. This flooring also turned out quite well and had the added advantage that we could sprinkle it with water to keep the rooms cool. The concrete floors were, however, extremely hard, so hard in fact that the pounding on the bottom of my shoes made me slightly footsore.

The veranda we prepared with special care as it was to be our living room. It was very roomy, the ceiling was high, and we used stones for the flooring. My husband painted the inside of the veranda blue with various kinds of flowers as decoration. Much of our day was spent out there. We took our meals on the veranda, and often, after the day was done, sat in the light of an oil lamp until the wee hours of morning. To further beautify the exterior of our house we decided to construct a gabled roof over the veranda. My husband painted it. In one corner, he painted Pegasus, symbolizing spiritual strength, and in the other corner he painted a centaur, symbolizing physical strength. Between these two symbols my husband hand painted the lettering of my favorite verse from Horace: *Caelum non animum mutant qui trans mare curru* (For those who cross the sea, the sky above changes but the heart remains the same). With the symbols and quotation in place, we were ready for our struggles in this inhospitable land.

We had brought most of our furniture with us on the voyage from Germany. We owned a lovely dining room set of dark oak, a living room suite of Viennese furniture that was easily assembled, and bedroom furniture of solid walnut. Anything else we needed we were obliged to make for ourselves from the wood of packing crates. Old Ertmann and Sergeant Scholz gave us a good deal of help. Our homemade furniture looked a bit rustic, yet it admirably fulfilled its purpose. It fell to me to paint these pieces, as well as the doors and windows of our house.

The wood-burning stove in our kitchen was small, but it served us well. I spent many a challenging hour at that stove, but in time I learned to cook on it. Baking was not as difficult for me. I rather liked it, and to make me happy, my husband built a beautiful brick oven in our yard. It was plain and simple, of course, but well suited for bread baking. We packed the oven with wood and lit the fire. Once the wood was burned through, a servant cleaned out the ashes, wiped the inside with a special wet cloth, and placed the breadpans filled with dough inside the oven. As the oven had no door, we set a piece of corrugated metal across the opening, then placed heavy stones against the metal to keep it in place. Over the chimney I laid a metal lid with a single brick to hold it fast, and in this manner the heat was kept inside.

My husband's old bachelor hut stood behind our house. One of its rooms became the chicken coop; the other room served as our tool shed. We also added a wooden stable to the original structure, as well as a small room for the lambs. Every evening we carried the lambs inside to protect them from the cold, the rain, and from beasts of prey. We also wanted to prevent them from drinking their dams' milk all night long. Behind these buildings stood the servants' pontoks (huts). We simply could not keep them in our house because of differing standards of personal cleanliness and the possibility of vermin.

Our home was very pleasant. Everything was new and comfortable, and we had built it all by ourselves with great care and love. Our pride and joy was the dining room. My husband had built a fireplace with a well-functioning flue in one corner, and on cold evenings we sat beside it in our comfortable chairs and watched the crackling fire. The tea kettle hummed, and old Ertmann—our constant guest and helper—told us again and again, in his own melancholy way, of the experiences and disappointments of his life, always ending his reminiscences in the same way, "I hate people. I love the wilderness with its silence and grandeur. It conveys quietude and helps me forget all the suffering. Someday, when I pass on, my only wish is that I not be buried in a settlement, and under no circumstances is some missionary to recount my good and bad deeds to a group of curious people. I wish to die by myself that my body may become one with the wilderness." And oddly enough, that is precisely what happened. Many months later, the strange old man went hunting in the area surrounding the village of Outjo and never returned. Several search parties were organized to find him, but their efforts went unrewarded. The wilderness, his only true friend, finally took him and refused to give him back. Perhaps he was shot by fleeing Herero; I simply do not know. But Old Ertmann was a true visionary, a teller of tales, and we were deeply fond of him.

We had only been in our new home for three days when our first son was born. In giving birth, I depended solely upon the care and experience of the natives who did their utmost to help—though in a manner not particularly pleasant for a European woman.

The birth of our child seemed to be the most important event in the settlement. Not half an hour after giving birth I had to endure the congratulatory visits of the captain and subcaptain of the local Damara, as well as twenty to thirty Damara women. This scene was repeated again and again over the next few days. I hired a half-white young woman as a nanny, and except for her deep love of alcohol, I had no complaints about her. I was completely inexperienced and depended on her advice and support. I am greatly indebted to her for the splendid development of our little one. When I bathed and dressed him, numerous onlookers appeared. Our nanny was frightened of the bath water: "Madam, do not use water, surely the child will 'go dead' if you put him in the water. You must rub him down with grease." Yet when they saw that the baby was delighted to be in the water, their astonishment rose to a fevered pitch.

During the first few weeks, our child was given many, many presents—far more than he ever again received in his life. He was given a calf from the captain, goats, lambs, a chicken, meat, eggs, tobacco, wooden spoons, pumpkins, sour milk, etc. Of course, we were obliged to give unending presents in return!

The baptism took place six weeks later, and we decided to celebrate the dedication of our house at the same time. It was truly a festive occasion, and the entire settlement joined in the merriment. We had decorated our house, both inside and out, with palm fronds. We built an altar in my room and the young people came with flowers in hand. We had made Chinese lanterns from colored paper and hung them around the veranda. Everything was as festive as we could possibly have made it. The local missionary baptized the child; his wife and a soldier from the colonial forces were the godparents.

The natives very much enjoyed the festivities. Two springboks and a sheep cooked away in a large washtub in our yard; in a second washtub, several pails of rice with a few liters of raison sauce hissed and steamed. Coffee was served to everyone, and the captain and the chieftains received as much sugar beer as they desired. The celebrations climaxed before sundown, though at midnight large fires were lit and the natives fired three hundred shots into the air. The newborn baby, in whose honor all these festivities were conducted, slept peacefully in his little basket. No one had any idea of the very difficult times that would soon face us.

The celebrations made such a deep impression on the people roundabout that for months on end it was the subject of conversation around their fires. Often I was asked, "Frau von Eckenbrecher, when will you have another baby so we can celebrate another joyous baptism?"

6. Future Plans and Disappointments

Our idyllic situation in Okombahe had, of course, to come to an end. All along we had planned to find a suitable farm, one we could afford, and this is precisely what we did. However, making the right decision proved most difficult. There were plenty of farms available, but to find the exact one that met all our expectations was no easy task.

The first farms we considered were those owned by the government, though it was quite bothersome to actually purchase one of these. One must submit an application with proof that one possessed all the necessary capital to begin a farm, a statement detailing the specific reasons one desired the particular farm in question, and so on. Next, the government sends a notification setting forth the conditions of sale:

1. A house with at least two rooms must be on the property.
2. A well must be dug.
3. Trees may only be felled with the permission of the government.
4. Proper roads have to be provided.
5. If a large public road traverses the farmland it must be well maintained.
6. Within two years, a garden must be planted; if not, the property rights revert to the government—even if the farm has been paid in full.
7. One must have a certain number of cattle.
8. A special permit is required for mining.
9. The farm must be fully fenced.
10. The farm must be paid for.

If one knows the countryside of South West fairly well, then one is aware that most of the so-called government farms are of poor quality. The best land is primarily in the hands of private land syndicates.

There was no point in applying with the South West Africa Company to purchase farmland. It owns excellent pieces of land but does not sell them.

We wrote the Company and offered to pay cash for a piece of land in the Otavi region, but the Company responded with regret that it was unable to offer us any land at that time. Of course the farms were slated for purchase, just not at this precise moment. We should exercise patience, and when the Company felt the time was right, it would consider our request. Also to be born in mind was that mineral rights over the land were retained by the Company, and if at some time in the future the Company desired to create a mine, it reserved the right to buy back the land. At that time a hectare (2.47 acres) sold for three to four German marks. Consequently, the most agriculturally productive lands in the northern portion of the colony were settled by Boers—very few Germans began farming there.

If a single settler purchases government land, he is strictly forbidden to speculate on it should he be in the process of losing the farm. I personally know of no company here that adheres to the governmental regulations we farmers are expected to follow, even though these regulations are to be strictly enforced across the board. Neither the South West Africa Company, nor the Kaoko Land and Mining Company, the Settlement Company for German South West Africa, and the German Colonial Company for South West Africa observes these regulations. Also, to my knowledge, none of the fallow farms belonging to these companies adds anything to the economy of the colony. So what is the use of these companies? Every settler asks himself this question, especially since he can purchase farmland from the government for fifty cents a hectare, whereas these companies charge between three and four German marks for comparable land.

After we had been rejected by the colonial government of South West, we turned to the Colonial Company for South West Africa and offered to buy Aubinhouis farm along the banks of the Omaruru River. However, the Colonial Company was unwilling to sell any of its farms as the Company needed to retain them for its own interests. The Company did offer to sell us one of their properties, a farm called Gaynasib—a notorious cliff-nest in the region of Franzfontein that Old Sabatta, an Englishman from the Cape, was forced to sell to pay down some of his debts. For his part, Sabatta had purchased the farm from Lazarus Swartbooi, the chieftain of the Franzfontein Hottentots, yet had only repaid a fraction of the purchase price. Hence, Lazarus had the right to repossess the farm any time he wished to do so, and he himself told us as much. Then the old crook laughed and said, "But I don't like that farm because the cattle hurt their hooves on all the rocks and stones. My cattle always got terriblyfootsore, but it serves the stupid Germans right that their cattle become footsore!"

The Shepherd Company of the southern part of the colony offered us part of their grazing land. They had purchased too much land from Hendrik

Witbooi, though we were rather doubtful as to whether or not old Hendrik had actually signed over the very best land under his control. And since we wanted to remain in the north, we did not investigate the matter further.

We eventually decided to write the government and explain to the relevant people that since my husband had served in the military in South West, he still had a valid claim to a farm, as farms had been promised to all members of the Colonial Forces who took part in the 1898 campaign against the Topnaar-Hottentots and Northwest Hereros. We requested a farm owned by the government situated along the border of Kaokoland. Unfortunately, the governor informed us that the farm in question had only recently been sold to someone else.

We quickly recovered from this disappointment and requested a farm between Okombahe and Omaruru situated along the banks of the river. The farm was close by and could easily be managed from the Okombahe settlement where we lived, especially since we had planned to begin with cattle. Later on we wished to cultivate extensive gardens in Okombahe, mainly fruit orchards and vineyards. We had already contacted German companies and had made arrangements for a windmill, along with various pumps and water pipes, to be shipped to us. The area manager at Omaruru, a most capable and responsible man, fully supported our request. However, the negotiations, in typical African fashion, lasted forever and were not concluded before the outbreak of the Herero War.

7. Cattle Raising

We had spent time and expended effort on our cattle while building our house, but with that undertaking behind us, circumstances were right for us to study more intensely and follow the methods of cattle herding and husbandry practiced in the colony. The kraals (enclosures) in which our cattle were housed for the night stood behind the chicken coop, just a short distance from the house though close enough for us to be able to see them from our bedroom window. My husband and several natives had spent a number of days in the wilderness cutting one-and-a-half to two-meter high thorn bushes. The cut bushes, called tackies, were laid in piles, then pulled by oxen to our property. We marked a large circle on the ground and then set the thorn bushes along its periphery, stacking one bush upon another until the desired height was reached. In this, we had to be very careful not to leave any openings through which a jackal, the largest and most aggressive of the cattle thieves, could crawl. Two poles were rammed into the ground close enough to each other to create an entrance that could easily be closed with a sturdy bush.

The natives always throw a handful of dung into the kraal before their herds enter to protect the animals from simple illness, epidemics, and wild animals. Rarely will a leopard or hyena leap over the high thorn barricade, so the cattle are quite safe inside. The manure that accumulates in the kraal is used as fertilizer in the spring when the gardens are prepared. Natives, as well as many of us settlers, use this kind of kraal. Others prefer kraals surrounded by bricks which are, of course, far more durable, but also very expensive. Should an epidemic break out among the cattle, they are difficult to disinfect. A thorn bush kraal, on the other hand, can simply be burned and another one built in a different area.

The cattle are gathered into one kraal, while the sheep and goats are led to another. The herdsmen sleep in pontoks close to the kraals so they can quickly respond should the livestock be disturbed by wild animals. Dogs are either chained nearby or run freely in the yard. The local dogs have mated so often with imported European dogs that one rarely sees a purebred native dog anymore. Native dogs are of medium height, in looks resemble wolves, have a highly developed sense of smell, and will bravely attack even large predators. As soon as a native dog smells a leopard or hyena, it announces the predator's approach by its characteristically long howls. The next dog hears it, joins in, and warns the dog in the neighboring yard. Soon the entire neighborhood is awake. Then, with burning twigs for torches in their hands, the natives position themselves in front of their huts and scream; if the dogs fail to calm down and think the animal is still close by, large fires are lit on three sides of the kraals and guards are posted. Early in the morning, as soon as there is sufficient light to see, they search for spoors in the area where the first dog began howling. Once evening has fallen, they set out an iron trap—usually one that belongs to both the natives and a white trader. The trap must be very carefully placed if one wishes to catch an animal. Old Ertmann taught us how to do this one day when we made the rounds with him, walking for hours to inspect all the traps.

All traps must be securely anchored in the ground to prevent the trapped animal from running off with it. The anchor itself is buried deep in the ground and its accompanying chain weighted down with heavy stones. The trap is cranked with a handle and then covered with a light coating of sand to hide it. Animals have such a keen sense of smell that they can detect a trap prepared by human hands. Hence, cow dung is smeared on the trap and then strewn about to cover any human smells. Next, a small lamb is tethered to a peg near the trap and surrounded by a barricade of thorn bushes, with only a narrow opening that leads to the trap. The leopard or hyena attempts to reach the lamb by jumping over the barricade. In this, the animal succeeds; however, the high, thick hedge proves too difficult to jump from inside the barricade, so the animal looks for an escape route. The creature soon discovers the opening, steps through it, and is caught in the trap. An angry howl echoes in the mountains and alerts the trappers who were hidden nearby—the beast has been caught! A well-aimed shot and the predator lies dead at the trapper's feet.

Things do not always go according to plan though. We, along with Old Ertmann, had set a trap, but the anchor had come loose, and the leopard—whose front paw was caught in the trap—jumped forward with the weight of the iron, snarling and yowling. Accompanied by several natives, we ran after it, keeping, of course, a healthy distance between it and us. We were uncertain whether or not the beast could free itself of the trap and turn and attack us with its desperate ferocity. In the darkness of night, all we heard was its angry howling and the rattling of the trap. We simply could not see a thing. The leopard ran farther and farther away from us, and we only spotted it when the anchor it was dragging dashed against rounded stones, causing sparks to fly in the air. Finally, the anchor wedged itself between two boulders and the leopard came to an abrupt halt. There it lay hissing and glaring wickedly at us with its green, shining eyes. Cautiously, we approached. Shooting from such a distance in the pitch black is unwise, for if a bullet strays from its target and hits a rock, it could easily ricochet and injure one of us or one of our servants. I personally had no desire to see the animal at close range, and when Ertmann and my husband closed in, they fired simultaneously and put the animal out of its misery. My youngest son now plays on its skin.

We had few problems with predators in Okombahe, perhaps because we were surrounded by natives' homesteads. There was one night, however, when I was truly frightened. My husband was out hunting and I was at home alone. I heard the dogs barking and peered out the window and noticed that the cattle were greatly agitated. Hardly dressed, I walked out the door toward the kraal to see if jackals were scaring them. My European hunting dog, Lump, crept along close to me and began whining. Suddenly, a creature right next to me—and in the faint moonlight I could not tell what it was—jumped into the kraal. Bleating loudly, the animals moved close to one another and a single goat cried out pitifully. I called the herdsmen, the dog barked, and in the confusion the animal jumped out of the kraal and disappeared in the night. Together with the natives, we searched the kraal and soon found a dead goat, its neck broken. "What kind of animal kills like that?" I asked one of the locals. "That is the leopard, Madam. Tomorrow, you will find its spoors."

There was nothing else to be done, so we returned to our houses. Early in the morning, the natives returned and woke me. "Madam," they said, "come and see. The leopard walked into the yard and right up to the oven. Several times it crept around the kraal. Look here at the spoors!" But the spoors did not help me—the leopard was long gone and my goat was dead, leaving behind three kids I now had to tend and bottle-feed. The following night, the leopard killed a sheep in the neighboring homestead, then a calf farther away. Finally, the natives were successful in trapping the beast; angrily, they beat the old sinner to death.

Though audacious, jackals are rather jolly creatures. They enter a kraal through small breaks and holes; they breech a chicken coop by digging under the threshold; they attack livestock, particularly lambs, in broad daylight; they eat ox-hide straps, animal skins, and nearly anything else available. Nothing is safe from a jackal. On cold nights, they often come in packs to human settlements and howl impertinently, leaving only when we make a racket in return and allow the dogs to chase after them. They are difficult to trap as they are prone to gnaw off the crushed paw, as do foxes, and flee on three legs. It is very rare indeed that one catches an old jackal, for they are so clever that they will dig their way under a trap to safely get to the meat that serves as bait.

The basic care of livestock is fairly simple. Before sunrise, activity begins in the kraals with the ewes calling for their lambs and kids, the cows for their calves. Usually the ox kraal is opened before the others, and with their stately gait, the oxen step out and walk to the watering hole, after which they are left to themselves. Oxen are accustomed to one another, as well as to their kraal, and have little problem defending themselves against wild animals. Rarely do they get lost, and if good pasture is not too far away, they return every evening to the watering hole, then march themselves into their kraal. If their grazing leads them farther afield, they may spend the night in the rough and return home when they are thirsty. If oxen are needed, a herdsman will track them down and lead them home.

Cows remain in the kraal until their calves are brought to them. Only then does the milking begin—a process not without its difficulties. The native herdsman tightens a strap around the two hind legs of a cow so that it cannot move. He allows the calf to take a few drinks, to establish the flow of milk, then drives it away with a stick and begins milking. As soon as the cow notices that the calf is no longer drinking, the dam becomes uneasy and the milk flow ceases. The calf must start nursing again, but as soon as the milk flow is established, the calf is pushed away and milking begins anew. It is a drawn-out process, and one must be careful not to take too much milk as the calf must be fully sustained by what its dam provides. In Africa, it is rare to raise calves with artificial food. Many cows are so wild that they simply cannot be milked at all and are merely kept for breeding. So long as the calves are very young, a specially hired herdsman looks after them; once they have been fully weaned, they accompany the cows to the pasture. Cows also walk to pasture by themselves and every evening return in time to feed their calves. Occasionally a cow gives birth in the veld. When this happens, she remains in the veld for several days, returning only when the calf is strong enough to walk beside its mother to the kraal.

Settlers' livestock is branded to distinguish one person's animals from the animals of other settlers and natives. We had the large iron letters, *v E*, at the end of a long iron rod with a handle on the other end. The iron is heated in a fire of dried cattle dung. The animal to be branded is caught by means of a strap with a large sling on the end. The strap is wound about a flexible stick and the sling is thrown around the hind legs of the beast, just above the hooves. Two or three men pull the strap to one side while another, who has hold of the tail, pulls it to the other. When the animal's backside falls to the ground, a native immediately pushes down its head in such a way that the animal's forehead is pressed flat to the ground, the tips of its horns boring into the earth. At this point, the beast is completely incapable of moving, and the brand is quickly applied with the red-hot iron. The sling is easily undone and, gently mumbling to itself, the creature with a strange decoration on its backside trots away. The wound quickly heals. The branding, however, must be repeated every five or six years, and the specific mark is registered with the government and then protected so that no one else may use the same mark.

Natives also brand their cattle using marks of all kinds—a sun, a cross, a star, a circle—to prevent them from mixing and mingling with other herds. Should an animal get lost, the brand is made public, and if the animal is found, it is safely returned to its owner.

The situation in the sheep or goat pen is a good deal livelier. There always seems to be a deafening racket, but it is delightful to watch the tiny kids scurrying into the pen, making comical leaps as they search for their mothers. Sometimes they make a mistake and try to drink from the wrong mother—which seems to bother the doe—and the kid is unpityingly brushed aside. Motherhood seems rather unnatural for many goats, as they must be tightly held by their horns just so their kids may drink; it is up to the owner to see that this is done. Sheep, on the other hand, are more patient.

Quite often, grown animals, both sheep and goats, have the unfortunate habit of drinking a doe's or ewe's milk, which results in the starvation of the young lambs and kids. It is difficult to break the animals of this habit, and for this reason natives push a sharp, thorny stick through the offending animals' noses so when the sheep or goat approaches another animal to suckle, the stick pricks the intended victim, and the "milk thief" is unable to get close enough to drink. Our herdsmen, of course, pay no attention to this at all—if a small animal starves to death it is of no concern to them. Every morning and evening found us in the kraal checking to see that each lamb and kid had

had sufficient milk to drink before allowing their mothers to go to pasture. This was only possible because our flocks were small; if one has a thousand lambs, it is completely impractical. Sheep and goats are driven to pasture and then watched. Whether or not they have enough to eat depends on the herdsman and where he has led them. Pasture here is limited, and if the animals eat bitterbush, for example, the milk has such an intensively bitter taste that it cannot be consumed.

I often marveled that native herdsmen were so familiar with their herds that they noticed at once if even one animal was not present—and this despite the fact that they could not count to ten. A careless herdsman often loses an animal to a jackal, and those who are truly unscrupulous will slaughter one of the best animals, callously devour the creature with their friends, and then accept punishment in the form of a lashing without batting an eye. Since herdsmen are normally given an animal that has died, some will resort to despicable trickery to let the chosen animal perish. Typically, they pull out the creature's rectum, tie it closed, and let it snap back inside. With agonizing pain, the tortured animal dies a few days later. Another trick they play is to choke an animal for a long enough time that its neck gradually swells. In this condition, it can no longer eat and eventually dies. Later on, when I became aware of these horrible tricks, I had each "sick" animal slaughtered in my presence and investigated the cause of death myself. If it was foul play, I refused to give the animal to the herdsmen and instead had it prepared for the dogs and chickens. I stood and watched as they ate the meat. In the beginning, the natives murmured their dissatisfaction, yet as time wore on, the herdsmen became accustomed to the strange ways of the white woman. Once in a while an animal was killed by a snake, but even the best of herdsmen can do little to prevent such a death.

Lambs and kids spend their days close to the house and are carefully watched. Shortly before sundown, they become restless and bleat, then suddenly jump away in all directions only to return disappointed. As soon as they notice a growing dust cloud in the distance, signaling the imminent arrival of the flock, all the young ones race in that direction, paying no heed to any obstacle in their path. Then the flock begins to run, and the mother sheep and goats hurry to reach their lambs and kids. After the mothers have had enough water to drink, they march single file into the kraal. I always positioned myself at the opening of the kraal so I could count them as they entered one at a time. After the headcount, the herdsmen gave me their reports, then I examined any sick or injured animals. I examined the hooves of an animal that had stepped on thorns, and washed and disinfected the sore. I bandaged another's wound, massaged and applied wagon grease to a third animal suffering from a sore hoof; my fourth and fifth cases had colic, which I treated with a couple of spoonfuls of cognac. (Had they been severe cases, I would have applied warm compresses.) A sixth animal appeared to have been bitten by a snake and was treated with vitriol. I often worked more than two hours treating animals, sometimes caring for the last few under lamplight in our kitchen.

Unlike cattle, sheep and goats cannot be branded. One either notches a certain mark into their horns, makes a slit or a cutting in one ear, or marks them with clips, as we did, using hard nickel buttons that fastened onto their left ears. Our flocks looked quite smart with shiny buttons on their ears. The buttons, however, were short-lived; they eventually fell off from all the head shaking and we had to resort to painting a wide, bright-red stripe across their backs. The rain soon washed away the paint, and my husband decided to carve a small *v E* in their horns. It turned out to be the best choice.

As for bovine, we purchased ours from the Herero, and deliberately chose short, heavy, long-horned Damara cattle. We never planned to improve the breed by crossing the cows with Simmentaler or other European-bred bulls. Swiss cows need Swiss food, and it is obvious that in regions where the animals depend on laboriously foraging for the dry grass that sprouts alongside pathways, among rocks, or on sandy soil, Swiss cattle would not thrive. Under such conditions not even the best Swiss cow can give milk. Besides, it would soon become lame in our rocky terrain, and when lame, would no longer be able to walk the necessary distances to find adequate grazing; hence, the poor beast would rapidly weaken and then perish. The Herero had splendid cattle, and it seemed reasonable to us to improve the quality of our animals with a good Damara bull instead of experimenting with imported cattle. Before we arrived on the scene, the now deceased Daniel Cloete, from Okombahe, had used local bulls to increase the quality of his cattle, and his livestock were known throughout the entire region for their excellence.

We had a flock of fat-tailed sheep native to southern Africa. They are an excellent source of meat though not of wool. Our initial experiments to improve wool production failed, yet we were not discouraged. And since we did not have sufficient funds for expensive cross-breeding experiments, we decided, for the time being, to raise the fat-tailed sheep. For the same financial reasons, we raised common Damara goats—though it was our plan to eventually begin improving the breed by crossing it with Angora goats.

In all of this, we had anticipated the government as the primary market for our livestock, and immediately after our arrival we made application to be a supplier. During the first year in the colony, we also applied to be a supplier to the Damaraland Guano Company located at Cape Cross (on the north-central coast). A year later, the company declared bankruptcy as the guano supply was finished, and our only possibility then was to supply the Colonial troops. Every now and then, my husband delivered some livestock for sale to Karibib where he had business connections with various merchants. Most of the time, though, he ended up trading the animals for various goods. Naturally, we attempted to sell our livestock as advantageously as possible, and since our operation was rather small we simply did not have the quantities necessary to export animals to the Cape Colony as other farmers and traders did.

I personally undertook poultry raising, though my ambitious plans were eventually cancelled because of the German-Herero War. I began with four pigeons I had purchased for one German mark each. I also bought six chickens from a soldier in the Colonial Forces. At first, I had absolutely no idea how old they were or how to take care of them. But one gradually learns, particularly in Africa.

My favorite chicken was an old, stocky Boer hen that was both tame and trusting. A steep ladder afforded access to our chicken coop, but it was quite difficult for the hen to climb the thing when it was ready to lay an egg. At a certain time every morning, the hen appeared on our veranda; it cackled, flapped its wings, pecked at our feet, and made excited attempts to get us to follow it. We watched the thing until it finally dawned on us that it wanted us to follow along. As soon as we got out of bed, the hen cackled excitedly, quickly waddled on ahead of us, kept looking back to make certain we were following, and finally led us to the chicken coop. There it began the same act again, and it became clear that the hen wanted to get inside the coop. We opened the door, and the bird majestically entered. Nearly an hour had passed when it made such a racket that we came to investigate the matter. As soon as we opened the door she contentedly marched outside. Every day the hen repeated the exercise. Finally, my husband arose as soon as the hen appeared on the veranda, accompanied it to the ladder, opened the door to the coop, bowed deeply, and addressed her, "You are most welcome, your grace!" She rewarded us with eggs that weighed three or more ounces.

I purchased a few additional chickens from the natives and fed our flock water-soaked rice, Ovambo grain (millet), maize, meat scraps, and lots of greens. Fortunately, I discovered an instruction manual on poultry raising in my husband's library. I read the book and was eventually successful. Perhaps the climate was responsible for our chickens only laying eggs for three or four weeks. When they ceased laying, they started clucking and nothing I did brought the racket to an end. I gave them heavy doses of Ricinus oil, placed them in a dark chamber, withheld their food, set their bodies under water; yet nothing helped. When I took away all their eggs, they found stones on which to sit. Finally, I allowed those chickens that insisted on sitting on eggs to do so, but I locked them in crates that were barely large enough for them to turn around. Every other day, I pulled them off their nests and made them walk for fifteen minutes or so before I locked them up again. Food and water were within easy reach, and I had ten crates full of chickens in the tool shed.

My unending troubles were finally crowned with success. I was delighted when the first chicks were peeping and could not take my eyes off of them. Shortly before the Herero War began, I had more than eighty chickens, one hundred and five chicks, and innumerable pigeons. The chicks required additional work, for when they ran about the yard, birds of prey gathered in the air above them and it was not unusual to lose three or four of them a day. The natives' dogs also went after them, so I engaged a young boy to look after the chickens—but he often was powerless against these thieves. We decided to purchase chicken wire, and my husband built a small but airy enclosure. The wire mesh proved to be too wide so that the larger birds hung themselves in it while the smaller ones slipped right through. Despite these and other difficulties, I was proud to have raised one hundred and five birds, and to have had so many eggs that I could fairly squander as well as sell them.

In time, our pigeons also grew in number and became quite bothersome. Nama pigeons—small, blue-grey birds with forked tails that are native to South West Africa—joined our pigeons, and I constantly battled them as they flew into my garden, devouring the fresh seeds, the fresh lettuce, and pulling the buds from many of my plants. Still, they did present us a welcome change in our otherwise monotonous meals.

All the animals living around our house were remarkably tame. As soon as the chickens caught sight of us, they ran in our direction. Audaciously, the pigeons settled down on the table on our veranda—even when we were enjoying a meal. Lambs by the dozen jumped through open windows. A small dog, given to me by Old Ertmann, was suckled by a goat and always sat in the kraal howling when the goat remained out too long. The guard dogs, a hunting dog, and a Great Dane obeyed perfectly. A small baboon performed its callisthenics in front of our house. The creature was the sworn enemy of the natives and played outrageous tricks on them, but it was also a great friend of all the animals—including two baby jackals we were raising by bottle. When the jackals grew older, their violent natures manifested themselves as they stole and killed chickens and lambs, and bit anyone who came too close. One day they disappeared into the desert.

8. Gui-Gams

I have already mentioned the business relationship we established our first year in Africa with the Damara-Guano-Company at Cape Cross. Many farmers were anxious to sell their extra livestock to this company, as its manager,

Mr. Elers, was known to pay the best prices. This could not be said about most traders and distributors who typically paid between twenty-five and thirty-five cents per unit of living weight, while Mr. Elers paid forty cents. This, of course, made a substantial difference for us and was the reason we were very pleased that he wished to purchase our livestock. Through correspondence, we agreed to take our animals about two-thirds of the distance from Cape Cross to his cattle post at Gui-Gams, where he intended to meet us. The road leading toward Gui-Gams mostly followed along the contours of the Omaruru River. I had been told the countryside was exceptionally beautiful and, as I wished to become better acquainted with the Omaruru area, I decided to accompany my husband on this difficult journey. Had he gone by himself, he would have traveled by horseback; since I was going with him, we took the oxcart—slower, but far more comfortable.

I had little time to make preparations for the journey. While my husband had the cart greased, saw to all the provisions, and had a new Kudu skin whip made, I portioned out flour, rice, salt, coffee, and tea, then baked bread, put the other provisions in cans, and packed the needed utensils in a crate that was set at the front of the cart. Together we cleaned the guns, checked the bullets, and made some fresh buckshot.

We certainly did not look fashionable in our travel garb. My husband wore a dark blue linen shirt, corduroy trousers (with nineteen patches I had lovingly applied to them), field shoes of untanned leather, a colorful silk scarf around his waist, a care-worn, sun-bleached hat from yesteryear, and, to complete the ensemble, a short pipe that he held between his teeth. For my part, I wore a very long skirt, a faded blouse, field shoes, and a Helgoländer (a hat resembling a fisherman's hood). Our better clothes made the journey in a suitcase.

We stored toiletries, my watch, pens and pencils, Goethe's *Faust*, Themis's sketchbook and paints (which he took on all his travels) in side pockets within the cart. We also brought along piles of blankets, pelts, bedding, pillows, tools, wheel grease, and anything else that might come in handy. When we were ready to leave, the oxen had to be taken to the military station for a health inspection. A commission, consisting of a white man and two natives, confirmed in writing the number of animals, the fact that they had been vaccinated against lung disease and rinderpest, and that the team was in good health at the time we left the station. This so-called ox-pass was the only document in South West Africa issued free of charge. If one is not in possession of a valid ox-pass, a fine is assessed; also, before leaving on a journey, one must not forget to register at the military post.

We locked our house and nailed the entrances shut. Inside, there was a window we could not completely close, so we set an old, broken, unloaded rifle that appeared ready to fire at any intruder. We only meant to scare off the natives, but in one of his sermons, the missionary took us to task because of our apparent mistrust of the "dear, sweet people of Okombahe."

Our driver was a Cape boy named Manuel Timbo, an excellent animal handler, but otherwise more than just casually suspect. He had once been accused of pilfering from human corpses, but for lack of direct evidence linking him to the murder of the same, he could not be properly accused of facilitating their deaths. My husband liked him because he was circumspect, though I never cared for him at all. Two other young servants, as well as an older boy to drive the cattle, rounded out our crew. Travel clothes of the first two were simple—they merely asked for a piece of cloth and a string with which to tie the cloth about their waists. Hanging from the waist string was a fearful weapon—a traveler's knife—made from a piece of strip iron sharpened diagonally on one side. Karl, the herdsman, left a day before we did, driving the cattle and resting them for a time at Dawitsaub, our first stop.

Finally all was ready, and I climbed aboard the cart and was handed our two dogs, Khaki and Schmolli. As the young dogs were not used to traveling, I had to hold them fast, and they cried and howled so pitifully that the natives rushed out of their pontoks to see what the matter was. The road we traveled was a very difficult one. It went steeply up and down, leading across mighty boulders and through thorny bushes and ana trees. The several times we crossed the Omaruru riverbed, the wheels sank deep into the sand. The oxen were also difficult. It had been some time since they had last trekked and they kept pulling on the yokes, one this way and one the other way, to avoid the soft sections of the road.

Close to the road lay several stone piles, the stones set on top of one another in a regular fashion. These were graves of non-Christian Damaras and reminders of the cruel, perfidious wars fought between Herero and Nama over land and livestock. Well beyond them, we arrived at a strangely level, wavelike plain where lay the buried remains of hundreds, nay thousands, of cattle that had perished from rinderpest. The road continued along a natural avenue of camelthorn trees, on top of which flourished thick bunches of reddish blooming mistle plants. We plucked a few sprigs to adorn the portable shrine of our patron saint—an old, bleached-out picture of the Madonna that Themis took with him on all his African travels and adventures, whether by wagon or by saddle. The camelthorn trees were loaded with yellow blossoms that filled the air with a fragrance similar to that of a lupine field in Germany.

The day was hot, and around four o'clock in the afternoon we reached Dawitsaub, a watering hole beside the river that offered fertile grazing for small livestock. All along the riverbed in this area grew huge, beautiful ana trees, whose pods, called anapills, are a particular delicacy for cattle, sheep, and goats. During years of drought, when proper grazing is difficult to find, these pods are a welcome supply of food and the animals are led to areas where they are likely to find anapills. The pods lie either on the ground in large piles or are shaken off the trees by herdsmen. In times of severe drought, when the cattle are starving and too weak to continue on, entire ana trees are cut down—though this is now forbidden—and the animals rush to eat the pods. Soon enough they recover to the point that they can be driven on short, slow treks.

Because of the abundance of anapills, the chieftains of Okombahe kept their cattle at stations on these plains. The station "masters" were mostly Mountain Damara of the absolute worst sort. They wrapped themselves in filthy, stinking, vermin-infested animal skins, and so lazy were these people that they did not even bother to build themselves huts to live in, but huddled under leafy branches. The men, who apathetically lay by the fire dozing, were entirely maintained by their women who not only had to care for the livestock but also had to forage for food to feed their large families.

Their food consisted mainly of a kind of grass seed that ants gather and store in small piles beneath the ground, augmented with roots, herbs, and berries. Locating a cache of seeds is no easy task, and a woman who is specially adept at it is highly regarded by men. This is, one supposes, just another chapter in the "History of Woman" who support their useless husbands. In the evening they pour the grass seed onto a skin and beat it with a wooden club to separate earth, sand, and small fibers from the grain. The wretched grain is then boiled in an old pot or tin filled with water, and when finished, eagerly consumed—though without benefit of salt the porridge is absolutely tasteless. It also has a curious effect on the human organism, for it causes large, hanging bellies, especially in young children. Mountain Damara also eat many an *unkjie*, a type of onion that grows abundantly in the savannah and tastes like floury paste. It also causes their chewing muscles to swell unduly and makes their already unattractive faces look all the more frightening and brutal.

The watering hole at Dawitsaub was small, which made drawing water for each animal a difficult task. The ground was covered with sharp prickly grass that stuck to us like burs. Taking branches, our servants laboriously swept a spot so we could spread out the skins on which to rest. We took a bit of a walk and suddenly found ourselves standing in front of the old Hartebeest House. The roof had collapsed, all the windows were broken, and the walls were crumbling to nothing. The old house gave an impression of utter desolation.

The name, Hartebeest House, originates from the time of the old Jäger (hunter), though I do not know why it was named after the Hartebeest antelope. The house was constructed like a pontok—though it had four corners—with the frame made of tree trunks and tree branches. The gaps between the trunks and branches were filled in with reeds and rushes, and on both the interior and exterior of the walls a plaster made of cattle dung and clay was applied. The roof was covered with bunches of reeds or maize straw and resembled the thatched roofs of German barns. The doors and windows were made of various kinds of packing crates. A house of this type usually has but two rooms: a combined living and bedroom, plus a storage room. The kitchen is housed in a nearby pontok.

Hartebeest House in Dawitsaub had originally been built by Theophilus Cloete, one of the richest Basters. But the wastefulness and slovenliness that typify the chief character flaws of the men of his tribe also contributed to Cloete's downfall. He sold hundreds of his cattle for practically nothing, then moved to Okombahe where he proved too indifferent and lazy to work for his own maintenance. His sole possession was an emaciated cow whose unborn calf already belonged to a trader.

All of South West is characterized by this obstinacy and rigidity. It seems as if even nature rebels against the culture we white people wish to impose—forcefully, if necessary. For once the white man has created something great, something that seemed almost impossible to accomplish, a year or two of sitting back and resting is enough to nearly erase all traces of our civilization. That which remains is but a sorrowful witness of the transitory nature of the things of the world.

Cold and windy, the night was uncomfortable. At a considerable remove from us, the driver had started his fire—most likely to avoid any disturbance while he ate his stolen meat. Themis sat at our fire and stared into the flames. I looked into the deep blue sky with its large, twinkling stars. Each of us pondered, and there was no overt communication between us. Early the next morning we continued our drive, leaving the riverbed and laboriously wending our way along a dreadful road that led between rocky valleys and steep cliffs. Stiff and ugly stood the aloe plants and candelabra cacti with their deadly poisonous sap. Their defiant and inflexible nature added a charm and ambiance to this wild landscape.

It was a dreary and lonesome stretch of land. Not a single soul did we meet, neither did we encounter even a solitary animal roaming about the veld. At times we had to stop and search for the road, often recognizing it only because the grass was shorter and had even less color on the tracks than on the surrounding land. Only an abandoned fire pit under a tall camelthorn tree reminded us that people had once passed this way. Ten or so small piles of ash lay in a circle, indicating to us that Ovambos had spent the night there. It is their custom to seek warmth during the night by having each member of the party make his own fire, which he uses to warm the front of his body while relying on his neighbor's fire to warm his back.

The most distinguished member of the group sleeps in the center where he enjoys the heat of all the fires. Herero and Hottentots are satisfied to stretch out around one large common fire. If the pile of ash is particularly large with thick, unburned branches nearby, it was the fire of a white man. Thus, it is possible to read from the remains of a fire, with a high degree of accuracy, precisely who had spent the night there.

The landscape gradually changed as we neared the river. Leaving behind this desolation, the road now opened onto a large savannah dotted with bushes. After a time, we decided to stop for lunch, unhitched the oxen, and ate beneath the heavy branches of a very old omborombonga tree. The wood of the omborombonga is so hard it is useless for construction. Indeed, it is nearly impossible to fell such a tree with an ax, and like all other trees and plants in South West, omborombonga trees should be left to grow into old age. The Herero actually trace their origin to this tree and it is, after a manner of speaking, sacred to them. Herero set small shavings from this tree in their milk gourds so their cows will always give milk. During illness and epidemics, Herero fumigate their pontoks with wood from this tree and carry omborombonga amulets to protect them from death, love sickness, the wrath of god, and whatever else might afflict them.

By this point in our journey, the livestock were so accustomed to the movements of the cart that we no longer had to watch them. As soon as we stopped and spread our animal skin blankets on the ground, they lay down on them as if it were the most obvious thing in the world for them to do. The servants fought a constant battle for their pots and pans against the curiosity of the goats.

Late in the afternoon, we resumed our trek beneath the shady trees that grew along the riverbed. It was evening by the time we reached Aubinhonis. During the previous year, several Herero settlements had stood in this area, but after the Herero had moved on, Mountain Damara drifted in and occupied the houses for a time. Because of this, all the dry wood close to the watering hole had been gathered up and used. As it was difficult to find any in the dark, we made a huge fire from desiccated cattle dung and prepared our evening meal: a porridge made from flour paste then mixed with goat's milk.

Shortly before sundown, we left the camp to take a few photographs. While we were gone, our driver, "Friend Manuel," stole some of our flour and rice, visited the local natives, and traded his ill-gotten provisions for a lamb. Late that night, assuming we were fast asleep, he crept away from the fire smoldering close to our cart and cautiously built his own quite some distance from our camp that he might roast and consume his lamb in complete privacy and without disturbance. We duly noted the theft, of course, but said nothing about it. The roads ahead of us were treacherous and we were completely dependent on the driver's skill to see us safely through; to confront him now could mean losing several of our oxen.

Very early the next morning, we arose and continued our journey. We were frozen, due in part to the cold, damp fog that rose from the river. As we rolled along, the landscape began to look more and more like the area around the Namib Desert. Along the roadside lay masses of bitter yellow melons. We had no meat with us and were very hungry, so we kept a constant lookout for game of any sort. We were in possession of a valid hunting license, which had to be renewed every year at a cost of thirty German marks but allowed us to hunt as much game as we needed for our daily maintenance. We were prohibited from hunting female ostriches, kudus, and oryx. For an additional four hundred marks, the government issued a hunting license that allowed one to hunt everything other than elephants and hippos. One who purchased a thousand-mark license was allowed to hunt any kind of animal. Licenses were not required for hunting springbok as they were plentiful and even considered a public nuisance.

Hunting licenses were, naturally, a welcome source of revenue for the government—especially since the hunting laws were strictly enforced on white people and violations were punished by disciplinary measures. It is questionable, however, if the laws actually protected African wildlife. The natives were not obliged to pay any such tax and could shoot any game animals within their territories, which usually had no precise borders. I doubt they had any concept of a "closed season," as they could not even name the days of the week. Several times I purchased female ostrich feathers and the skins of female kudus from them, and when I enquired after the origin of these items (since it was forbidden to shoot such animals), I received the typical answers: "Why does madam ask such strange questions? We have been in possession of these things for a long time," or, "Wild dogs bit the kudu."

The road led through an undulating landscape. At the top of a hill, the silhouette of a springbok suddenly appeared against the bright morning sky. Immediately, we stopped the cart; Themis and the driver jumped down and very cautiously began approaching the animal. In the inordinate clarity of the morning light it is easy to underestimate the distances between things, and as Themis and the driver ran toward the animal, the springbok stood absolutely still, its gaze continually upon us. The only available cover was milkbush. These appear as thick, stiff reeds; the dreaded sap of the milkbush causes blisters on the skin and can even blind cattle if the sap gets into their eyes. Like cats, Themis and the driver jumped from bush to bush. Two shots rang out simultaneously. But the springbok jumped away, its white tail visible in the distance; when it jumped again, we saw its hair bristling across its back. The creature gradually slowed—it had been hit. Two more shots were fired, then I noticed Themis waving a piece of white cloth. The springbok was down! I sent on a servant to carry Themis's rifle while I took charge of the ox team myself. Soon Manuel returned with the dead, bleeding springbok draped across his shoulders.

Immediately, the mood of the men around the cart lifted. The prospect of a proper midday meal—with opportunity to eat as much meat as they desired—continued to whet their already happy anticipations. Moments later, we encountered an entire herd of springbok and shot a female. A wild hunting frenzy took hold of Themis and the driver, and the two of them lay behind a rock formation ready to fire. A second buck was killed. The animal had fallen quite a distance from the road, but we had a good idea of where it lay. Crows and predatory birds suddenly appeared in the sky, circling the kill and cawing as they did so, greedily expecting to enjoy their spoils.

After a third animal had been shot, we left the road and rested under the blistering midday heat. We disemboweled the three bucks and carved out great pieces of meat. I cannot say who was the greedier, our hirlings or the dogs. Much meat was wasted. Thick pools of blood soaked into the ground, attracting innumerable flies. I found the entire exercise disgusting, and when the servants fried a liver for us to eat, I was fully unable to even countenance the thought—my appetite had been completely undone.

All the remaining meat we set into bags; the bloody skins were rolled into bundles and loaded into the cart. We were anxious to continue, but the men had stuffed themselves to such a degree that we were at pains to get them moving. After some groaning and stretching, we were finally able to get them to work. The two dogs lay motionless on the ground, their sagging bellies filled to capacity. The hunt had cost us a good deal of time, and we were keen to reach Achas, a place set along the bank of a seasonal river. Passing a night in the Namib Desert is hardly enticing. There is no wood to be found, the nights are bitterly cold, there is nothing to eat. On that particular evening, darkness settled in by six o'clock; no moon stood on the horizon. A chilling wind rose; the sky had grown so dark that one of the servants had to take the guide rope to lead the oxen along. Searching our way forward, we progressed but slowly, all the time hoping not to veer off course. Because the day had been so very hot, the animals were especially thirsty, and even the normally resilient sheep and goats appeared thoroughly spent.

Eventually, we turned out of the Namib and the road led us across some cliffs. It must have been ten o'clock at night when suddenly we noticed a bit of a rocky precipice, several meters high, rising ominously along one side of the road. Along the other side, sharp, smooth cliffs soared above us. We had found our way into a deep ravine and had to make our way through it. We recalled how dangerous this pass was even during daylight hours and chided ourselves for the sheer recklessness of attempting to traverse it on a very dark night, how easily our two-wheeled cart could topple over; then we worried about what might break first, how expensive it would be to replace a wheel, and so on. Without warning, the dark figure holding the guide rope in front suddenly started. He must have been sleeping when he led us onto the precipice. But he was awake now and attempted to draw us back on the road, hoping we had failed to notice what had happened. Unfortunately, the lead oxen had already followed him onto the precipice. The second yoke of oxen had sensed the danger and turned away; the cart, however, began rocking precariously to and fro. Should we jump off? Darkness and the steep cliffs made this inadvisable. A loud crack, and the cart tumbled. I was the first to fall. Then the crate, the mattresses, Themis, and the rifles all toppled down on me. I vaguely heard screaming and cursing. The driver was yelling. Themis beat the servant who howled for mercy. I slowly untangled myself. Sensing danger, the livestock bleated loudly, pressed close together, and were on the verge of a panicked rush into the darkness. Our men stood by not knowing what they should do. Immediately, I ordered them to unload the cart so that we might right it and check for damage. I feared for the wheel. Our dogs had been sleeping inside the cart and had not moved from it. Pinned between bags and blankets, we assumed they had suffocated. As the men offloaded, we finally uncovered them—sleeping soundly, then snarling angrily when we woke them. All our belongings were strewn about the rocks.

Fortunately, several turpentine bushes (called such because of their pungent aroma) grew close by. We tore them out of the ground and started a fire whose crackling flame burned a pale green. After tying leather straps to the cart's raised wheel, all of us tugged and pulled on the cart until we brought it into its proper, upright position. We examined the critical side and determined, Heaven be praised, that all was in order—not even a scratch in the paint. We checked the roof, the bars, and all the rest of the cart, and everything was in good condition except for a small piece of wood that had fallen from the very front of the cart. Calmed and relieved, we decided to stay the night there; the livestock would have to go without water, but there was nothing else we could reasonably do. We sat around a tiny fire, tired and exhausted. Willi, one of the servants, served us hot tea, which helped to revive our good spirits. I was the first to struggle to my feet and, after taking care of the cart, I prepared the evening meal.

We had no wood for a proper fire so we decided not to sleep on the ground but in the cart—nights in the Namib are terribly cold. Soon we fell into the deep, dreamless sleep of deathly tired people. When we arose the next morning, the sun was already high overhead. Our limbs were as heavy as lead. My left hand was sprained and swollen, my right hand shone in all the colors of the rainbow, and my entire body ached. While the servants lay sleeping, the oxen had run off to find water, and it took quite some time to round them up and securely hitch them to the cart. We finally broke camp and proceeded very carefully, knowing as we did that more treacherous stretches of road lay ahead.

The Namib presented us a splendid view. A blue haze enveloped the desert plane, and rising above it in the rose-colored shimmer of a full sun was the massive Brandberg (mountain of fire). It is one of the most beautiful sights I have ever seen in my life—grand and noble, yet also desolate and lonesome. What is man with his struggles and problems, his pomposity and nothingness, in contrast to the sublimity of nature? If someone asked me to describe its beauty, I could not do so. I do not know precisely what creates this wondrous and beautiful scene: is it the endless yellow desert with its long sandy waves or the magical blue haze or the sharp jutting cliffs? I do not know, but it piqued my senses; the beauty and grandeur of the moment seized my whole being. Tears ran down my face. I felt like clapping my hands together and saying: "Dear Lord, I thank Thee that Thou allowest me to see Thy beauty in this lonesome landscape. I do not mind hunger, thirst, and difficulties, but allow me to enjoy the miracles of Thy creative powers, far away from the hustle and bustle of man. Let me grow up and become good, so that I may become worthy of Thy handiwork."

Again we set our course toward the river, but Klein-Achas held no water so we trekked on to Gross-Achas. Themis was well known in this area from the times when he had had a cattle station there. Joyfully, the natives came to meet us. News that my husband was returning with a white woman and a cart en route to Cape Cross had already reached the people living along the river. In droves they came to greet us and looked me over, observing me from all sides and angles as if I were a creature from another world.

The reunion between Themis and the man who had been his stationmaster was cordial, though not without humorous overtones. This man had been given a wonderful nickname, Urib (louse), and struggled mightily with his Dutch.

"Good day, my honored Baas, do you still know me?"

"Of course, you are Urib, the Louse. Well do I know you."

"Eio!"

"How are you?"

"Not well, not well, my dear Baas."

"What's the matter?"

"My stomach is empty, only hunger lives in it."

"Where are all your cattle?"

"Look, Herr, now you are asking about the saddest thing. Rinderpest was in the country. All my oxen and cows died of rinderpest. I was very hungry and it was pitiful that all that meat should go to waste. So I ate some of the sick meat and it gave me rinderpest. It was my responsibility to herd the sheep and goats—which I did though I was very ill. I infected the animals and they also fell sick. Now everything is gone and I am still hungry."

"I am terribly sorry, Urib. Come to our cart tonight and eat together with my people—then you will get nice and full. Where is your father?"

"Look, Herr, again you are asking about a very sad thing. My father was very weak and old and sick. I had to leave soon with the cattle and take them elsewhere, so I went to my father and told him all of this. So my father says, 'I am sick and cannot come with you. I am old. Old people must die. Leave me here and go without me.' I began crying and lamenting. I cleaned out the pontok and only left the skins for the old man. I laid him on the skins and placed a bowl with unkjies and a jug of water next to him. And so that he could die in peace, I closed up the entrance to his hut."

My husband answered, "Urib, think. Perhaps your old man is not so sick after all and in the end may recover."

"Oh no, Baas, he is already dead. Every morning I returned to the pontok and listened to hear whether or not he was still moving. But since the day before yesterday all is quiet; besides, the jackals came during the night and tried to dig their way in. I also found the track of a hyena. You know how they follow the scent of dead people and can smell the dead person inside his house. Nothing can be done. I am sorry, very sorry he had to die, but he was old and I really could not use him anymore."

That was the story of our sentimental Urib.

We rested for one day in Achas and used the time to dry our game meat. The rascals cut small strips of meat with their dirty, unwashed hands and pocketknives of questionable cleanliness. The strips were salted and hung on a tree to dry. Endless swarms of flies tormented us, and the stench of game was most unpleasant. My aversion to meat grew to the point that I only had roasted bone marrow to eat.

From Gross-Achas we followed the river course for a while. Samuel and Karl took turns driving the sheep and goats, which, along this stretch of our journey, was quite difficult work. So many bushes grow along the riverbanks that it is quite impossible to see ahead, and if an animal stops to browse, as often happens, the herdsman may not take notice. We impressed upon the servants the need to be extra careful. As we made our way, we noticed oxen that were to be sold for slaughter leisurely sauntering along by themselves with the occasional sheep here and goat there. Samuel the herdsman was nowhere to be seen. We finally discovered him lying under a tree sleeping. My husband walked up to him, turned him around, and walloped him. The chap awoke and screamed his head off while Themis left him there and returned to the cart. Defiant, Samuel followed us at a distance, though by mealtime he had shed his surliness and joined us as though nothing had happened.

We made haste wishing to reach Leewater, the next watering hole, before darkness fell. The trail we followed defies all description. It leads along the edge of the Namib through canyons and valleys. The trail is so dangerously positioned along steep cliffs that everyone is well warned of the risks involved in following it. Indeed, horror stories are told throughout the country of broken axles and wheels and tired oxen. Needless to say, we were more than a little apprehensive about it. It was not too long before our cart began swaying with surprising force. Now and then, a wheel scraped horribly as it ground along a rocky outcropping or clattered as it turned in the air when the cart balanced on a single wheel. It was a perilous track, and given that we had just experienced an accident that left us hurting, we decided to walk behind the cart. This too was difficult as very sharp stones pricked right through the soles of our shoes. Strangely, these stones were beautiful—sparkling white, pink, and bright red crystals, a huge area filled with rose quartz. We set a number of them in a small pile to collect on our return journey.

At last beautiful green trees appeared in this desert wasteland; we had reached Leewater. The draught animals were unyoked by early afternoon and set to rest. We wanted to take a few photographs and were preparing to do so when Manuel approached and told us he was sick, that he had eaten too much springbok meat and now had violent stomach pains. He groaned miserably and, like a hedgehog, rolled himself into a ball on the sand, certain he was about to die. I gave him a large dose of calomel, had him drink a cup of strong tea, and covered him in blankets so that he would perspire. Generously, I offered two heaping spoonfuls of pork fat that Willi employed in massaging his stomach. He soon recovered.

That evening, a native from a nearby village appeared and reported that a leopard had killed one of his goats the night before and that the same leopard had been taking an animal a week from his kraal. We were grateful for the news and kept our animals close to the cart. It was too dark to gather thorn bushes for building a temporary kraal, so we decided to make a huge fire that could be seen from a great distance. We felt completely safe, and Themis played his harmonica while the boys danced around the red flames. Warmly wrapped in blankets, I slept in the cart. Suddenly, a hyena jumped through the fire into the midst of our animals. In a panic, they darted in all directions. Immediately we were up and began searching for them in darkness, fanning out in all directions. Now and then we heard an animal bleating and followed the sound. Sometimes we reached out to grab at a moving shadow, other times we detected something running before us. All around us howled jackals and hyenas, and in the distance, wild dogs barked. I prayed not to step on a snake and to find as many of the animals as possible.

By about four o'clock in the morning we had gathered all but five animals, and these we would seek at daybreak. We had just fallen asleep when we heard the most horrible screams. Confused, we quickly rose but immediately stopped with a sense of embarrassment. The cries were those of the Damara Guano Company donkeys stabled here at Leewater. Their dreadful braying echoed a hundredfold off the rocky walls that surrounded us and nearly scared us to death.

Our terrified servants came running. "Mister, Madam, do you hear?" they asked. "What is that? Oh, dear Lord, what kind of evil creature screams with such horror?"

We tried to convince them that the cries came from the tame long-eared donkeys, which were not at all frightful animals, but we simply could not calm them.

Soon Manual arrived and cried out, "Mister, let's quickly hitch the team and hurry away. This place is haunted by evil spirits. The bones of dead rattle snakes and ghost-hyenas lie all about. No one spending a night here ever finds peace. We may all be Christians, but God has no power over the evil spirits of the heathens!"

Under the circumstances, we decided it was probably best to get on our way. A short while later the team had been hitched and the cart rolled forward. A few minutes' drive brought us to the water where all five of our lost sheep stood peacefully. Two had been bitten by jackals, but even they were generally in good shape. With Leewater behind us, we soon reached the actual desert—an endless expanse of yellow sand and blue sky. We felt the bright sun and saw the air quaver from the heat. For an entire day we trekked across the desert and saw neither tree nor bush, not even a solitary blade of grass, just desolate waves of yellow sand. The cart moved slowly, and its wheels screeched and moaned as they cut deeply into the sand. The poor oxen had a difficult time moving forward, their big heads bowing under the wooden yoke, their long tongues hanging out of their mouths, their flanks quivering, their eyes protruding beyond their sockets. Again and again, the driver encouraged them forward, "Trek, trek," and then the crack of the whip. The long whip left thick wales in the hair of their skins. We lay quite listlessly in the cart following the principle that one felt the heat less if one moved as little as possible.

Far in the distance we could see the Spitzkoppies. At last the trail turned and we journeyed to the top of a tall sand dune. Standing before us was a black cone of a mountain and the green plains of Gui-Gams. We arrived there after two more hours of travel. It was about noon, and we had decided to rest near the water so the animals could drink their fill and recover from the exhausting march for several hours before trekking onward to Gross-Gui-Gams, where we would stay the night. We had been told that Gui-Gams was a very beautiful place—a wide plain covered with green, rustling reeds situated between tall mountains. I had been very much looking forward to seeing it. We wanted to take photographs, to paint, and make sketches that we might capture all the beauty of that area. As we drew closer, however, Themis shook his head and said, "This is not the same Gui-Gams that I remember." Manuel jumped off the cart to search for another trail as he feared we had somehow lost our way and now stood in a completely different place. A few native pontoks still stood in a semicircle upon a nearby hillside, but they appeared to have been abandoned quite some time ago and had fallen into a state of disrepair. We alighted the cart and began to look for those magnificent reeds whose magically beautiful rustling sounds are so much a part of the sagas and stories of Damaraland—stories of the heroic efforts of the Herero, who ruled over Xoudaman (dirt-natives), to finally repulse the malicious incursions of the Hottentots, as well as other natives who fought but finally had to submit to the hated whites.

We stood and looked. A great, wide valley lay in front of us, but there was no sign of the lush, green reeds. Instead we saw an ugly, stinking swamp with the occasional charred tree stump. A baboon quickly fled from between several stumps. The natives had burned the entire area because they believed the place had become home to an enormous, legendary snake. In vain we searched for fresh, clear water, but two wretched, stinking pools were all we found. The water level was so low that tree trunks had been tossed into pools and our servants had to climb down the trunks to fetch the water, the one on the bottom handing the pail to the second, who passed it to a third man kneeling at the edge of the hole.

We stood there surveying the devastation a group of foolish people had wrought in their stupidity and superstition when one of the servants came running and breathlessly told us that a white man with three horses was riding toward us. In our dirty and unkempt condition we were a bit embarrassed that Mr. Elers was nearly upon us. We had anticipated having some time to make ourselves presentable, but there was nothing to be done. Mr. Elers approached us, truly astonished to see that a white woman had ventured into this wilderness. He was further amazed that the woman, I, spoke English, his native tongue. He knew my husband understood English, but when I answered his enquiries in perfect English and told him that I had lived for a time in Merry Old England, he became my lifelong friend. We three decided to spend the evening together, and with Manuel's help, I prepared a fine supper. As ever, we spread the skins on the ground, this time beneath a beautiful old tree, served the meal, and relaxed as best we could. The evening wind picked up, but the tree sheltered us from it.

Soon the sun dropped below the horizon and African darkness fell about us. Elers and my husband quickly agreed on business matters, and the rest of the evening was spent in lively conversation while lying around the fire. By midnight, things drew to a close.

There were leopards in the area, so our animals were kept close to the cart. In fact, on the morning of this same day, one of Elers's servants, a Boer named Du Toit, had shot and killed a leopard not far from Gui-Gams. Elers had met him while the leopard's body was still warm, and Du Toit was dealing with the skin. The manner in which Du Toit had taken the leopard demonstrates how cool-headed these Boers are. Accompanied by his two children, ages four and six, he inspected the various cattle pens belonging to the Damaraland Guano Company. A huge leopard jumped in front of him and fled into a tree. Du Toit fired but missed, and it was his last bullet. The leopard remained in the tree, snarling and baring its fangs. Unwilling to let the old thief escape—it had stolen innumerable animals—Du Toit remained beneath the tree while his children hurried home to fetch him some bullets. For an hour or so he awaited their return, making a terrible racket to keep the leopard from leaving the tree. Luck was with him. When the children finally returned, the agitated beast was still perched on a limb. Du Toit put a single bullet in its head. The leopard fell from the tree, bringing a good deal of debris with it, and nearly landed on top of the children who were standing a bit too close to the tree.

The night passed quietly. Fortunately, the animals remained close to the cart, and the silver half moon looked down upon us peacefully. After breakfast, we separated. It was unfortunate, but Elers had to return; we, on the other hand, desired to stay on a few days in the area between Achas and Leewater. We had discovered a great many animal tracks in that area and wished to do some hunting.

It was such a splendidly cool morning that we took a long walk. We inspected the ruined pontoks on the hill near Gui-Gams. Samuel had entreated us to go with him to visit his father's grave. Samuel's father had died two years previous, and we had difficulty locating the grave. Smiling broadly, the rascal finally led us to a half-meter-deep hole in which a few strips of animal skin lay. This was the grave. There was no trace of the corpse and nothing but several shreds from the skins in which it had been wrapped. Hyenas had dug up the body and had had themselves a gruesome meal. As we tried to picture this scene in our minds, I felt as though ants were crawling up my spine. But such is the custom of the heathen natives. The deceased is wrapped and sewn into skins, then buried deep beneath the sand. A noise is suddenly heard coming from the gravesite. As the noise grows louder and louder, the weight of the sand covering the body grows lighter and lighter. Night winds whistle and howl across the desert, and the sandy covering grows ever thinner until the greenish moonlight shines upon the body. The noise changes to a frenzied scratching and scraping, and with greedy, bloodshot eyes the beasts stare at the corpse. They tear into it, ripping it here and there. They pull it from the sandy grave and quickly drag it across the desert. One of the beasts starts to eat; the second follows her example; the third and fourth are jealous. Finally, they join the struggle for the favored pieces and rip the corpse asunder with their sharp teeth and muscular jaws. The human bones are scattered to and fro, and when one rides across the desert sands, the bleached skull lies there grinning—the open grave yawns and tells its sad story.

We white people always take care to dig the grave much deeper, set the body within it, return the soil, weigh it down heavily with rocks, and finally to scatter dense thorn bushes all around the burial mound. The silent, slumbering body down there must be protected from gruesome beasts. I was glad to leave the remains of that grave. Samuel, on the other hand, seemed to accept these events as just the given manner in which things must happen.

The return trek through the desert seemed interminable to us. The heat nearly melted us, and the yellow desert sands blinded us so that tears stood in our eyes. We did not overnight at Leewater as Manuel and the other servants had fervently begged us to continue on. At the edge of the Namib Desert we finally unhitched the team. After enduring the intense heat of the day, we prepared for bed, but our sleep was broken and distressed. A damp, icy fog rose from the river. Our blankets were drenched, and we froze the whole night through. Needless to say, we got a very early start the next morning! Along the way we gathered the rose quartz we had set aside. Curious, Manuel asked, "What does the gracious lady wish to do with these stones? They can't be eaten and they don't contain gold."

At a good clip, we traveled past Achas and spent the night in the desert. There was, however, no wood to be found, not even enough to build a small fire to prepare supper or boil water for tea. In the morning we were met by a large herd of springbok, though we had no interest in shooting any of them, neither did we take aim at the quaggas. All we really wanted to do was get home. Themis fired a salvo of buckshot at a pair of bustards but only hit one of them in the wing. Cheering loudly, the servants snatched it up, and before I realized what was going on, had it under a cartwheel to remove its head. Later in the evening we spotted a flock of ostriches marching like soldiers on parade. They were too far away to shoot, and besides this, they were downwind from us so we had no cover at all. We spotted a good deal of other game animals, but our temporary distaste for meat was still stronger than our passion for hunting. By the time we reached Aubinhonis we had yet to fire a single shot. Early the next morning we continued on. In Dawitsaub, we only stopped for the oxen to drink and reached Okombahe after having been gone for twelve days. Everything was in good order.

The next day, we started to prepare our garden; it was, after all, the middle of September and high time for setting seeds in the ground.

9. Cultivating a Garden

During our first year, we limited our activities to a piece of land Chief Cornelius had assigned us, situated along the river. Themis first plowed the land, then fifteen women worked for three days pulling weeds, couch grass, and wild cotton bushes. We covered the soil with ox and goat manure and plowed it once more. We now considered the ground well prepared and began to set out beds for the plants. In one corner, we prepared beds for lettuce, cabbage, tomatoes, and eggplant. Close to these beds we sank a water hole. A second hole was placed in the center of the garden, and around it lay the beds for melons, pumpkins, and cucumbers. A third hole was dug on the other side of the garden that watered the beans, peas, radishes, carrots, and kitchen herbs, as well as the plants that had been grown in the seed beds. We dug the beds deep and mounded soil like a frame around the beds to conserve water by preventing it from running off. Daily watering would have meant a lot of work, and for that reason we connected all the beds with irrigation ditches so we could distribute water when and where it was needed.

We experimented with different methods of cultivating potatoes. We heaped soil onto one row, had very little extra soil on the second, and on the third the soil level was somewhere between that of the first and second. The row that was not heaped with soil produced the most potatoes. We did not grow corn or wheat because we could easily trade with the natives for them. Maintaining the garden was an enormous amount of work, and we absolutely had to stay on top of everything lest the plants should not receive enough water and so on. But it was wonderful to have fresh vegetables to eat. Vegetables we did not immediately consume I cut into strips and dried them in the warm air. I always guarded the drying vegetables from the bright sun as I did not want them to lose their color. When the vegetables were completely dry, I set them into thin linen bags and hung these from the ceiling of our pantry. During the cold season we had such an abundance of vegetables that we were able to share them with other settlers. With a thriving garden, we no longer needed to eat monotonous meals, and when we traveled, we were spared the expense of buying canned goods.

Pumpkins, melons, and cucumbers grew luxuriantly in huge numbers—Okombahe is known for its excellent varieties of these. We had first-rate seed plants and produced in such abundance that we had far more than we needed for ourselves. We gave bucketfuls to our people, fed our plump chickens on them, and did not punish our dogs when they ate the most beautiful watermelons in the patch. I tried to bottle things but did not have bottles in sufficient numbers to make the effort worthwhile. I pickled eighty prickly cucumbers in a large earthen pot. I checked them two weeks later and found that Old Ertmann and my husband had already devoured every last one! Pickled cucumbers were such a new and delicious treat in Africa that the two men could not even wait for the cucumbers to pickle all the way through!

We had big plans for growing produce as the land on the banks of the Omaruru River, all of it alluvial soil, was considered to be unusually fertile. The garden of Herr Merker, a settler, was certainly proof of that. The settlement of Omaruru also had beautiful gardens that had been established long before those in Okombahe. Worthy of special mention were the gardens of the mission stations, particularly the garden that belonged to the missionary, Dannert. His gardens included date palms, fig trees, and a vineyard. The grape skins were tough, but the flesh of grapes was deliciously sweet. There is no need to mention our experiments with palm trees, figs, and vines, only to affirm that I am convinced that large vineyards can be successfully cultivated in the Omaruru region.

Tobacco cultivation in our colony should not be underestimated, especially since the natives are keenly interested in this and already possess the basic skills. They very carefully begin a seed bed in the shade of a large tree and never neglect its watering. They also cover the seedbed with thorny branches to guard against the intense rays of the sun as well as to prevent trampling. As soon as the seedlings are about eight inches tall, they are transplanted. With unusual care, the natives construct a small, protective roof over the plants, resembling a miniature pontok though built of long branches. They do not irrigate by flood, but water each seedling individually. Each plant is set in a soup bowl–shaped recess in the soil, and as the plants become stronger, the thorny branches surrounding them are removed until the plant finally stands free. With the help of a good deal of water and the warmth of the subtropical sun, a plant reaches full development and is ready for harvesting within about four months. The leaves are then pressed together in large piles and covered with skins until they have ripened. The final product is sold on the market as rolled tobacco.

The finest tobacco in the colony is produced by the Boers in the Grootfontein area. In 1900, when my husband returned from South West for the first time, he brought several tobacco plants from that area with him and had them examined by experts in Berlin. The consensus was that the tobacco had an excellent aroma but that its taste was too strong, most likely due to improper fertilization. Under competent supervision, however, tobacco cultivated in South West could become the equal of tobacco from Mexico. Themis also learned how to roll cigars in Berlin, and when we became engaged, he delighted my father by presenting him a blackish-looking imported cigar from the colonies. With hopeful enthusiasm and a degree of courage the cigars were lit; yet by the time the cigars were half smoked, the enthusiasm and courage had dwindled to nothing.

One must never forget that cultivating gardens is but of secondary importance in South West. Given the huge, expansive, grassy plains of our colony, the true priority is cattle farming. Cultivating fruits and vegetables is difficult because water is scarce, and to create a large irrigation system requires a heavy investment.

We indulged in the luxury of planting a flower garden at the front of our home. Since the house had been built on rocky ground, we had to dynamite the ground in order to add humus to the soil. But for all this bother we were richly rewarded. In due course, sunflowers, mignonettes, petunias, and portulaka were in full bloom, and a type of vine popular in South West had begun climbing the walls of our house. Its clusters of whitish flowers spread a delicious, heliotropical aroma. We brought a tall aloe from the veld in our cart and planted it in the flower garden. We also planted eucalyptus trees, agave, oleander, cactus, and wild tobacco (which grows very fast and very tall). The wild tobacco blossoms look very much like the blossoms of genuine tobacco, and for this reason the plant mistakenly received its name.

Thus it was that our quaint home with the blooming bushes, the flowerbeds, and the several flower boxes tucked into the niches of our veranda left a singular impression of geniality and comfort. A white settler, who had lived in the colony for a long time, once saw me standing in the doorway of our veranda, surrounded by all these beautiful plants. He was genuinely touched as he looked over this lovely spot in the otherwise lonesome wilderness and with tears in his eyes admitted that he felt a bit of envy in his heart.

10. Experiences and Observations

Although pecuniary difficulties were always at hand, we were very happy in Okombahe. Our day-to-day lives consisted of hard, often difficult, work; the routine could prove monotonous, yet at the same time, our lives were eventful. We typically arose at five o'clock in the morning. My husband went to the kraal, roused the animals, oversaw the milking of the cows, and hired workers to toil in the garden. I fed the chickens and pigeons, which were routinely joined by wild peewits, Nama pigeons, and a variety of finch that caused much damage to the fields of grain. After this, I set our baby in the pleasant morning sun and prepared our breakfast. Unfortunately, these were tasks I could not leave to the servants as they were forever stealing things from the household. They pinched coffee beans and chunks of bread; they wolfed down large pieces of meat and added water to the milk. In short, nothing was ever truly safe.

It was pleasant to be on the veranda early in the morning when the outside world was fresh and the plants were all dewy. The bright sun smiled, flowers and herbs had been refreshed by the coolness of the night, wonderfully scented flowers were in bloom, and insects hummed and buzzed about them. Across the river, fog stood on the mountains. Natives attended to their work while singing their repetitious melodies. Herds and flocks of animals were led to pasture accompanied by their herdsmen's lively calls. We had not the luxury of spending too much time contemplating all this activity, for I had my work to do in the house and with the child, and Themis needed to head out to the gardens where his supervision and help were urgently needed.

Cooking in the heat of the day was an awful task. I could hardly bear to stand in front of the smoking stove, yet I had to be there to prevent the servants from stealing the food before it was even cooked. Though I tried everything including denying them food, coffee, and sugar, I was hardly ever able to catch a thief. With the most innocent looks on their faces they stood there and lied to me. To restore and maintain order, occasionally I resorted to my riding crop.

Midday was also the time everyone took a very long break from work. The entire area seemed dead at these hours. Everywhere, whether inside or outside their pontoks, people slept. It was utterly impossible to force anyone to work at this time of day. For our part, we usually lounged on our cool, shady veranda. Themis smoked and slept; I relaxed in my rocking chair, sometimes with a book from our well-stocked library. Studying at this time of day was not easy. The cookbook and the manual on poultry raising were too dull, Kant and Schopenhauer too challenging at this hour, and Nietzsche simply created pessimism. Thus I mainly read the classics and sometimes even indulged in entertainment literature—mostly French and English novels. We never lacked for reading material. The postman who traveled every two weeks from Omaruru to Okombahe brought plenty of this with him—in addition to the stacks of newspapers and magazines we regularly received.

If it were too hot or if I had had too much cooking and baking to do, exhaustion overtook me and I rested to my heart's content. When the air quivered with heat and I knew I had the time, I rested, for everything was quiet, yet everything around me was likewise festive. The sunflowers stood there, made dull and sleepy by the sun's heat. Brown, gray, green, and iridescently colorful lizards crawled out of the numerous crevices in the rocks. I knew them all and knew under which stones they lived, how cleverly they caught flies and bugs, and how they ran across the pebbles and rushed in between the leaves. House lizards lived on the outside walls of our veranda. They had big heads and flat stomachs, and we could have written volumes about them. Every now and then, when a hunt on the walls went too fast, one of the lizards would fall to the ground. I then snatched it up, set it on its back, and hypnotized the thing by stroking its belly. Sometimes more than a dozen lizards lay about hypnotized in this manner. It was very amusing to watch them as they began to waken and stretch, slowly turning their eyes, until they gradually regained consciousness. Then quick as the wind, they rushed up the wall, and once arriving at the top, began to chatter with one another—a sound similar to the chirping of sparrows.

I also found it charming to observe a female hummingbird. She had a plain gray coat and a long, pointed, and unspeakably thin little beak. For a time, she was a regular noon-hour visitor to our veranda. Searching for a place to build a future nest, she finally discovered just the right spot between the branches of a wild tobacco tree. With all diligence, she fetched wool, hairs, and small pieces of straw and artistically wove them into a round hanging nest. For two weeks, the little creature worked tirelessly. Every day, her husband—black-coated and red-chested—visited her. Very nobly, he flew to the nest, examined it, and praised or criticized it (occasionally, he pulled out something his little wife had so carefully woven into the nest). Finally, he made a few melodious sounds and flew away. He never helped. I never even once saw him bring a single bit of nesting material. He did, however, visit every day at the same hour. I also found it amusing to watch as the female tested the doorway of the nest. Eventually, everything was completed to her satisfaction. Then one day, I noticed three tiny eggs in the nest. The male still came every day as he had done before, but he no longer criticized things; instead, he sang his heart out. Such was our idyllic life in South West—and it was only the Herero uprising that brought it to an end.

Our flower garden was charming. To conserve the moisture in the soil we had left various fieldstones and porous rubble on the ground, but the moisture was also attractive to a variety of small animals. I was particularly terrified of the snakes, especially since it was during the rainy season that snakes were out in greatest abundance: poisonous mambas, horn vipers, puff adders, spitting cobras, and the African giant snake. The black mamba was very much at home in Okombahe and was often found in the pontoks of the natives. Many a restful midday break was interrupted by the loud cry, "Auba haa!" (The snake is here!). I too lived in constant fear of snakes. Many nights I was up searching the rooms because I thought I had heard a rustling from somewhere. I combed through my child's little bed and shone a light in every corner. There was a time when I almost fell victim to a mamba. I was standing beside the kitchen table preparing supper when one of the servants called, "Madam, stand still—do not move!" The servant then lifted the fire tongs and hit me so hard against the shins that I jumped with pain. Triumphantly, he held the dead mamba before my face. It had risen in front of me from beneath the table. If I had made the slightest movement, it would have bitten me.

Around three o'clock in the afternoon, work started again. I was up and preparing a "three o'clock tea" for visitors who usually came at this hour. Inevitably, the missionary came, as did Farmer Merker, and the soldiers stationed in Okombahe. In South West, as in any new country, I suppose, hospitality plays a major role. Every European settler is invited and hosted regardless of his or her social class. In larger towns like Windhuk, Swakopmund, Okahandja, and others, settlers are very much more class conscious. We had a great deal of contact with the settlers on the surrounding farms and with the officers of Omaruru. It was my pleasure to make our home as comfortable as possible, and it was mostly the lonesome bachelors of the area who so enjoyed our home and hospitality.

It is no simple undertaking to host guests. Preparing the guest bedroom and the meals and drinks is certainly easy enough, but the guest brings his own servants, as well as his wagon, his horses or oxen, and his dogs. Thus the entire entourage must also be hosted. Lodging for the servants has to be found, the wagon must be guarded to prevent theft, the draught animals need to be fed, and an extra kraal must be organized for keeping them safe. Nevertheless, I would much rather host guests than travel myself. It was especially difficult traveling with a small child. Water sources were infrequent and the nights very cold. When it was necessary to travel, we took a cow along so the little one would have milk. He was only a few months old when he slept near a campfire at night. His eyes were filled with dust from the ashes, his little nose was a bluish-red from the cold, and his limbs were stiff. Fortunately, no lasting harm resulted from the fact that I could not keep him immaculately clean while traveling.

It was also during the afternoon that the natives came requesting things. Captain Cornelius, the local leader, always seemed to be thirsty and was a frequent visitor. Our conversations with this strange old man were indeed interesting. He hated the white man just as much as all the other natives and did not hide his feelings about that. "Our time has passed," he remarked. "Now the white man has the power in Damaraland. I obey him as his servant because that is to my advantage. The future rests in his hands, and we must obey if we want to get along—but I do not love him!" The Captain's many affairs with the women of the area made him a thorn in the side of the local missionary, and he had been excommunicated from the church many years ago. For his part, Captain Cornelius chided the missionaries terribly. He reported that once at communion—wine drinking, as he called it—he had taken the cup from the missionary's hand so he could drink more comfortably and generously. The missionary, however, became very angry, yet from the Captain's perspective, it was quite natural that he should have done this—for he was, after all, the captain and therefore more deserving of wine than the others!

The Captain's mother was an ancient, desiccated little woman whom he allowed to starve. Every morning, the old woman crept about with the help of a cane, and with a beggar's sack hanging from her shoulder, she went from house to house begging for her wretched maintenance. "Captain," we asked, "You are rich, so why don't you give the old woman something to eat? She is your mother, isn't she?" He replied, "Amei, it is true that I own many cows and oxen, but the woman is old—very, very old. In the past I gave her food, but now her time is up. Why does she still exist? Do you see anyone else her age around here? No, just let her die. I've done enough for her, now she must take care of herself." We could not persuade him otherwise.

Old Rudolf was one of the few genuinely pious men in our area. He could initiate hours-long discussions over some or other question of his own design with the local missionary or with me. He was probably the one man of the area who truly felt the importance of rearing his children with Christian discipline. At the same time, he was an odd sort of man who inflicted punishment in the strangest manner: biting. Despite his advanced age, he still had beautiful white teeth. He himself was frightfully unattractive; indeed, when he smiled he reminded me very much of an old baboon. Whenever one of his numerous children committed some infraction, he pounced on the child, held him down, and bit wherever he was able to on the child's body. If he wanted to remind a child of something, he always slapped at his shoulder.

His son, Friedrich, served as our driver for a very long time. When he drove the oxen with the whip resting on his back, he philosophized. I well remember a discussion we had around the campfire in Omaruru. Deep in thought, he stared into the flames. I noticed his silence, called him to me, and enquired if anything were wrong.

"Not really, Madam, I am just thinking."

"What are you thinking about?"

"I am thinking about God—whom I do not understand. May I ask you your opinion? The people around here say that you are very learned."

"Certainly you may ask."

"Listen then. When the Baas goes into the garden he works. He plants and pulls weeds, he builds seedbeds and measures them to make sure that they are as straight and beautiful as possible. He works as hard as possible so that his work is good. And when it is not quite as he wants it, he improves upon it until he is satisfied."

"And what do you want to tell me with your observation?"

"Wait, Madam, I am not finished. You have a child, a son. I can see that you take care of him as best as you can. When he calls for you, you are always there. But you also punish him when he is naughty. Why do you do that?"

"Because I want him to grow up to be a good man someday."

"Later, you will send him to Germany, you will spend a lot of money so that he is able to learn everything necessary. Why do you do that?"

"Because I want to help him to become a good man."

"Now pay attention. Our dear God has created people—is that not correct? And he must have made them good because he was satisfied with his works. He wanted them to live happily in Paradise. Then, suddenly, he sends them temptation, and man falls. Man was not entirely good, not entirely perfect after all—otherwise he would have resisted temptation. Our dear God himself must not be perfect since his creation is not perfect. Yet it would have been so easy for God to have made man perfect. God made man from a lump of clay just like this (Friedrich formed some earth into a ball). First, he gave form to the head, then the eyes, the ears, and so forth. Since God is supposed to be omniscient, he must have known that man would be tempted. Why did God not create man so that he could withstand temptation? Man could still waver, still struggle, but why did God not give man the willpower to win in the end? All God needed to do was to take the lump of clay and give him a tiny kink or bobble somewhere. The pressure of God's little finger would have been enough—perhaps on the head, on the forehead, but where precisely I do not know—and this bit of pressure could have changed man so that he could lord it over evil. Then sin would have no power to this day and we would be happy. You see, Frau von Eckenbrecher, I do not understand why the missionary preaches that God loves us. We are poor, hungry, ill, miserable, and sinful. Why does God not make us happy? He is omniscient; he knows temptation—He sends it to us, after all; He knows that we are powerless against it, and yet He does not help us. And He is all-powerful!? Then why does He not banish misery from the earth? The missionary says we are God's children—just like our white brethren. God loves everyone equally, but look at us, what are we? Dogs, slaves—we are even worse than the baboons on the cliffs."

I recalled the lectures of Professor Harnack and Professor Seeberg, but found no answer to help this struggling native in his anguish. Here in South West, like everywhere else in the world, people wrestle in their attempts to understand the workings and motivations of God. I was frequently asked these and similar questions. Why do we maintain that missionaries, who receive the barest of instruction and training at the missionary station, are suddenly capable of carrying Christianity to these "poor" heathens on the "dark" continent? Heathens are also capable of thinking—not all of them are dull-witted!

It was mainly in the afternoon that natives would arrive to trade livestock for various goods such as flour, rice, sugar, coffee, clothing, and shoes. To emphasize the festive nature of trading, they brought along a translator. Many of the natives had no concept of abstract numbers and used pebbles or coffee beans to help them. My husband spread out a handful of pebbles before the trader and took as many as he felt the animal was worth (one pebble equaled one German mark). Once they reached agreement, the seller pushed a number of pebbles or beans toward my husband, which signified that he wished to purchase various goods to this amount. If he intended to spend only fifty pennies, he cut a bean in half or placed a pebble on his palm and drew a line across it with his finger. When the money represented by the pile of coffee beans or pebbles matched the value of the goods wanted, my husband was expected to add a gift, usually of tobacco or matches, and the transaction was complete.

Trading sessions such as these required time, and the higher the value of the animal being traded, the greater the number of complications wanting resolution. People almost never paid in cash in South West. Servants, herdsmen, and messengers were paid in the same manner described here. Almost every farm had a messenger whose job it was to collect ammunition from the regional government office and take news and letters to neighboring farmers. The messenger was always given food for the days he traveled. Similar to the American Indian forest-runners, the letters carried by messengers were placed on a notched stick and tied in place with a string or plant fibers. During the entire trip, which could last for days, the messenger carried this stick in his hand. This arrangement, it was found, was far safer than permitting the messenger to place letters in his pocket, provided, of course, that he even owned a garment with pockets.

I rarely had much time to rest in the afternoon. I helped with the garden work, looked after our child, and mended clothing. Of all the household chores, I disliked the weekly laundering of our clothes most of all. Even under normal circumstances washing is not exactly regarded as a pleasure by German housewives. My two washer women squatted on the ground in front of the wash tub to do the laundry. They had short pipes in their mouths, used water sparingly, and took their time. If I did not keep my eye on them constantly, they would rub the clothes with handfuls of pebbles—which did nothing for the longevity of the clothing. They never actually got used to a clothesline and clothespins. What good were these things when one could use the nearby thorn bushes as drying racks? The clothes stuck to the bushes making it difficult to get them off when they were dry. One had to delicately lift the clothing from each individual thorn—which took time and patience—or each thorn ripped a hole in the clothes. These washer women had not the slightest inkling of how to gather the clothing properly. A large bottle filled with water was used to flatten them. I myself had to starch and iron the clothes. The wooden fire in the kitchen stove was not hot enough to properly heat the iron, so a special "ironing fire" had to be set in the yard, using cow dung as fuel. The kitchen became terribly hot and filled with steam. The laundry hanging outside on the line dried almost instantly in the warm, dry wind, and was immediately covered in a sheathing of sand.

Emma, one of our washerwomen, and a Hottentot, was especially devoted to me. She loved my little son very much and sang him lullabies that she herself had composed. She was particularly proud of her favorite song, which in German goes something like this: "You white child, born of white parents, are now a Bushman in Damaraland. You are lonely since your home is in a strange land. Your family (i.e., extended family) is far away, and your grandparents long for you. But some day you will visit them and their joy will be great. Then you will be lonely no more, you white child. Be quiet therefore and do not cry."

It seemed that whenever I thought I had a moment to rest, I was called to visit the sick. I was not a volunteer nurse or anything of that sort; people simply came and fetched me. Our drug and medical supplies were limited, but in his kindness Old Ertmann gave me his, and thus I became a famous "doctor" in the area. People came to me with the most unbelievable illnesses and requests. My main drugs were quinine, calomel, and opium, and with them, one can cure most diseases. Occasionally, I was able to help with internal illnesses, but most people were simply starving and only needed sufficient food to regain their strength, or they had a fever—which any white person can successfully treat over here. What I dreaded was cutting, washing, and dressing external wounds, and treating them with antiseptics. It was most unpleasant for me to treat people who had festering sores or who had smeared the universal remedy—cow dung—all over their wounds to prevent air from "disturbing" them. Countless flies and maggots infested these wounds, and often it required all my emotional strength to simply touch them. I helped three women deliver their babies, and in these instances I was happy to help when my services were needed. I experienced much of mankind's misery within my small circle of influence and I learned to overcome reluctance, though not without great effort. There were times when I was about to decline a cry for help but then felt ashamed of my selfishness.

One of the Auderlings (elders) came to me on a very hot Sunday afternoon and said, "Madam, come and help." He described the nature of the illness, and from what I could gather a child had begun experiencing violent seizures during a St. Vitus dance. I honestly felt there was nothing I could do to help in this instance, so I refused to go. The old rascal pleaded vehemently, "If you think you cannot help, then just come along anyway and allow the child simply to look at you. This alone will cure the child." What was I to do? I was deeply touched and accompanied him in the heat of the day. It took us an hour to get there. Incidentally, I did heal the child—after three months of bromine treatment. I received no thanks, of course. Often my cured patients later visited me. "Madam has healed me," they would say, "so now she owes me a present."

Herero also came from afar to seek my help. After church on Sundays, I held involuntary visiting hours for sufferers of various ailments. It was understood that I was obligated to console people who had sorrows, lovesickness, disappointments, and diseased cattle. Often I became a godmother for a newborn child. People also came to ask me to translate, to read, and even to write letters.

I was deeply distressed about the superstition and outright stupidity of the people. One day, for example, subcaptain Josua came and asked me to help his five-year-old grandchild. The little girl had fallen into a fire and was badly burned. Immediately, I went with Josua and found the poor thing in a most gruesome condition. Over half of her body had been burned. I did my best to help her, and was greatly assisted by Captain Franke, who happened to be visiting us. He promised the services of his staff physician in Omaruru and requested the child be taken there post haste. After a good deal of debate among the natives, our cart was made ready so the child could comfortably lie suspended in the cart. We were also prepared to lend them our team of oxen. Josua asked me to write a letter detailing the condition of the child. Preparations were finished when suddenly the mother refused to come along or to let the child go by herself; so Josua went alone to Omaruru to obtain clean, proper dressing material as well as written instructions. While he was underway, the child gradually improved. Miraculously, for want of a better word, the wounds began to heal, and I fully hoped the child would make a complete recovery.

I visited the girl every day. One morning shortly after Josua left for Omaruru, I found to my amazement that the pontok was closed. I knocked on the door and could hear the little creature moaning inside, but nobody opened to let me in. It was the same situation at noon. In the afternoon, I met Josua's wife who said, "You needn't come any longer. You know that the mother of the child is a heathen; she is afraid of your magic and fears that you could harm the child. You must stay away." I respected her wishes. During the following night, that same woman suddenly awakened me and with tears in her eyes pled, "Come quickly, the child is dying." At first I was reluctant to do so and told her, "If I come now and the child dies, everyone will blame me." She begged and cried until I finally gave in. The door of the pontok was closed. I bowed down and entered. A fire burned at the center, and its biting smoke filled the pontok. The smoke brought tears to my eyes and at first it was impossible to see anything at all. Then, six or seven women, squatting on the ground, came into view. The little girl lay close to the fire; I could tell immediately that she would not survive the night. I opened the door widely to allow the smoke to clear out. At once, an ancient, completely naked, beastly old hag jumped toward me, stretched out her leather-like hands, and screamed curses at me. Her neck, chest, arms, legs, and hips were covered with strange amulets. I told her to be silent, that later we would go outside and she could tell me what she wanted. Lydia, Josua's wife, stood in front of me wringing her hands.

> "Dear Lord, Madam, be careful," she admonished me. "You are insulting our greatest magician, the wise woman who came from far away. She was paid a cow to heal the child."
>
> "Well, has she been able to do it?"
>
> "No, she says another power is working against her and because of that, the child will die. This other power is your healing art."
>
> "Then I will leave at once."
>
> "No, don't do that. Your art is more powerful than hers and you must help."
>
> "There is nothing I can do. Before the moon reaches the corner of the river, the child will be dead."
>
> "You must try to help—come, have a look at the child."
>
> "Only if you do what I ask. You must throw out that old hag."
>
> "She'll curse us if we do!"
>
> "Let her curse you. In the morning, I'll give her coffee and sugar—then she'll bless you."

They did as I asked. The old hag sneered at me and offered her dirty hand as she went outside, her amulets clinking. The first thing I dealt with was the fire. The smoke was caused by a series of magic wooden sticks piled crosswise on top of each other, herbs, and tree bark packed between the sticks. I had them light the lamp I had brought, then I examined the child. They had taken her from the bed I had carefully prepared; now she lay on the ground atop magical sticks and herbs. My meticulously prepared bandage had been removed and covered with rancid butter, cow dung, and some kind of mashed root. She had an amulet around her neck, with blades of grass twisted into the curls on her forehead. There was really nothing to be done at this point.

"Have you given the child anything to eat?"

"Yes."

"The food I sent?"

"No. The magician ate that herself."

"What did you feed the child?"

"A maize cob and goat meat roasted in the ashes."

"Did she eat it?"

"She did not want to, she just called for you. We forced her to eat until she fainted."

"Lydia, help me to set the girl back on her bed."

Before excusing herself, she asked, "Madam, you did not hear me. Must the child die?"

"Yes, she will die tonight."

"Please try and wash off the magic salve."

The poor little creature whimpered quietly. Why should I compound her agony? I decided to do what I could to make her as comfortable as possible, and before the moon set that night, the little girl had expired. But my work was by no means done. I had cared for the sick child, and now it was up to me to provide a coffin. Normally, the body is sewn into animal skins and buried, but in Okombahe such a burial would have been disgraceful. "After all, we are Christians, and like Christians everywhere, we must be buried in a coffin." Wooden planks were a rarity in Okombahe, so the family came to me and said, "Just give us a wooden box instead of a casket, and provide us with linen for burial cloth." I did.

Like a pauper's bell, the church bell tolled pitifully shrill, and the little creature was returned to earth. Soon, the mourners came to me again. "Madam, you took care of the child," they said. "You furnished the casket and the linen, now give us coffee and sugar for the funeral repast." On occasions like these, one can well imagine that often I reached the end of my tether.

I will never forget another tragic occurrence either. One of our female garden workers, a woman named Christine, was a young mother of four children. During the morning's work in the garden, she had chattered away and returned to her home happy and well. About half an hour later her husband arrived at our house, out of breath and ashen-faced. He begged me for linen to cover his wife who had just died. When a native dies and loses his coloring, he looks positively frightful—I think even more so than when a white man loses his coloring. We tried to comfort this husband by telling him that his wife must have fainted, but he insisted that she was dead. He was right. It was only later that we found out he had killed her during a quarrel. We gave him four meters of cotton cloth that he used to cover his wife. A while later, he returned and asked if he might be allowed to gather old planks for a coffin. He found what he needed, and soon we heard him hammering during the heat of the day.

I watched as he worked under the shade of a camelthorn tree before our veranda. At first he seemed very sad as he worked away, his stubby pipe between his lips. I felt sorry for this poor man who had lost his wife so suddenly. We found a few better planks and gave these to him along with nails. His task became easier and his mood improved markedly. I went over and inspected the coffin—a short and unusually narrow box, so narrow, in fact, that I had no idea how a human being could possibly fit inside it. I told the gathered people of my concern, but Traugott, the deceased woman's husband, answered my worries by saying, "Don't fret, it will be all right. If she doesn't fit, we'll press down on her body until she does fit!" An hour and a half later, the coffin was finished.

Christine was to be buried the next afternoon (but not among the cliffs, of course, where the godless, evil heathens are scratched into the ground!) in the graveyard where all the pious Christians rest, where one is buried with the blessings of the missionary and to the tolling of the bell. Proudly, everyone admired the coffin, and everyone seemed satisfied. The coffin, however, was not carried to the pontok where the dead woman lay, but remained among the unused boards beneath the tree. The next morning at nine o'clock, Josua appeared and asked for all workers to be allowed to attend the funeral of poor Christine. Men and women filed past us in their most colorful Sunday attire.

"I thought Christine was to be buried this afternoon," I said.

"You're right, but that's no longer possible, her body is already decaying. I am on my way to find the missionary so we can get the burial started."

Soon after that, four chaps arrived with large, smoking pipes hanging from their mouths, and bore off poor Christine's body, setting it beside the coffin.

As predicted, the body did not fit inside the coffin, so they decided to chop off Christine's feet, which they matter-of-factly did, and set the severed members inside the coffin. Then with great effort and a mighty push, they squeezed the corpse sideways into the box. A few loose planks served as a lid, ox hide straps were tied about the coffin so that it could be carried, and the men heaved forward, moving as quickly as they could toward their cemetery. Arriving at the church, two shrill bells were rung in honor of the Christian soul's homecoming. Dressed in his cap and gown, the missionary stepped from the church, and the small procession advanced to the cemetery, everyone smoking like chimneys. Twenty minutes later, the exercise was finished, and an hour after that, the sorrowing Traugott came by and begged for sugar that he might prepare sugar beer for the celebration. Not quite three months passed before he remarried.

A further note on burial: The pious zeal of the good Christians in Okombahe went so far that heathens were not allowed to be buried in a coffin. It sometimes happened that a wealthy heathen Herero or Hottentot went to the store to buy wooden planks for building a coffin in which to bury his deceased relative, normally paying cash or trading a fat sheep for the wood. When this happened, immediately a group of Christians assembled and mobbed the heathen. The wildest scenes imaginable occurred in that store. Often fierce clashes broke out to prevent the sale of wood. Like wolves, they fell upon the buyer, and if he fought back and succeeded in getting away with his lumber and actually made a coffin, the mob destroyed it during the night. Occasionally, a mob would tear the heathen corpse from the coffin or even dig it out of a grave. He, the heathen, had no right to be buried in a coffin!

Late in the afternoons, before the livestock came home, we often went bird hunting. Sometimes we searched for wild honey, but only once did we actually find some. Old Ertmann went off work with dynamite, a stone drill, and other tools in order to break open the rock in which the bees had made their nest. After a lot of hard work, Ertmann finally succeeded. Yet just before he actually got to the hive, a very tall Herero man appeared and claimed the honey was his. He asserted that he had discovered the hive before we did, that it belonged to him, that if we wanted the honey it was only right that we buy it from him. We did not agree. The produce of the bush is free to everyone, so we refused and continued working, placing the last stick of dynamite in the rock. The Herero man gave no heed to our urgent warnings to step away. He refused to leave and stood there defiantly. With a deafening roar, the explosion sent stones and other debris flying about his head. Not harmed in the least, he fled as fast as his legs would carry him. We walked back toward the hive, thousands of frightened bees swarming about us. Inside the rock we discovered numerous honeycombs; however, they were old and completely devoid of honey. The entire week of hard work had been for naught.

Just as we participated in both the happy and sad events of our "neighborhood," the natives also came to celebrate with us settlers on special occasions. As soon as they discovered that visitors were expected, or festivities were being planned, they eagerly offered their services. Shortly before my birthday in September, we checked our finances and noted that we had made a considerable profit. After some deliberation over what we should do with the extra money, we finally decided to purchase a hamper filled with bottles of champagne and delightedly anticipated our celebration. But as it was September, the season of greatest drought, and all the local water holes were completely dried up, and the ground water so low that five men were needed to bail water—one above the other to form a vertical chain—we began to think twice about our plans. The animals found precious little to graze upon in the fields. Those who could afford it drove their herds to areas with fresher grazing. Each day, several animals from every herd died of hunger and thirst. Nearly all the lambs that had been born in late summer were dead. The cows were terribly scrawny and hardly had milk enough for their calves. It had been weeks since we had seen any butter.

The people of the area suffered terribly from the drought; many of the older ones actually died. I visited several of the heathen pontoks on the other side of the river and found people lying on their worn-out animal skins, looking for all the world like skeletons, too weak to even sit up. A few made the attempt, but then fell back. So spent was their strength that their heads just rolled to one side. I remember once having seen a very old dog too weak to even stand; this is exactly how these people looked to me. The ones who still could walk dragged themselves to our house where they squatted or lay in front of the animal enclosures, in the yard, or in the kitchen. They stole the bones from the dogs, and small children who could barely walk stole my chickens and lambs.

Under these circumstances, we could not drink the champagne; we were not in a celebratory mood. Themis went and talked to Captain Josua, telling him, "Tomorrow at noon, whoever wishes to, may come to our yard; and everyone who does will have something to eat." At dawn, the first ones had already arrived. We were deeply moved by the sight of all these hungry, miserable, and sickly figures. Friedrich, our driver, had to protect the two huge washtubs with his whip as the sweet maize-meal porridge cooked. I personally supervised the portions and the distribution. First the children were served, then the old people, followed by the women. Last of all the men were given their due. From that point on, we helped the people as much as we could. But what we were able to offer them was not much—like a drop of water on a hot stone. The missionary also helped. We petitioned the government for a good sum of money, and Captain Franke, in Omaruru, was indeed issued relief funds that were passed along to the missionary for him to disburse in ways that helped the starving people.

Shortly before Christmas, the landowner, Merker, lost his eldest child, a darling four-year-old girl. She had been ill for a long time and the Merkers had sent for the doctor; however, five days hence was the earliest he could come. Because of this, Captain Franke sent a nurse while another doctor from Karibib was being summoned. Waiting, hoping, then slowly losing courage is a terrible thing. We so much wanted to help them, but all our good will could do nothing. The little girl's condition worsened, and she died before a doctor was able to come. The poor Merkers blamed themselves over and over again: "If only we had sent for the doctor sooner, if only we had done this or that, perhaps the child would have lived." The exhausted mother was absolutely inconsolable. I can still picture her sleep-deprived, care-worn face, and how, with trembling fingers, she cut a lock of her daughter's hair. Again and again, she caressed that pale little head.

I believe that losing a loved one in South West is more difficult than losing someone at home. In Germany, the body can lie in state for several days before burial, allowing one more time to gradually accept the fact that one must bid farewell to a person whom one has loved above all others. In Africa, on the other hand, the very warm climate necessitates the almost immediate burial of a deceased person. The Merkers' child had died at noon; by five o'clock in the afternoon it was time to place her in a coffin. At seven o'clock on the following morning, she was buried. All of us were present. The missionary's wife and I prepared a garland and several wreaths for the simple coffin. We arrived at the cemetery—an austere place with sandy soil. The grave mounds were weighted down with heavy stones, and the entire grave is covered with thorn bushes. Nowhere was there any sign of loving care or loyal remembrance; nothing can be planted and grown in this wasteland.

Our Christmas celebrations were muted by these tragic events. While walking in the surrounding area we had often looked for a suitable Christmas tree. But not finding one, we had a collapsible, artificial, ghastly green fir tree delivered from Karibib. In comparison to the beautiful trees that graced our parents' homes, this one was modest (to say the least), but just the same, we decorated it with loving care. December 24th was probably the hottest day of the year, and when we lit the candles that evening, the branches bent low, the candles burned upside down, the chocolates slowly melted and dripped steadily on the floor. Still, it was very festive, the first tree in our own home, a sweet child who cheered when the candles were lit, and the peaceful but delighted faces of the natives. We were healthy, vigorous, and looking forward to the future. Our only guest was Sergeant Müller, for whom we had placed cigarettes, cigars, and a few other small items under the tree. The settlement was full of life. Standing in the church was an artificial tree, decorated with many candles and little cherubs. Together we sang Christmas carols, after which gifts were distributed.

Outside, the grain was ripe, and plenty of watermelons lay ready to harvest. The difficult period of sickness and starvation was now forgotten, and the natives enthusiastically sang, "Christmas is wonderful. It is the celebration of the whites, therefore they must give us presents. Now we eat sweet bread, we celebrate, and we dance. It's nice that so many white people are here—because that way we get more presents." During these bright, moonlit nights, accordion music was heard along with the rhythmic stamping of dancing feet and the wearisome melodies of the natives' songs. We settlers were happy and satisfied. Of course, the natives seemed not to know songs of gratitude, but we did not give to receive their gratitude in return. We gave because giving is joyous and delightful. Still, we often feel a bitter strain: "The white man is stupid. He has no reason to give us presents. We give him nothing in return for his generosity. But come, let us beg from him. The main thing is that he gives us things."

11. Spitzkoppies

On horseback from Okombahe, one can ride a southwesterly course and reach the Spitzkoppe in about six hours. It is a giant massif of compact, red granite peaks. Its name, Spitzkoppe (Pointed Heads), derives from the sharp, conical shape of the peaks. As to formation, they are reminiscent of the Dolomite Mountains in Italy, and in clear weather, the highest peak is visible from the ocean and is even marked on ocean charts. The Spitzkoppe is part of the final large mountain range that stretches into the Namib Desert. These mighty peaks rise from an immense, endless plain, and in the brilliant sunlight appear reddish in color. The foothills are draped in a mysterious blue haze. Like huge watchmen, the mountains stand in the quiet desert sands enchanting many a person with their mystical charm, sometimes causing a fatal desire to rise in their minds. "I wish to go there," they think, "I must go there—see how they beckon me, how they entice me! Surely, they cannot be too far off!" Once these unfortunates are charmed, they walk and walk and walk toward the shining red granite peaks *just* in front of them. In their desire, they march onward in the clear, sunlit air unaware of how little progress they are actually making. Night comes, then the golden sun of morning rises above the pointed cones and beckons them forward. Another day, another night, and with great difficulty they drag themselves along. Their feet are sore, their legs hurt, their throats are parched with thirst, but they finally arrive.

Aimlessly, they wander about the mountains. There is only the blinding sun, craggy peaks, and an unforgiving loneliness. They recognize the beauty, but are too exhausted to partake of it. They are being punished for having dared to penetrate the heart of this wilderness without nature's permission. Now thirst becomes more and more acute, their weary limbs become heavier and heavier, and the desire for water is unbearable. But there is not a single drop of water to quench their thirst. The futile search continues, and their footsteps become aimless. Finally, death comes and saves them from their agony and provides them with eternal rest. Their bodies and bones are later found by those who know the area, by those who know where water is to be found.

As we approached the Spitzkoppe, we saw the chain of small hills lying before the massif proper. Many dead and nameless people rest there after having been embraced by this cruel sphinx. We found the skeletons of those who had not yet been laid to rest. As bedazzled and enraptured as we were by the sheer unrestrained wildness of these massive rocky cliffs, in which leopards and other wild cats roam, we could not help but shudder at the price some people had paid to explore them.

My husband proved an excellent guide through the maze of rocks and stones. In 1895, he had visited the Spitzkoppe with Herr Merker, an acquaintance of his then, and had hiked through the mountains and valleys. One day, after having climbed for many hours, they reached a valley between the cliffs that was so beautiful they stopped in their tracks, spellbound, marveling at the wonders of nature. They named it Paradise, and that name has been retained to this day. Together, Themis and I visited this valley. The climb up was extremely difficult, the descent even more so. But climbing and sliding, we reached the graded terraces. Huge callabra cacti, aloe plants, ancient wild fig trees, and other trees with white shiny trunks (which I had never seen before and whose designation I do not know) stood before us. All kinds of brightly colored flowers, mainly lilies, bloomed among them. Astonished by the scenery, we ambled about in Paradise. Ahead of us, a leopard jumped up and sauntered off hissing. Several wild dogs and klipspringers darted across the smooth rock surfaces. Rock dassies eyed us suspiciously from their holes and hiding places. Innumerable birds, mainly a small variety of parrot, sang their hearts out. Above all us shone the hot African sun.

We retreated into the canyons and grottos formed by the overhanging rocks. In our imagination, we discovered dragons, monsters, giants, and dwarfs in the rocky formations. By all appearances, people in former times had made their homes here. Pottery shards lay strewn about, but it was beyond us to identify them. By appearance, they seem to have been made of darkly burned clay. In several of the caves we discovered sketches or paintings made by Bushmen. Similar sketches are found all throughout Africa, and we succeeded in taking photographs of them. Bushman art possesses a charm all its own. Certainly it is but a remnant of a people that no longer exists. The sketches include human figures, always marching in diagonal rows, carrying either a musical instrument—now completely unknown among the tribes of South West—or a bow and arrows. On their heads, the archers wear strange ornaments, the likes of which not even the oldest natives remember ever having heard of or seen before. A type of antelope dominates the sketches, whether single animals or entire herds. Leopards, lions, and hyenas are also depicted. All these figures are sketched on the rocks with red pigments, probably lead based, so deeply embedded in the stone that it cannot be scratched off, as if the pigments themselves had been etched into the stone.

I am not certain why these drawings are called Bushman sketches. In fact, they seem to have nothing at all in common with current Bushmen whose lives are so elemental that to this day they are incapable of drawing even a very simple figure. Some time ago, my husband asked several Bushmen in the Kalahari to make a few sketches for him. They laughed at him. They had never before seen paints, or anything similar to paint, and were unbelievably clumsy in their attempts to use a paintbrush or even a wisp of hair that could substitute for a brush. What they did produce showed no talent or character at all, but made a childlike impression that could in no conceivable way be compared to the discoveries in the caves. Themis's many questions and researches always led to a negative response from the Bushmen.

It is peculiar that these drawings are found all over Africa, even as far away as the Zambezi River. Peters mentions them in his publication, *Ophir, das Goldland des Altertums*, and published photographs that square exactly with the ones we saw. Many more are to be found in Rhodesia (Zimbabwe). We saw some of these in Ameib, where, unfortunately, lesser individuals had added their drawings to the originals. Many such drawings are found in Spitzkoppe and even more on the Brandberg. Besides these sketches, prints of almost all the African animals are found cut into a stone plate. The etched spoors of elephants, giraffes, rhinos, ostriches, lions, leopards, all types of antelopes, hyenas, jackals, and so on, are so magnificently lifelike that they have true artistic value.

The Mountain Damara living near the Brandberg also identify this rock art as the creation of Bushmen who, supposedly, had lived in that region many, many years ago. Others say that Elob, the great God, chiseled them into the rock as a kind of template for him to ascertain whether or not the creatures he was about to create would have the proper shape.

It may prove worthwhile to research the relationships between all of these drawings in order to determine if, in prehistoric times, all of Africa might have been ruled by a people that wished to document its origin and its history. I would like to assume that there once existed a particular culture in South West that eventually vanished. Certain stories of the Bushmen are also peculiar in this vein, for they tell of temples and ruins located far away in the Kalahari, built by a people that long ago faded from existence. These Bushmen had been given the task of guarding these great secrets from the eyes of intruders. Several Europeans demanded to be taken there, but to show them the way would have meant to die a horrible death and to be taken by evil spirits. Those who went by themselves to search for the hidden temples never returned. The temples themselves were described by the natives as being large, gleaming, white, and beautiful to such a degree that one would lose one's mind by simply gazing at them. Of course, I do not know how much of this is true. Perhaps this great oral legend refers to the ruins in Matabeleland (the great Zimbabwe).

A while ago, the imposing beauty of the Spitzkoppe misled the South West Africa Colonial Company to create a model farm there. Its construction was remarkably expensive. A concrete basin for storing water was built, but it quickly proved to be inadequate. Five or so wells were dug, but they yielded nothing. Long before the insurrection began, the material remains of this expensive misadventure fell to nothing, rusting and wasting away. The glory of the Spitzkoppe was short-lived indeed. For lack of good pasture, the cattle did not thrive. Rain in and about the Spitzkoppe was never good, but during the past few years it has been even worse. The grass dried out, died, and was quickly consumed by animals. Eventually, the livestock had to trek great distances to find pasture, and when nothing was left for them to eat, the herds had so weakened they had to be loaded into wagons and taken to an area by the Omaruru River. Animals died by the dozen, virtually no lambs had been born, and because of these miscalculations the Colonial Company was not able to determine whether raising Angora sheep could be successful. The scarcity of water entirely ruled out the cultivation of trees and other crops.

A store had also been established on the Spitzkoppe farm and was favored by the natives of Okombahe—at least so long as the Colonial Company had a license for selling liquor. At that time, the store turned a profit; however, by and by business declined. At the time of our visit in 1903, the farm as a whole was in a state of disrepair. One year later, in the midst of the uprising, it was completely destroyed, the manager murdered, and any remaining livestock stolen.

INSURRECTION

1. Foreboding Weather

A peaceful Christmas was followed by a tranquil New Year. The cloudy days were hot and humid; the nights were uncomfortable and gloomy. Banks of thick, black clouds formed above the Okombahe Mountain—after which the settlement is named—and darkly brooded there above the desolate colossus, threatening at any moment to break into treacherous storms. Like glaring rockets, flashes of yellow lightning shot along the horizon. It was all quite unbearable, and the sultriness of nature seemed to influence the mood of the people. Everyone felt uneasy. Something strange was in the air, though we could not quite put our fingers on it.

The Hereros living in Kawab, half an hour's distance from us, were very busy. Often their leaders came to Okambahe to trade with the various stores, but now they came more often than ever before. If one enquired after the increased frequency of their visits they simply answered, "We fear the grasses in Kawab will not be any good at all. We want to move the livestock sooner, so we are busy preparing everything." They often dropped in on us at our home. We had always had friendly relations with them and were surprised when one day they enquired with great interest about when I would be returning to Germany. Themis was astonished and answered that we had never considered leaving South West. Then Justus Hongera, the son of our foreman, said, "Herr von Eckenbrecher, it would be a good thing if your wife were to leave. She must leave soon and stay away for a long, long time."

"Why are you telling me this?"

"I am telling you because it would be best for her to leave. Africa does not suit her. This is a land for men who love war, men who are rough and ready. You may stay—you have a gun and a saddle. But if you love your wife, you will send her away!"

Themis listened with astonishment, not quite knowing what to make of it.

On another occasion, when we had gone to visit Captain Josua to discuss a trip we were planning, he answered us in a very strange manner.

"I feel," he said, "that you won't get far, that you ought to stay here."

"Why on earth are you saying this?"

"You whites must not think that you're safe here in the land of the blacks. I've read in the clouds that God, the Lord, will send a terrible war that will devastate the countryside and kill all the whites."

At the time, we considered his words as boasting, just idle chatter.

A song sung by my poetically gifted washerwoman, Emma, startled me. "You poor white folks," she began, "you will perish in this land. Little child, you are not to be blamed for having been born here. But you must also die. I am crying, my son also cries, but there is nothing to be done."

"Emma," I asked, "what are you singing?"

"Oh, nothing, Madam. I'm just singing what's on my silly mind. Don't pay it any attention."

Another time, she asked, "Madam, are you sleeping well?"

"Yes, thank you, very well."

"That makes me sad."

"Why?"

"Because you would be better off not to sleep so well. Set pebbles between your ribs then, you won't sleep so soundly."

One day, Emma's son, Samuel (who several months previous had accompanied us to Gui-Gams), came onto the veranda and stared at me.

"Why are you staring at me so?"

"Oh, Frau von Eckenbrecher, I'm looking at those who are marked by death."

I took note of this and discussed the matter with Ertmann and my husband, who were preparing for a journey.

"Wouldn't you rather stay home?" I asked them.

"Heavens no!" they answered.

"But it seems to me that their words are veiled warnings—though I can't interpret them properly."

"Nonsense," they replied, "it's just that the insurrection of the Bondelswarts occupies the imagination of these people. As soon as news arrives of its suppression, things will quiet down. They're just a gang of cowards."

We did not worry ourselves about the insurrection in the south. We had become accustomed to the small incidents of resistance that occurred often enough. Even if the insurrection in the south became more serious, we in Hereroland had no need to worry. The Bondelswarts, a Hottentot tribe, were enemies of the Herero. The insurrection was far away, and the governor of South West, in his familiar gentle manner, would settle the matter. As it turned out, we were completely mistaken. Captain Christian of the Bondelswarts was not appeased; indeed, he entirely rejected the peace for rum and tobacco settlement. Troops became necessary. And when soldiers from the northern district of Omaruru were transferred south, and when Captain Franke, who was both feared and loved by the natives, also went south, the old experienced Afrikaners became worried.

We remembered that during the previous year, several Herero homesteads had been searched for weapons and ammunition—and that the search had been successful. The Herero, who were by nature brutal and imperious, had become even haughtier. Individual Herero behaved in an insubordinate manner toward us whites. Where before we had been greeted warmly in their homesteads, now we were met with an icy coldness. The possibility of an insurrection was something we now anticipated, especially as very few soldiers remained stationed in the north. If the Herero were contemplating a rebellion, it would likely come soon. At least these were the conclusions of our many discussions; we even proposed several possible scenarios for the insurrection.

The government had also received warnings but remained optimistic. It was fortunate for us that the troops that had been scheduled to return home had not yet boarded the ships when the Bondelswarts rebellion began. At this point, only four soldiers remained in Omaruru, and to us this seemed an insufficient number for maintaining order. It was decided that reservists be called up to support the small unit of soldiers. A few settlers, traders, and merchants who were not too busy at the moment volunteered their time. These were men who had become old and gray in their struggles against the sun and the wind. They had worked hard in the colony, but now the Fatherland was calling and they were ready to leave their own affairs behind them. The districts of Karibib and Omaruru were combined under the administration of First Lieutenant D. von Kuhn. After Captain Franke had been called to the south, there was little activity at the Omaruru post. The army horses were with the Field Company or quartered at Soris-Soris. When the old reservists reported for duty, the post became busy again and the daily routine more befitting the military. The transition from civilian to military life was not easy for these men.

At the beginning of January, my husband and Ertmann traveled via Spitzkoppe to Swakopmund to buy fresh supplies. Our existing supplies were almost entirely spent, and we were in need. I was at ease when they left; whatever should come could not be averted, for one cannot avert the course of fate. They had been gone for half a day when a messenger on horseback arrived at our house. He was dripping with sweat and had brought orders for every reservist to report for active duty. Themis was ordered to report to the military post at Okombahe at once and not to leave there until further notice. The poor rider was completely exhausted but told me in confidence that the government had serious concerns about the Herero. The Herero captains from the different settlements, as well as many men capable of bearing arms, had disappeared. Rumors were circulating of large assemblies, secret messages, and heavy traffic between the various villages. I had no choice but to send for my husband to return. My most trustworthy messenger reached him, and by the evening of the second day, Themis was back. He had just enough time to barricade the bedroom window and to move the furniture around so that I could not be shot while sleeping at night. He was to spend twenty-four hours a day at the post so he could defend it if that became necessary.

I was left to my own fate, all alone in the house with my little child. The induction order was very hard for me, but we women of the colony, no matter what our social class, shared the fate of our husbands. Together, we likewise suffered from hunger and thirst, and just as our husbands did, we also placed all our strength in the service of the colony. Our children were born under the most difficult circumstances. We women endured more hardship than the people in Germany could possibly imagine; and now, in time of danger, my husband was taken from me. For ten days and nights I was alone with only my nine-year-old servant, Isaak, whom I quartered in the kitchen so that he could help with the baby from time to time. During the day, I had all my regular work to do plus my husband's. At night, I was left to myself. Several times my dogs barked angrily, and I arose thinking a hyena or some beast of prey was bothering the livestock. I found nothing and returned to the house. The next morning Isaak showed me fresh footprints in the sand around our house and in our garden. The local natives or the Herero had attempted to steal livestock. Some of my lambs, and even a calf, disappeared in a mysterious way. But I was powerless to do anything about it. My husband was only twenty minutes from me, but could not help.

Things only got worse. About noon on January 13th, two Herero, whom I knew well, arrived with their servants, wanting to sell me a sheep. I duly noted that all six men were armed. Two of them had rifles; the servants carried kirris (clubs). In times of peace, armed visitors occasionally dropped in, but they left their weapons outside the house to indicate their peaceful intentions. This time they did not. I had no use for another sheep, and without cash or anything else to trade with them, when they asked for two shirts I consented, just to get rid of the scoundrels. Since I did not suspect foul play, I did not prevent them from following me into the house. We quickly agreed on the matter, and as is customary after even a small deal, we carried on a conversation centering on the impending insurrection.

Geert Afrika, the eldest Herero, asked me, "Do you think it is right that the Herero should want to start a war with the whites?" I answered truthfully that it would be most unwise, in my estimation, since the white man would be victorious in any case.

"Well," they asked, "what do you think will happen?"

"If you are so unwise as to start a conflict," I replied, "much blood will be shed. You will lose your land and livestock, and your lives will be at the mercy of the white man!"

One of them answered, saying, "Don't you worry. We are the servants of the white man. But the sleeping lion doesn't know when danger may strike."

My visitors sneered. Geert Afrika uttered a quick word, and two of his men stepped before the two doors in the room in an attempt to inconspicuously block the exits. Instantly, I sensed the danger and the seriousness of my situation. Cackling, Geert Afrika walked up to me, then screamed in my face, "Now we'll have a proper conversation about the war—which you'll soon see!" My heart was pounding, but I forced myself to remain calm. There was no one to help me, so I had to help myself. I breathed deeply and responded, "You've come to bargain with me yet you haven't a clue about how to act in the presence of a white woman? Step aside, you scoundrel!" I brushed past him to the door, where I expected resistance, but the two rascals were so taken aback that they simply stepped aside to avoid the cuff I meant for them. The door flew open, and I was outside. To my great relief, I saw a good number of Mountain Damara milling about. I was safe, for the Herero would not dare to confront me in public view. They would not have dared to fire a shot in the house for the same reason, but a blow from one of their clubs makes no sound. It was such a blow that I had expected.

I stopped in front of the store. Grinning, our Gerhard stepped outside and doffed his hat. "This time Madam was clever," he said. "She was cleverer than we were. But that's all right—the time is close at hand when she will hear from me!" He then disappeared with his comrades. My young servant, Isaak, who was looking after my child, came running at full speed and reported, "I don't know what the Herero were up to, but they came to me and the child, and asked me whether it was a boy or a girl. When I told them, they left, laughing and calling back to me, 'Just stay here, we'll be back in a little while!'"

I was deeply unnerved by his words. The Herero wanted to kill my child as well. Never in my life had I felt so completely left to my own devices as at this very moment—though I was relieved that I had been alone when the Herero had made their call. Had my husband been home, they would likely have mercilessly beaten him to death. I believe they were a bit awed by my behavior and had decided not to carry out their plans. Perhaps they also remembered the many meals they had received from me, the many times I had allowed them to grind their grain with my mill, and the many Sundays when I had provided them medical services.

My existence was now one of misery, but for the sake of the child I collected myself: I had to—for his safety and well-being rested entirely upon my shoulders.

2. The Insurrection Begins

Early on January 14, 1904, an express patrol of two riders arrived. They were ordered to round up the horses kept at the horse post in Soris-Soris. Soris-Soris is one of the few places where the dreaded horse disease does not occur, and for this reason, horses belonging to the Second Outjo Field Company and the Fourth Omaruru Field Company are kept there during the rainy season, but only under the protection of a corporal and several natives. The patrol was also charged with closing our military post in Okombahe, consisting of all active and reserve personnel, all of whom were to participate in this critically important transfer of horses. These orders only complicated things for us in Okombahe. Our military post would now be completely abandoned and, in a manner of speaking, handed over to the insurgents. All that remained was a solitary soldier who was not only ill but also had a broken hand. He would hardly be very effective in defending the post against an attack by the Herero. We had no clear idea of the true intentions of the Mountain Damara either. The two troopers also told us of massive gatherings of Herero, mainly near Okahandja and at the Waterberg. They reported livestock theft, the burning of farms, and sundry cruelties against whites. As my husband now needed more supplies than he had taken with him, I sent my experienced servant, Gottfried, ahead to Omaruru with these necessities. Gottfried had been with my husband at the post, and I hoped he was wily enough to deliver the items successfully.

On January 15th, at four o'clock in the morning, the patrol departed Okombahe. During the night, my washerwoman, Emma, came to see me. With tears in her eyes she begged me to do all I could to persuade the patrol to stay. Of course, it was not for me to rescind the orders from Omaruru. I resigned myself to the fact that our troopers would never reach their destination. They laughed at my warnings and told me I was overly frightened. I could hardly even convince them to keep their side arms cocked and at the ready. I later learned that because I had sent provisions with them, thus abrogating the necessity of stopping to prepare a meal, they took only a short noontime rest and were able to evade the Herero who were pursuing them in order to rob them of their horses.

After the patrol left, I walked over to the post and enquired if it might not be best to warn the settlers in the surrounding area. Surely they had no definite idea of the real situation. My request was ignored. The soldier stated that he had not yet received any instructions from Omaruru, and without instructions he had no authority to do anything at all. I grew impatient and felt that I had to take matters into my own hands. Fortunately, after much discussion, I received permission to deliver fifty rounds of ammunition to each settler. The ammunition was stored at the post but belonged to the settlers. At my own risk, I added twenty English bullets that fit our own model 71 so that each farmer could use at least two rifles for his defense. I carefully set the bullets, one hundred twenty in all, inside a small milk pail. I planned to take them to the two settlers who lived in Ais. I spread a layer of carrots and tomatoes over the bullets and explained to the messenger—one of our herdsmen who had no idea of the important contents of the pail—that the pail was quite heavy because I had packed the vegetables in wet sand to keep them fresh. As a precaution, I sent along two letters: an innocuous one tied to the lid, and a second one—the important one—placed at the bottom of the container. I promised the messenger a very fine reward if he safely delivered the container, and to my joy, he trotted off immediately.

It was my good fortune to have acted correctly, for the two settlers, Mr. Bowe and Mr. Crighton, were in dire want of ammunition. Isolated, as they were, from all regular communication, they would not have had opportunity to lay hold of ammunition in any other way. Immediately after they received my letter, they combined their livestock and drove their herds to Okombahe, and in so doing they managed to evade the Herero who were coming from the direction of Vogelkranz and paid their farms a "visit" just a few hours later. Shortly before they arrived at Okombahe, a Baster brought them news that the Basters in Okombahe loathed both the blacks and the whites, and that they were as much a threat to the whites as were the Herero. Mr. Bowe and Mr. Crighton therefore marched their livestock into the Namib Desert and safely led the animals to the Brandberg. For days on end, a band of Herero followed them but finally gave up their pursuit. It was only much later that I learned the two gentlemen had safely reached the coast—though not without great difficulty. The few bullets they had with them had saved their lives. Without ammunition they would have been unable to defend themselves or hunt game to sustain themselves on their long trek.

My former servant, Samuel, took it upon himself to look up a white man who lived in the veld among the Herero. The man gave no heed to Samuel's warning and paid for his recklessness with his own life.

At noon on the same day, a Herero from the veld brought news that Daniel Kariko, the chief of Kawab, wished to send me word that I had no need to worry about my husband—because I no longer had one, that he lay dead behind the cliffs near Okarundu, that every trooper of that patrol had died with him and that the Herero had captured all the horses. Now, Daniel Kariko, Gerhard Afrika, and two hundred armed men were on their way to burn down my house. Nothing would happen to me, for Kariko would take me captive. His report rang false to me, though just a little while later two other natives arrived, and breathing heavily, reported having heard very heavy shooting near Okarundu. The native mail carrier, who traveled between Omaruru and

Okombahe every fortnight, brought identical news. Even the missionary came and reported the same horrible events.

Around six in the evening, my servant Gottfried arrived back, exhausted and ashen-faced, his clothes torn to shreds. He reported that he had made it just beyond Okarundu when he was met by hoards of armed Herero. They stripped him of the goods he was to deliver to Themis. Delighted with their spoils, they immediately divided it among themselves, and what they could not use, they destroyed. They had found the letters Gottfried was carrying, tore them up, and at last let him go. Gottfried was to tell my husband that they possessed many wonderful things now, and that on the following day they would plunder his house and gather the rest of his belongings. Gottfried was traveling with a native policeman who wore a red sash as a sign of his authority. The Herero tore it from him. Then, magnanimously, they gave my husband's English riding trousers, his pride and joy, to Gottfried, saying, "These are for your troubles."

Hastening back to Okombahe, Gottfried met our patrol on its way to Omaruru and described in detail what had happened. These events were serious and called for extreme caution. Themis sent me a note and asked me to be very careful. It was simple enough for him to talk about prudence, but I was alone with our eleven-month-old child: where on earth was I to seek protection? There was none to be found. I was at least relieved to know that Themis had not been killed near Okombahe, and it was my fervent prayer that he arrive safely in Omaruru. Yet what was to become of me? As the most dreadful things we could have imagined were now a reality, I had to think very clearly and make plans for our safety. There was no time to allow myself to be arrested by fear. I wanted to make no rash decisions I would later regret! I must have made a thousand plans and rejected them all. If need be, I could hide myself or gradually make my way to Omaruru, a distance of seventy kilometers. I had the stamina; I was tough. But I could not bring myself to expose my nursing child to these hardships and thus gave up the thought of escaping. I hid food supplies in an old dilapidated lime kiln, including boiled water, milk kept fresh by being wrapped in wet cloths and blankets, and so forth. I buckled my revolver about my waist (thank goodness I knew how to fire a gun!). I found it curious that this small, shiny object could calm me down, but it did. And I was fully prepared to defend my own life and that of my child with everything I possessed. Indeed, I would have first shot my child to death in cold blood, and then myself, before allowing the two of us to fall into the hands of these monsters.

After a great deal of thought, I finally decided to go to the missionary and ask if I could stay the night. I did not wish the enemies to take me by surprise in my own house, and a mission station is, so to speak, neutral ground. I was granted shelter and packed some children's clothing, our silver, blankets, underwear, and more critically, all our important documents. Preparing for the worst, I dressed in an ankle-length skirt and sturdy boots so that I could flee more easily with my baby in the event it became necessary. I emptied all containers of gasoline so the scoundrels would find none with which to set the house ablaze. I took along our axes and hatchets to make the destruction of our home as difficult as possible for them.

We sent for the wife of the estate owner, Merkel, who lived on the other side of the Omaruru River. Her husband was in the furthest reaches of their farm and had no idea of the dangers surrounding him. Several messengers had been dispatched to bring him the news, but returned without having made contact. I offered to share my room at the mission with Frau Merkel. I also had enough milk for her children. She arrived with great haste, having had to wade across the swollen Omaruru River carrying her fourteen-day-old child in her arms. Again, I visited the lone soldier at the post. Apparently, he had not a single care in the world, for I found him in the pontok of a Baster's homestead loving a Baster woman. I asked him how he planned to protect himself and others. He did not know. He had seventy bullets that fit his 88 rifle; the post was still under construction—two of the four exterior walls had yet to be erected, and one side of the guard tower had been torn down. The direction of any attack upon the post was obvious, and it was equally obvious that this one soldier was incapable of defending the post. I suggested we send for Captain Cornelius and ascertain where he stood in regard to the insurrection and how well he could be trusted. He came, and as we wanted to avoid any possibility of misunderstanding one another, the missionary was at hand to translate.

"Will you support the whites?" we asked the Captain.

"Yes, I shall try to protect them."

"Are you contemplating the possibility of working together with the Herero?"

"No."

"What are your reasons for siding with the Germans?"

"My people and I well remember working as slaves for the Herero. If we joined them, we would certainly become their servants and would be allowed little more than to bear their guns and clubs, and to fetch them their waterbags during battle. We wouldn't receive any rewards from them."

"Do you think the Germans will reward you for loyalty?"

"This I don't know, but since they are our masters, they'd punish us if we were disloyal."

"Does everyone feel the way you do?"

"That I don't know."

"Do you have many dependable people?"

"Some of them I do trust."

"Would you be willing to send out some of your men on a night patrol?"

"I don't have enough guns and ammunition for that."

We ordered all the men of the village who possessed firearms to assemble in front of the church. I sent to the post for my husband's four rifles and all his bullets, a total of 120, all of which he had deposited at the post as a matter of caution. I took gunpowder, lead, and percussion caps from my own house so that we could make additional bullets. The rifles I gave to men whom I considered to be especially trustworthy. We distributed bullets so that every man received five and the captain, ten. The name of every man who received ammunition was dutifully recorded. They were given instructions and sent off on their various rounds with explicit orders to immediately report any suspicious activity. At sunrise, they were to return to the church and present their bullets. This was our plan for conducting the patrols in a manner that allowed us to keep track of both men and ammunition.

We had done what was reasonable and what was within our power to do. Now we had to cope with the most unpleasant part: just sitting still and waiting for something to happen. It had been dark for quite a while, and I put my little son to bed, wondering in my heart if he would still be alive the next morning. I returned to our house once more to bid it farewell. We had toiled with all our might to build it, and, for a final time, I wished to look at all those things I loved, things that reminded me of my distant homeland. Along the walls in our dining room, shelves stood full of all types of glasses. Some of them, particularly the blue-green crystal glasses, had been in my family's possession for more than a hundred years. On many a happy occasion they had adorned the table of my forefathers. My heart ached. Was all this destined to be plundered and destroyed before sunrise by a pack of wild men? I took a single glass with me as a remembrance.

I entered the bedroom and looked through the medicine cabinet. Perhaps I would find a fast-acting poison I could use in case the revolver failed. I only found a glass container of sublimate pills. Then I did something ugly and unchristian, something for which I have no excuse—and for which I feel no regret whatsoever. I took the last bottle of cognac from our pantry, which was nearly empty, and opened it, dissolved several sublimate pills in the liquid, then set the bottle on the counter. For whites, I penciled three crosses on the label to mark it as poison. For the Herero it would be a welcome drink. Just let them come!

I locked the door behind me, visited the chicken coop, and then the livestock resting in the kraal. I had no choice but to abandon them to their fate. I picked flowers from our garden for the last time—a small patch of earth we had tended with so much love. Then I walked into the dark, mild, African night. The silvery crescent moon shone above the Okombahe Mountain. What on earth would the next day bring?

During the night there was a good deal of unrest. Letters addressed to Cornelius arrived from Captain Daniel Kariko of Kawab, and Cornelius handed them over to us as proof of his loyalty. The first letter contained a statement that war had broken out in Hereroland and a serious question for Cornelius: "What do you Mountain Damara intend to do?" In the second letter, Kariko asked Cornelius to accompany him and help kill all of the white settlers. In a third letter, Kariko wrote that he was camped with two hundred men in the cliffs of Okombahe Mountain. If the cliff-dwelling natives refused to join his side, he would regard them as his enemies. On the following morning, he would come and not only plunder and then burn to the ground the houses of the whites, but would also capture the livestock and other belongings of the Damara at Okombahe.

Diplomatically, Cornelius responded that he had sent for his chiefs who lived out in the bush to come to him so they could counsel together about what to do. Cornelius knew that Daniel Kariko would mount an attack if he decided not to wait for a final decision, and Cornelius and his men were prepared. Finally, another letter from Kariko arrived in which he summoned Cornelius to meet with his chiefs in Kawab. Cornelius replied that it was just as far from Okombahe to Kawab as was from Kawab to Okombahe, and he invited the Herero to visit him. During this same night, several Herero arrived bringing promises from Captian Daniel, while kindly seeking permission to destroy our houses and drive off our livestock. Courageously, Cornelius firmly denied their request. Showing their true face as cowards, the Herero left, postponing their plans for another day.

At the mission home, the unrest continued through the night. Messengers came and went, as did numerous other natives, and many an exciting tale was told. The now legendary shooting spree in Okarundu was also clarified. A native told me he had been working in Kawab during the morning of January 15th. When he arrived, he had found the place in great commotion. A large assembly had been called. The old people, among whom was the current captain, Johannes, plus one other, the influential Old Martin, had tried to hold back the young men. Under the command of Gerhard Afrika, former captain, Daniel Kariko, and his son, evangelist and schoolmaster, Samuel Kariko, the young men planned to attack the patrol escorting the military horses to Omaruru. They very much wanted to steal the horses. Finally, Old Johannes agreed and encouraged them, "Go about it cleverly and allow the first transport to pass by peacefully—for that is the smaller one and consists of female horses and their foals. Don't let the troopers see you, but sneak away to Okarundu. There you will wait for the second and much larger transport. You can expect that there will be shooting, and later, when the settlers in Omaruru hear about it, they'll believe that it was the Herero of Okarundu who mounted the attack and nothing will happen to us in Kawab."

Rejoicing, the young men went on their way. When the patrol rode at full gallop through Kawab, everyone remained quiet, and Kawab appeared to be slumbering nicely. As the soldiers rode through Okarundu, they spotted a few armed men, yet these did not harm them. In the meantime, the young men of Kawab were ready for action and hid themselves in the bushes and behind the cliffs of Okarundu, waiting in vain for the second transport to pass through. When it did not arrive, their patience gave way and they decided to walk and meet the transport. Later, back in Kawab, Johannes could only admit that he had made a mistake: all the horses had been in the first transport after all. But the young men did not believe him and decided to march onto Okombahe to attack the military post and steal the horses they assumed were still there. On their way, they were met by other natives who confirmed that Johannes had been mistaken. Thankfully, they turned around. If they had gone on and stormed Okombahe, all of us would have been killed. This was not the only time our lives hung in the balance.

The Herero divided into two groups. One group was on horseback and followed the patrol—the one that had entered my home to visit with me—and would certainly have caught up with it if the men of the patrol had taken a long noon rest. And this they most certainly would have done had I not prepared plenty of rolls and filled their canteens with tea. The other group, an angry, howling mob, rushed down the path that led to the estate of farmer Merkel. "All right," they said, "we lost the military horses, but we'll be able to find horses and plenty of other booty at Herr Merkel's place."

I thought a good deal about the action of Old Johannes and wondered why he had decided to protect our troopers. As time went on, he became the greatest criminal, one who never shied away from any opportunity to murder a white man. He and his trusted men knew full well that no second squadron of horses would follow the first. And it would have been to his distinct advantage to have had his men riding horses. The Herero living in Okarundu had also intended to shoot the troopers and steal their horses but later explained that the patrol was moving so quickly they simply had had insufficient time to act. Thus they had intended to shoot them as they passed but ultimately allowed them to ride on in peace.

At the beginning of the insurrection, I noticed time and again the indecision of the Herero. I remain convinced that many of them did not want war but were forced to follow their captains who sought freedom *and* power.

Samuel Kariko, the previously mentioned schoolmaster, evangelist, and son of Daniel Kariko, was a revolutionary. His father had been demoted by the government because of his civil conviction for smuggling arms and munitions. He was also suspected of giving revolutionary speeches and had, years before, encouraged his people to start an insurrection. He had been banished from German South West Africa and fled to the British enclave of Walfisch Bay. A few months before the current uprising, his banishment expired and he was allowed to return to Hereroland. Immediately, he expended his time—and the considerable influence he had among his people—on soothing their misgivings and uniting them with the insurgents under the command of Captain Michael of Omaruru.

Samuel Kariko always had aristocratic pretensions. Once he visited me wearing a white suit and starched shirt, yellow riding boots, a pocket watch, a dainty handkerchief, and riding stick. He clapped his heels in military fashion and, with his arm stiffly held against his side, shook my hand, petitioning, "Madam, will you permit me to sit down?"

"Please do," I replied.

"I like shaking your hand," he graciously added, "for I can tell that you come from an aristocratic family. I do not care to shake hands with the other whites, you know." He then asked me for a favor.

Samuel Kariko spoke excellent German and Dutch, and fair English and Namaqua. He had also mastered writing in Herero and Dutch, and wrote quite well in German. As he had charge of the Christian parish in Kawab, he preached, conducted services, and taught the children. He played both the violin and the harmonium, and served under the supervision of the missionary in Okombahe, who performed the most important ordinances in Kawab such as baptisms, confirmations, and marriages. The missionary regarded Samuel as one of his truly pious and most devoted servants.

During the week preceding the uprising, Samuel had visited the missionary several times, conversing with him as a true Christian, and had fervently invited him to journey to Kawab as soon as possible to perform confirmations and to serve the Lord's Supper to everyone. The parishioners desired the sacrament with all their heart. The missionary was unable to come immediately, but promised to fulfill the wishes of the people. The pious of

Kawab were forever supplicating him, sending messengers nearly every day bearing the same request. But as the missionary's oxen were lame, he told them he was simply in no position to undertake the journey. Immediately, they proposed a solution, offering to send him a team of oxen to pull his wagon. The missionary replied that at the moment the river could not be forded by an ox team and wagon. Again, messengers were dispatched to invite him to come without his wife and children. They would send him an escort and an ox he could ride. The missionary was quite delighted with the religious fervor of these pious, Christian people and often mentioned how wrong it was to condemn them as hypocrites.

Then, on January 15th, everything suddenly changed. We were puzzled. Why had the Herero repeatedly sent these urgent invitations to the missionary and then abruptly stopped? The missionary continued to believe the matter was not an omen of impending harm but the consequence of Samuel being among the people again. Samuel, he thought, could certainly be counted upon to admonish them to do the right thing, to avoid evil, and to be loyal to the whites. Because of his confidence in Samuel, the missionary had asked him to stay with his tribesmen. Personally, I think the natives wanted the missionary in their midst so that they could more freely engage their evil designs. Perhaps they intended to pressure Cornelius by threatening to murder the missionary if the natives of Okombahe continued in their reluctance to support the Herero cause. Perhaps they hoped to pacify Cornelius's conscience if he knew the missionary were safe. The rest of us would be easy prey, and I do not think that even a single native would have lifted a hand to defend us. They would have stayed away, acting as if they knew nothing, yet fully confident the Herero would leave behind sufficient booty for them.

The missionary was utterly speechless when I shared my candid thoughts with him. Entirely dumbfounded, he murmured, "But Samuel is with them!"

"Samuel is as much a crook as the rest of that lot—perhaps even worse."

"No," he retorted, "it's impossible that I could be so deceived."

"My dear Missionary," I replied, "errare humanum est (to err is human)."

Later we learned that Samuel had served as the secret secretary of the head captain of Omaruru for a very long time, that he knew of the plan for insurrection in exact detail. During the night of January 16th, Samuel visited Cornelius, asking his permission to plunder our house. He also stole the livestock from Merker's farm. Worse still, he and his father, Daniel, murdered in the most appalling and cold-blooded manner a settler named Johr, who was a retired petty officer. Johr was an odd duck, to put it mildly, who lived exactly like a native among the Herero. He spoke their picturesque and melodious language perfectly well, and was allowed to sit in their councils and to vote in all their deliberations. Following Herero custom, he married a Herero girl. Because of his counterlifestyle, he became estranged from the white community, yet despite all this, I knew of his basic goodness and decency, and wished very much to help him improve and regain the respect of the whites. In this, however, I was not successful.

Johr was also called up as a reservist, but failed to report. Old Ertmann, who was a fatherly figure to him, urgently wrote to Johr, but Johr answered that even though his skin was white, he felt perfectly comfortable as a Herero, that he feared nothing and planned to remain with the Herero. He gave no heed to our warnings, and it cost him his life. Elly Bam, a Baster woman who lived in the same village as he, reported that the Herero began holding their meetings without him. When he enquired as to why he was being excluded, he was told they were discussing matters concerning their own race, matters he was not in a position to understand. From that point onward, he noticed that his every step was being observed by armed Herero. One night, he was awakened and summoned to appear before the assembly, and was told that the war had begun. What, they wanted to know, was he going to do? Johr is alleged to have said, "Like you, I am Herero, and come what may I will remain with you." He was then dismissed. Johr may have contemplated fleeing, but knew he could not get away with it.

Elly Bam reported further that at daybreak Daniel and Samuel Kariko came to him. "We are thinking," they told him, "that someday the time will come when you'll remember that you're white. Our cause is a war that pits race against race, and no man can betray his own blood. You have a great knowledge of us and could very well betray us. But this must not happen!" Immediately, they seized upon him, tied him up, and tortured him in a most horrible manner. Finally, they left him lying on the ground, bleeding from innumerable wounds. He struggled to get to his feet, but those dogs had gouged out both of his eyes. Justus Hongera, Johr's friend, came to help him and beat those two Herero monsters with his bullwhip. "What you have done is filthy and despicable. Leave or I'll forget that you belong to my tribe! If he were truly our enemy because his skin is white, I would not have objected had you simply killed him. But one does not torture one's enemy." Having said this, Justus took his rifle and, with a well-aimed shot, ended the agony of poor old Johr.

When I received this distressing news, I could not resist asking the missionary, "Do you still believe in Samuel and in the sincerity of his Christianity? Will he influence the parish for good?" He answered with a dangerous logic, "If Samuel does not prevent his tribesmen from going to war, it's because he can't help himself—he is a good patriot after all." I did not answer but thought about the passage in Romans concerning obedience to authority.

While all this was going on, the midnight hour had come. Time surely flies when one resigns oneself to death, and these hours passed quickly. Now and then the three children in the small room stirred. Frau Merker lay crying and praying on the hard mattress. I asked her to calm down for the sake of the little child she nursed. The room soon became quiet, yet we could not sleep. Too many thoughts whirled about that night, perhaps our final night! Both of us were worried sick about our husbands. What fate had befallen them? Were they dead or alive? Where were they? Lying wounded and helpless somewhere in the veld? Or were they alive and well?

3. A Dangerous Ride

The firmament was a huge, black mass. Around midnight, heavy rain began. The first drops hit hard, booming on the tin roof. Before long, it came down in sheets, pouring as though it would never end. The rain trickled off the rocks and formed little streamlets in the sand that grew into miniature torrents. The tributaries emptied huge quantities of water into the Omaruru River. With a sublime melody, mighty waves formed in the river, roaring down the bed toward the ocean. Along the banks, water tore away everything that was not firmly anchored—trees, bushes, animals, and people. For a small group of five men, it was a matter of life and death as they urged their exhausted horses onward beside the rising yellow river, trying to reach the river ford in time to bring assistance to our otherwise hopeless situation. This group consisted of Themis, the corpulent Sergeant Josesowsky, and three other troopers who had accomplished the impossible: they had delivered the horses to Omaruru and were now returning, physically exhausted, but in possession of a sizeable cache of ammunition.

They had hurried to reach the post in Omaruru before dark and were met by our servant, Gottfried, who reported that he had been robbed. Admonished to exercise the utmost caution, they did not even stop for a minute's rest. As each of the riders raced along, he loosened the saddle beneath him, held it with one hand while taking hold of the horse galloping next to him, then threw the saddle on its back, jumped on the fresh horse, tied down the straps as securely as possible, and raced ahead. This was their exchange of horses. Groups of armed Herero, many of whom they knew personally, met them along the way and greeted them in a friendly manner and with smiling faces. "Moro, moro (good day), mein Herr," the Herero called out, one hand removing their hats as a greeting, the other hand cocking the rifles. "Moro, moro, you fiends!" came the reply as our men tightened the reins and searched for the snap-link in their rifle scabbards. The rascals yelled at my husband, "Be glad that you're gone, tomorrow we attack Okombahe!"

Shortly before they reached Omaruru, rogue Herero appeared wearing large, white armbands, imprinted with the seal of the Omaruru Regional Office, as a symbol of their neutrality. How splendidly they proved themselves "neutral," for just several hours later the staff physician, Dr. Kuhn, the defender of Omaruru, observed the countryside from the guard tower and noted that the Herero trenches were occupied by their warriors, whose armbands signifying neutrality were clearly visible! The people in Omaruru were very worried and feared for the safety of the patrol. After duly noting the threatening posture of the Herero, Dr. Kuhn sent nine mounted troopers to meet the patrol; thanks to this precaution all the riders and horses arrived safely in Omaruru.

Many whites had settled in and around Omaruru, and the people of the area were up in arms as they endeavored to salvage at least some of their cherished belongings and transport these things to the two military posts. Head Captain of the Herero, Michael, had visited Dr. Kuhn that morning, and together with Subcaptain Mutate and the chiefs, had enjoyed the hospitality of the doctor's charming wife. They had assured them of their most sacred devotion and loyalty. Suspiciously, however, neither Michael nor his men accepted a second invitation to stay at the post. Moreover, an Englishman arrived at the post reporting that his Herero wife had told him Herero warriors were planning a huge blood bath in Omaruru during the coming night. The lives of Englishmen, Boers, Bastars, and Missionary Dannert were to be spared, but not the lives of the Germans. Taking this information seriously, Dr. Kuhn invited all Germans to come to the post for protection but asked the foreigners to voluntarily remain in their homes. Both posts were fortified and, with no time to spare, barricaded with bags of provisions.

At first the Herero were baffled by the sudden shift in the white settlers' activities. But when they noticed that the whites were all coming to the post, they realized that someone must have betrayed the secrecy of their plan. The next day, the plundering of houses and stores, as well as the battle for the posts proper, began in earnest.

As soon as Themis reached the post at Omaruru, he went to Dr. Kuhn to explain the desperate situation at Okombahe. Dr. Kuhn, whom Themis knew very well, immediately responded that if Themis wished to scout up several other men and attempt to ride to Okombahe, he was certainly welcome to do so—though it would be a very risky undertaking. Themis, of course, was not about to abandon his plan and asked for volunteers to accompany him. At once, Sergeant Josesowsky volunteered, as did Komanns, our employee, Herr

Merkel's inspector, a man named Sander, and a former railroad employee named Bruder (for whom this plan was especially difficult as he had never before ridden a horse and would now be required to go seventy kilometers on horseback in the dark of night). It would not be easy for the others either, just having finished the ride from Okombahe to Omaruru and being quite exhausted from all the strain, excitement, and speed of that journey. From among the horses they had brought to the post they selected the very best. The animals were brought forward and shoed. They waited for darkness to settle in before taking flight.

Thick, heavy clouds covered the sky and rain seemed imminent. The men began to wonder if they could cover the entire distance in one night; they also wondered if we were already dead and the efforts would be for naught. Some of the men made light of the situation, calling the patrol "a funeral commission," and when my husband urged haste, one of the men encouragingly said, "No need to hurry, you'll arrive in good time for the funeral." All five members of the patrol were issued new uniforms. They filled the ammunition belts and stashed as many bullets into their saddlebags as they felt the horses could carry: two hundred and fifty bullets for each man. And here one must bear in mind that the horses were small, and that both horse and rider had already endured an exhausting day.

Night fell and it began to rain, making the darkness even darker. The riders could not see their own hands holding the reins. Breathlessly, they stole out of the post, straining to see through the dark, carbines across their saddles. Taking a longer, looping path in the direction of Okombahe, they carefully rode around the Herero trenches or bravely jumped over them. They saw silhouettes of various men, smouldering fires, and they heard Herero words. At any second they might be recognized as the enemy and stopped, but the rain favored them and they rode on quite unnoticed. The direct path to Okombahe had to be avoided as they were likely to encounter Herero sitting along the way as it wove through Otjimpaue, Okarundu, and Kawab. One careless sound and they would be lost. They even had to sacrifice smoking their beloved pipes—which normally provided some consolation in difficult situations. Smouldering tobacco or smoke of any kind might have roused suspicion. Death awaited them at every turn.

Riding through the brush, thorns scratched at their skin and tore their clothing to shreds. They rode over slippery rocks and often had to dismount and lead their horses on foot. The rain beat against their faces, growing stronger and stronger until it came down in torrents. In their hearts, they harbored frightful doubts: Will we arrive in time? They also began to worry about the flooding. If the driving rain continued, the river would rise and flood the surrounding fields. That would make crossing the river impossible, and all their efforts would be for naught. They urged their brave horses onward, and whenever the terrain permitted, they galloped as hard as possible that they might reach the river crossing in time. Soon they heard the roar of the river and for a while rode alongside it. They reached the shallow part before the flood and were able to wade across to the other side. Relieved, they stopped, spotted a tiny spark, then the rest of the fire. One of their horses almost trampled a sleeping "guard" who lay covered in blankets. Once more they summoned their courage. Taking turns, they led Bruder's horse. Besides not being able to ride, the poor man had practically lost his wits from fear and the superhuman strains he had endured. Bruder pleaded with them to shoot him dead, that he felt unable to go on. But on they went, often becoming entangled in thorn bushes that left them no option other than stopping and cutting themselves free. Under these conditions they came in sight of Kawab and hurried past the sleeping village. It was still very dark, droplets of rain continued falling from the branches of trees that lined the riverbank, and the temperature was falling. Morning would soon break.

At first it was only a silhouette, then the broad back of the Okombahe Mountain appeared. Now the men had to be doubly careful if they were to reach their destination. Knowing the Herero planned to attack at daybreak, and that they were most likely camping along the cliffs close to Okombahe, the men decided to play a trick. Riding in pairs bedside one another, they made their horses trot, made their horses do a slow "dance." By so doing, the tracks left behind on the path would appear to have been made by many, many horses—as if a large military force had arrived in Okombahe to protect the settlement.

Under Daniel Kariko's command, the Herero had slept well hidden among the cliffs and overhanging rocks, awaiting the morning in eager anticipation of the rich booty and the great numbers of livestock that awaited them on the farms of the whites. Yet because of the rain, they failed to post guards, following the old Herero adage that when it rains, one should go to sleep. When daylight came, the Herero noted the tracks along the path. The rain had somewhat lessened their sharpness, but after thoughtfully contemplating them, the Herero withdrew figuring they could not prevail against such a powerful force.

At daybreak, I listened carefully. The clap of horses' hooves interrupting the silence startled me. As they came nearer and nearer, I gripped the revolver tightly, believing the end had come! Then I recognized the uniforms! Help had arrived, and my husband was in the forefront! We embraced. "Themis, old boy," I thought, "you were very brave!" Behind the riders to the east, the sun rose brilliantly like a ball of fire. Fate had granted us another day, a new life. Herero do not attack after sunrise. But how did these brave men look?

Their bodies were steaming from the rain, yet they were scratched to pieces, their uniforms torn, and they were completely exhausted after the long, anxious ride of nearly one hundred fifty kilometers. During this one night Themis had aged several years, for it was he who had most to lose. The brave men had earned their rest, yet under these difficult circumstances how could they possibly sleep? We had to organize our defenses, then act accordingly. We had no way of knowing whether the Herero planned to attack in the evening or on the next morning.

4. In the Fortress

We all agreed that we should move to the army post, that we needed to bring plenty of wood, water, and other provisions, and that we should barricade ourselves there. At six o'clock in the morning, we hitched the oxen to the cart and loaded the most necessary and precious goods. The post was a long, brick building containing four rooms. One was used as the office; another served as guard and sleeping quarters for the men. A third room was occupied by Frau Merker, our three children, and myself. The fourth room housed the Bastars—one of the soldiers had a Bastar wife, hence, all her relatives required protection and help. The rooms were small with low ceilings and the windows in each room were bricked up two-thirds of the way to the top because the building was situated on an incline, and from the top of the hill, one could shoot into the windows. Our belongings were stacked from floor to ceiling with just a bit of space for mattresses, children's beds, and the baby carriage. Close to the building stood a livestock enclosure. These animals, together with horses tethered in the yard of the post, attracted hoards of flies.

Sweating profusely, we labored the entire afternoon until sundown repairing the walls around the yard. Since we had no bricks we used stones. It was the most physically demanding work I have ever done in my entire life. It was not long before the skin on my hands peeled away, but I could not stop, for time was of the essence! At least I knew I had done my part to fortify the post so it could be defended. I was much annoyed with the Bastars who considered themselves to be much too distinguished to perform such common work. Instead of helping us, they repeatedly asked in a highly astonished manner, "Why does the lady perform such dirty work? Why don't you make the natives or the soldiers do this?" They obviously had not considered the fact that the latter had been on horseback for an entire day and night. Our progress was slow. It was hot, the men were dead tired, but finally the walls were high enough to afford us some protection. We also repaired the guard tower that stood against the eastern side of the building, rising above the roof. In those places adjacent to the roof where the building's upper tower wall was missing, we set crates filled with sand upon the roof. When we ran out of crates, we stretched my clothesline several times across that empty space and hung blankets on it to serve as a bullet screen.

One of our numbers was ill. Trooper Bruder was so completely exhausted from the long ride that he lay there as though dead. It was not until the evening of the second day that he regained some of his strength. For several days I prepared him special foods, and with these, in addition to plenty of milk, he made a full recovery.

Shortly before twilight all our hasty preparations were finished. We longed to sleep, but under the circumstances, it was not really possible. A bit later that evening, a cart peacefully arrived from Spitzkoppe—it was Old Ertmann's cart. He was astounded to see so much commotion about the post. Old Ertmann had heard nothing about the insurrection and, fortunately, had met no one on the road. Immediately, he volunteered and was a considerable reinforcement to the six soldiers of the post. Since he had been ranging over a huge tract of land, we had had no means of warning him and were deeply concerned about his safety. After his reception, I hoped to be able to visit with him privately. He came to me right away and said, "If only I had known how bad the situation was, I would have come to fetch you immediately. We could have packed sufficient provisions, hitched the cart, tethered the other oxen beside us, and traversed the Namib quickly, not sparing the animals, leaving behind those that could not keep up. At least I could have gotten you safely to Swakopmund!" I knew he was sincere and that I could have depended on him, indeed, that he would have stood by me until his death.

Old Ertmann unhitched his cart. Our cart, along with his and the Merkers' carts and wagons, were positioned in a long row at the front of the post. This was not only protection, but also a ruse. For when the Herero spies saw this column of wagons, they returned and reported to their fellows that judging by the number of wagons in front of the post, there must be at least fifty white men inside. Perhaps this report contributed to their postponement of the attack.

During the early part of the evening, all of us remained together in the darkness of the yard and very carefully avoided the lamplight streaming from the open doors. Starting at about ten o'clock, two men were to stand watch until four o'clock in the morning, when everyone was to be awakened. I accompanied my husband during his midnight shift. The stillness of the night was occasionally interrupted by loud barking. We recognized the voices of our own dogs, and assumed the Herero were already plundering our home. The next morning we learned that some natives had broken into the pantry and the storage room. Dogs on the opposite side of the river also barked—the Merkers' property had also been burgled.

Although each of us was in dire need of sleep, no one got any real rest. Our nerves, the continuing strain of anxiety, and our exhaustion, now overwrought, prevented it. We still had no idea as to the fate of Herr Merker. The messengers we had sent out could report nothing of certainty, only that there had been shooting on Merker's farm. I believe all of us doubted that Merker was alive. Indeed, the missionary and the others asked me to gently inform Frau Merker about the death of her husband. I refused. Once again, we sent an armed native patrol to search for his body; they found nothing. Finally, during our second full day at the post, Herr Merker, assumed to have been dead for several days, appeared. We received him with great rejoicing. He had fought off a group of Herero who sought to drive off his cattle. He fired several shots at them; they did not return his fire, but stole nearly all of his cattle. Hunted by the Herero, and completely out of bullets, Merker had to hide. During his hurried flight he had injured his feet and was only able to progress but slowly. Now, God be thanked, he had returned.

Around four o'clock the next morning, my husband awakened me. So deeply had I slept during the last hour that I was at a loss as to where I was and what Themis wanted with me. Soon enough clarity returned that I was able to realize what I had to do. I went to the office to check the medicine box and organized all the bits and pieces so I could easily lay my hand on anything I needed. It was possible that men would be wounded, and I had to be ready to take my place. Since other firearms were simply too heavy for me to handle, my husband's hunting rifles and bullets lay ready for me to use. Should one of the defenders be wounded and no longer able to fight, I would take his place, but until such time, I asked that I might be at my husband's side. His position, along with the sergeant, was in the tower. The native policemen and servants lay around a smoldering fire in the yard. They quickly awakened, and a cup of steaming coffee, though of unpleasant quality, lifted their dampened spirits. No one had an appetite. The first wispy rays of morning light crept across the sky, and like the walls of our post, everyone stood in his assigned position. The sun rose and we anxiously looked through our field glasses, but nothing suspicious was to be seen, not a solitary Herero. Our expectations of battle had been so high that we actually felt disappointed. Well then, what had not taken place today could easily do so tomorrow!

As the day wore on, we became calmer and calmer. I entrusted the corpulent sergeant—who, I am certain, had shed a few pounds over the last several days—with the care of my baby. He was so natural and competent in caring for little children and did everything in such a loving manner. Indeed, it was almost surreal to watch the suntanned, weatherbeaten, bearded warrior gently tend my little blond boy. But Themis and I were now able to return to our home once again and retrieve a few more of our belongings. We fed the remaining chickens and pigeons, which had been decimated by jackals during the night as someone had left the stable door wide open. Only a few pumpkin runners remained in the vegetable garden; in the orchard, dates, vines, and our little fig plants had all disappeared. Because the Omaruru River had flooded, everything was covered with mud; nothing could be harvested. This was a pity as the cabbage had grown so beautifully! The floods had made an absolute disaster of the pumpkins, cucumbers, and tomatoes. After viewing the destruction and loss, we knew what to expect: meat and rice, rice and meat, and finally, only meat.

We were able to make ourselves more comfortable at the post, yet it was difficult to have to change so quickly from a peaceful, homey existence to the rough life of war. Somehow, though, we managed. Since the Merkers and we received plenty of milk from the cows that were still in our possession, we arranged a regular milk business. Due to a lack of space, milk was poured into bowls and set under the beds. Now and then, some stray mutt came in, but ran out howling as soon as we discovered it. We had no way of battling the insects attracted to the milk. Hundreds of flies, and a few bugs, drowned in the milk every day; still, we used it to make butter. There was very little fat at hand to use in cooking, hence, our slightly questionable butter was a most welcome ingredient. The curds were as well, and all faces shone with joy when I placed a washbasin filled with curds on the soldiers' table for the afternoon meal. At once, they diced onions, added salt and pepper, mixed it well, and spread it on a slice of bread. I also fried mutton tails for grease, then baked the most wonderful pound cake. I prepared "potato" pancakes from grated pumpkin. Everyone craved variety in our meals and enjoyed our creativity. We boiled mutton and goat meat to serve with the daily ration of noodles or rice. Once the man guarding our cattle shot a badger nearly the size of a roebuck. We decided to braise it in the oven. The corpulent sergeant was eager to help and smiled in anticipation of this delicacy. We added a bit of rancid bacon to the roast, and it tasted very nice.

Every evening, I distributed omaere to everyone—it was a refreshing favorite. All of us helped one another: one soldier baked bread, another was the chef, a third made sure that all containers were filled with water, a fourth kept everything clean and tidy. I assisted whenever and wherever I was needed. If the night was damp and the guns had rusted a bit and the bullets accumulated moisture, I sat with the bullets in my lap and carefully cleaned each one of them while the men polished the guns.

The sounds of Okombahe were quite peaceful, yet we knew the insurrection was threatening, that all around us lurked dangers of every kind. Each morning we stood at our posts prepared to repel an attack. Every day we held discussion with Captain Cornelius. In droves the local people came to report suspicious activities, and if we had believed all of their reports, then the Herero had been prepared for weeks to storm Okombahe. To be certain, there was no peace. Cornelius told us he received letters every day from the commanders of the Herero imploring him to join their ranks. The message of their supreme leader, Samuel Maharero, was characteristically to the point: "I fight, I kill!" Michael of Omaruru wrote, "Hurry, all the whites must die. Do your duty; come, join us." Mutate, the subchief, wrote, "Make haste, for the day has arrived. White and black can no longer coexist, the blacks must rule. If you choose not to follow the order, you and your tribe will perish along with the whites." Johannes of Kawab kept asking, "Aren't you finished yet with this handful of whites? How long will you wait? Come to your senses and take my advice—work with poison if you don't want to shoot them."

Coupled with all these threats was the simple fact that we were entirely isolated from the rest of the world. Herero lived all around us, and every messenger we tried to send was intercepted and often lost his life. To reach Omaruru was completely impossible; besides, the people of Omaruru were also surrounded and unable to send us news. Though we repeatedly made the attempt to contact them by various means, our efforts came to nothing. And because it was so deathly quiet and the air so dry, each day we actually heard the thunder of the large cannon in Omaruru. Needless to say, this gave us little solace.

A messenger sent to Swakopmund returned telling us he had seen Herr Loutsch, an employee of the Colonial Company in Spitzkoppe, dead, shot through the mouth by a Herero who had approached him in a friendly manner. Not far from our post, these brazen thieves trekked by with cattle and other goods stolen from Spitzkoppe. Unfortunately, we had too few men to successfully confront them. We assumed this might be the precise time the enemy planned to attack the post. If our men were returning from an encounter outside the post when the attack came, they might not be able to get back in, and divided in this manner, we could easily be overrun.

The tension gradually subsided, and we became quite bored with it all. The constant danger had become routine, the constant waiting annoyed us, and we wished for a decisive move. As we were unable to leave the post, we had no way of watching the natives. Yet since we assumed the locals still sided with us, we had some feeling of security. Then, quite out of the blue, Old Cornelius appeared with four armed men and told us we should not only consider the

Herero our enemies, but the Mountain Damara as well. We should, in fact, fear them all the more, as there was a strong movement among them in favor of the Herero and against the white settlers. Old Cornelius openly told us that he was constantly admonished to order us killed. Now, he said, we needed to decide for ourselves what we should do. He could no longer protect us and did not himself know how much longer his own life would be spared. Because of his friendship with us, he had made a legion of enemies among his own people and expected to be murdered at any time.

We knew that several of the most influential Mountain Damara leaders were related to the Herero, indeed, even considered themselves to be half Herero. It was these leaders who had fomented a change in loyalty among their people. If the natives actually planned to attack us, we stood no chance of survival. They were well acquainted with the weak points of our fortress and knew the walls would never withstand a concerted attack. The people were armed. They had even received their ammunition from us, and though we considered requesting them to return the weapons and bullets immediately, prudence forbade that course of action. We held a solemn war council, but arrived at no consensus. Several of our hot heads wished to take up arms straightaway, arrest the presumed leaders, and keep them as hostages at the post. Most agreed with this plan. The sergeant wavered, though, and I was strictly against it. I made clear to them what it meant to pour oil on the fire. These native leaders would certainly not come voluntarily. So either the soldiers would be forced to return without the hostages—a defeat that would prove our weakness and bring serious repercussions—or the shooting would begin, initiating a sequence of events we hoped, by all means, to prevent. I drew upon all my powers of persuasion to prevent them from acting too hastily. Themis and the sergeant agreed with me; finally, we succeeded in convincing the others as well. It was decided that we should remain passive and not irritate the natives at all. And since there was no immediate danger of attack, we sent an order to all the natives to leave their rifles at home in order to prevent accidents. We certainly did not want armed natives coming to the post.

Spies informed us that the natives living in the veld were gathering with the intent of joining the Herero. We enquired of the leaders about this. Evasively, they answered that they wanted to move to the Namib where it was easier for them to remain neutral. But we also learned they had informed the Herero, "You may come and murder the whites—and we will even help you do it." Late in the evening of this singularly eventful day, my husband approached the tent I had pitched in the garrison yard (as I could no longer tolerate the vermin in the post room) and found me awake, lying in a field chair. We discussed our situation and concluded it was desperate. He felt we would not be able to escape and that our final hour was near, and therefore asked my permission to shoot my child and then me should the post be overrun. He said he could only take his own life peacefully if he were certain that my child and I would not fall into the hands of these black devils. I felt sick to my stomach but immediately consented. We checked the revolver to ensure it was functioning correctly. "Be prepared. For when it comes, it will come quickly." I believe it was January the 20th.

The next day taught us how correct we were in deciding to wait. Some garrulous servant had spied on us during our deliberations, and like the wind, news spread among the natives that the two most important leaders were to be arrested. Immediately, they decided that the arrest would be the signal for storming the post. The whole attitude of the natives became more threatening. They awaited the arrival of one of us so they could begin the attack. However, we were careful and appeared to take no notice of the heightened tension, remaining peacefully at the post.

Miraculously, help arrived. Two messengers had journeyed to Spitzkoppe to deliver mail to Herr Loutsch. They had no idea he was dead, and for a week had traveled, quite by chance, on lonesome trails without encountering another soul. They had absolutely no knowledge that an insurrection was underway, and after failing to locate Herr Loutsch in the Spitzkoppe area, they spotted fresh wagon tracks and followed them, believing that Loutsch had trekked to Okombahe. When the tracks turned away shortly before Okambahe, they were uncertain where they should go. They chanced to meet a native who suggested they take the mail to Okombahe, and so they did. The sergeant officially opened the small bundle. He took out the newspapers, and cheering loudly, read the news that reinforcements had already arrived and that others were still arriving! Jubilant, we climbed on the roof and posted the black, white, and red colors, as well as our own blue flag with three roses on a white background.

The natives assembled in the square. First came the missionary, then the captain, the chiefs, and all the men, lastly, the women and children. They looked at us with bewilderment, doubting that we were in full possession of our senses. A man climbed atop a barrel (that just happened to be there) and with a loud voice read the reports of the reinforcements—adding a zero here and there to the tally of soldiers coming in order to suitably impress his audience. He concluded with a thunderous ovation for His Imperial Majesty, in which all of us participated. The natives looked at one another and gradually disappeared. It was interesting indeed to watch the facial expressions of some of them. The captain and the chiefs remained in place, as did those who, in their heart of hearts, had been solidly against us, feigning keen interest and deepest respect.

Old Ertmann turned to one of the chiefs and quietly said, "It's good that you cliff-dwelling natives remained peaceful rather than killing us as you had planned. I'm telling you, you'd have been in serious trouble then." On hearing Ertmann's statement, the old chief struggled to maintain his composure—even a native can get pale in the face. Horrified, he stammered, "What do you know about it?" and then departed in haste. Ertmann wished him a good night and pleasant dreams about the coming reinforcements, but the man was in such a hurry he probably failed to hear the last remarks.

Cornelius remarked to the sergeant, "Indeed, it was high time that message arrived, for I couldn't guarantee anything anymore. Now you can relax, at least you have nothing to fear from the Damara." We felt a great sense of relief knowing how fortunate we were to have escaped the jaws of impending danger. Yet we were also proud—and remain so. Thanks to our diplomacy, we succeeded in keeping the Mountain Damara as permanent friends of the government. I have no idea what might have happened had they broken away from us, and to this day, the Mountain Damara faithfully serve the military as guides, cattle drivers, servants, and so forth. Although no shots were fired and not a drop of blood was shed, it is to the credit of the Okombahe force that they served our colony with courage and prudence.

The following day we dispatched a messenger to Omaruru bearing this important news. He was later discovered with his throat cut.

The thunder of the heavy guns continued in the distance, so we decided to send out a few scouts to determine what the Herero were doing. Just as my husband and two others were about to ride off, two natives arrived. Gasping for want of breath, they called out, "Don't allow them to go, for they'll ride to their deaths!" "How do you know that?" we asked. One of them, a man called Keetman, replied, "Last night, I read it in the palm of my hand!" We were ready to laugh at him when word from Cornelius arrived that the patrol must turn back, that the entire area was blanketed with danger. Mountain Damara spies arrived shortly thereafter and reported that bands of Herero were swarming about the immediate countryside, lying in wait to ambush any patrol that dared leave the fortress. The Herero had every trail under surveillance, and if the patrol ventured out, the already weak force at Okombahe would be weakened further. The Herero planned to shoot as many men as possible outside the post so their eventual assault would prove easier.

Naturally, Keetman knew this very well, though he swore up and down that he had only learned of it through an oracle. He was an interesting man. Ancient, wrinkled, extremely thin, and dressed in an ensemble of animal skins topped with an old soldier's cap, he was greatly respected by all the people of the area because of his oracular gifts. People from all over the countryside came to consult with him. In answering their questions, he squatted on the ground, took three small pebbles, and balanced them on the back of his hand. Keetman divined his answers based on how these pebbles fell in relation to one another. For instance, he pointed out the direction in which the missing cattle had lost their way, and they were found. He said the Herero were not yet making their way toward us, and we had to believe him. Often, we came to him to ask whether or not our messengers had been successful in delivering our messages. Once he said, no, that one was dead, but the other was on his way back. His oracular ability proved correct. It all turned out as he had predicted. We asked if a relief force had reached Omaruru. Not yet, he answered, though one would arrive very soon, coming from the direction of Karibib, and much blood would be shed. We asked if Captain Franke was still in the south? No, he answered, Franke was on his way back, yet still had a long way to travel. Would we have to die? No, he said.

Of course all of his sooth-saying was utter nonsense, but the rascal was extremely good at his craft. He surrounded himself with an aura of mystery, and it was very effective. Like many others, he was better informed than we were about our situation; we were likewise confident that much of the local news was deliberately kept from us. The natives only told us what they considered to be necessary.

On January 22nd, several hundred natives from Omaruru had fled to our post. They reported heavy fighting in Omaruru and that many whites had been killed. One soldier shot during a breakout had literally been cut into pieces. Nothing, though, had happened to the non-Germans, and friendly relations between the Herero, the missionary, and the non-Germans continued. The natives complained that Michael and subcaptain Mutate were coercing them to join the insurrectionists, but that they had refused. Consequently, Michael called a large assembly of Herero and informed them that if the Damara continued in their neutrality, they, the Herero, would fall upon them and kill them. They also reported that at the very moment they were being threatened, a large piece of shrapnel, blown from a "large pipe," had landed in their midst and had caused a great confusion. During this moment of confusion they had made their escape. Since remaining in Omaruru was out of the question, they had come to us seeking protection. As it turned out, they were an aggravating lot—disobedient to the captain while pilfering our goods at every opportunity.

The Herero were a continuing worry, and since the local natives were now afraid of the Herero, we quickly fortified our positions and sent out patrols to determine the state of affairs. Every evening we saw the Mountain Damara squatting like baboons on rocky outcroppings situated around Okombahe. We had to admit, though, that they were most attentive in their observations of us. Around midnight on January 23rd, seven captured spies were brought to us.

Under the command of Captain Cornelius, these spies were encouraged by the whip and eventually confessed that the Herero had sent them to steal all our cattle during the night. Except for one of them, they were depraved and dull-witted scoundrels, for it is well known that the Herero treat their servants in the cruelest of manners and beat them with the *kirri* (a wooden club) until they have driven the last spark of sense from their brains.

After discussing the matter with Cornelius, it was decided that each would be punished with twenty-five lashes and then let go. We told them emphatically that should they ever return, they would be shot dead. The seventh spy was a very resourceful thief who had escaped from the prison in Omaruru. He was entirely naked, and we could see that welts on the noblest part of his body were a result of severe whipping. The man must have suffered a great deal as our natives transported him to the fortress. He appeared to be very weak, and when it was his turn to receive punishment, he just lay there in a despondent state, blood dripping from his mouth and nose. Of course, he was not punished but was carried to the shade by the horse enclosure. By our command, his fellow prisoners took hold of him, and with a dull thud, the head of the unconscious man fell against the stones lying here and there about in the yard. They took note of this, but seemed completely uncaring. I felt sorry for this poor, wretched man and dribbled some water over his head to revive him. When he failed to respond, I went to my tent to fetch some cognac. I returned to a gruesome sight. A tiny tot of a boy, a little black canaille no more than five years old, was kneeling on the dying man and pressing the latter's throat closed. His little eyes blazed with hatred and anger.

"What on earth do you think you are doing?" I roared at him.

"I'm killing this scoundrel," he replied. "He's my enemy!"

"Get away from him!" I yelled.

"No, not until he's dead."

At this I soundly boxed his ears. And with an angry look in his eyes, the little monster with a frog-like belly finally left his victim. Five minutes later the prisoner was dead. His fellow prisoners now had to carry his limp body from the yard and bury it some distance from the post. This they did in the same idiotic manner that they had carried the man to the shade. Again, there was the dull thud of his head hitting the ground. Several days later I chanced by the place where the man had been buried. I was struck by a foul, penetrating odor and followed it to the grave. The hole had not been dug deep enough, too few stones had been set upon the mound, and the unsecured thorn bushes had been blown away by the wind. Consequently, his body had been exhumed by hyenas and jackals and consumed as a gruesome feast.

On January 25th a native teamster named Galeb Goseb arrived in Okombahe. In his charge was an empty wagon as well as a half-loaded one. He had gone to Swakopmund to fetch a load of mineral water and beer for a merchant in Karibib. On the way to Swakopmund, he learned of the insurrection and determined that it would be unwise to transport the goods directly back to Karibib. He planned to reach his home in Okombahe by following little used paths, and from there he would push on to Karibib. However, on his return trip, he fell into the hands of the Herero who stole all of his freight. A carousing feast was held by the Herero, he told us, and the bottles of drink they were unable to consume were dashed against the rocks. They had stolen his oxen, but upon his fervent pleas the beasts were returned to him because he had been a friend of the Herero since his youth. As soon as they had disappeared, he and his men loaded the remaining goods on his wagon and made his way to Okombahe. Fifty cases of beer were left in our care at the post. It had been months since we had seen even a single drop of this delicious brown brew; yet as anxious as we were to drink some, we did not allow ourselves to touch it as duty had not yet been paid. We kept our thirst at bay and set the full cases along the wall—if we were attacked, at least they would serve as a barricade!

As time wore on, we noted less activity among the Herero and decided to send another patrol in the direction of Kawab. At ten o'clock in the morning the patrol left Okombahe so as to arrive in Kawab during the hottest time of day. We assumed that if there were people still living in Kawab they would most likely be resting at this hour. The riders approached carefully. In the heat, the place was deathly silent. It stood abandoned and desolate, even the gardens were dried up. The men burned one pontok after another as well as the tiny huts belonging to the leaders. From our post in Okombahe we were able to see the fires. The black smoke settled like a thick cloud above the ana trees along the river and lingered there for quite a while. The patrol returned, covered not with glory but blackened by acrid smoke. They were, however, in possession of spoils: a garden chair, a desk chair, a collapsible harmonium still in useable condition, and many other beautiful items.

We left the flags flying from our tower until His Majesty's birthday, and celebrated the occasion by pleading with the sergeant until he consented to open a case of beer. Cornelius received half a bottle of rum. And in the shade of a cart at the front of the fortress, Themis led three cheers for the Emperor. On this special day, our dinner was a bit grander than usual: meat with dumplings and pancakes fried in lard. It was indeed a day of feasting.

Somewhat later, we sent a second patrol that rode as far as Okarundu. There they found fresh tracks and an old woman who reported that the area was unsafe and filled with spies. According to her, the Herero still intended to attack us but were waiting as Omaruru had not yet fallen. Another patrol stayed out all night but returned without having gathered any useful information.

Negotiations with Captain Johannes of Kawab, whom I have previously mentioned, were amusing, to say the least. Several times he had sent written and oral reports that he was the most loyal friend of the whites. His intentions of receiving a gift of tobacco from us for his declarations of loyality yielded him nothing. A further letter arrived in which he wrote that he was a lone man in the village, left behind by all his people because he considered it a sin to join the evil insurrection. His woeful state had worsened still, for now he was threatened by his own people and was in danger of losing his life! To escape them, he wrote, he had gone into hiding on Merker's farm and requested that Herr Merker send an ox-drawn wagon, accompanied by a strong military unit, to bring him safely to the post. What a sly, old fox! But no one fell for his ruse. We sent him a reply: "We are deeply touched by your friendship, dearest Johannes, and believe everything you are saying. We are very sorry you find yourself in such a miserable situation. Naturally, the entire white force will leave Okombahe to bring you to safety. But until then, please remain patient."

Hardly had we sent our note when Cornelius appeared with a letter from Johannes addressed to him. How different the story was now: "Dear Cornelius! I am quite perturbed that you, our most loyal friend, you, the mighty king of the Mountain Damara, have persisted in staying so long among that handful of whites. I know you well enough to realize that you are cleverer than our great captains, Samuel and Michael. They kill with knives, bullets, and clubs. The two of us, however, have an understanding, do we not? When will your poison take effect and bring an end to these whites? To make your task easier, I have placed a small military force at your disposal. I am sure they will come. I will take care of the whites. I am Captain Johannes."

Our good, dear, and loyal Johannes! But thus it went back and forth, and it was taxing our nerves. How we longed for an end to all of this, one way or the other. On February 4th, we heard the cannon thundering loudly. It lasted until dusk. We worried, listened to the sound of the guns, and counted the volleys. Had the reinforcements arrived? Or were the defenders fighting desperately against these dastardly attackers? The next morning brought indications that the battle had been fierce. The river, swollen because of rainfall during the night, brought several dead Herero bodies. We watched them as the tide swiftly carried them through the churning waters. At noon, a number of excited natives appeared in the distance, shouting that reinforcements had arrived in Omaruru. The Franke Company, under the command of their brilliant leader, had arrived from the south after a series of forced marches, and relieved the besieged occupants of the fortress at Omaruru. Together with their supply train, the soldiers had marched sixty kilometers a day starting from Gibeon. In Windhuk, they had earned glory. Their defense of the railway lines, of Okahandja, of Karibib, and now of Omaruru were victories. These had been hard fought, and the cost was very heavy—especially at Omaruru where a mass soldiers' grave bears testimony of the hard-won battle. Detailed news we were not able to extract from these messengers. They knew only that Franke was there and that the Herero had fled.

My husband, along with two other men, volunteered to ride to Omaruru to report our situation and ask for reinforcements. An hour later, off they rode. During the afternoon, the messenger we had sent to Karibib returned, bringing with him a letter from the district head, second lieutenant Kuhn. The letter had been written on cloth and sewn into his coat. "Courage, hold out!" it read. "Cheers to the brave men and women! Perhaps reinforcements will arrive from Omaruru, but if not, send me another message, and I will take care of it from here." In the meantime, the crew of the *Habicht*, a navel vessel, and a battalion of marines had arrived in Karibib. How grateful we would have been had a small detachment of them been sent our way for one week or so. We had informed the people in Karibib of our desperate situation and our anxieties about the natives.

Sometime during the night of February 7th or early morning of February 8th, our patrol returned. The men reported the heroic deeds of the Franke Company with great satisfaction and pride. Through their now famous reenforcement and relief of Omaruru, they had won themselves glory for all time. Captain Franke, the commander, later told me he had been firmly convinced he could persuade the natives of Omaruru to give up without shedding a drop of blood. During times of peace, they had been completely obedient to him, and this gave him reason to believe he could still exert a bit of influence over them. Consequently, he donned his white uniform, and riding his well-known white horse, Blessbock, he led his column of men, firmly convinced they would either send a negotiator or flee as soon as they recognized him. Recognize him they did, and fired volley after angry volley at him. The rain of bullets aimed at him felled many of his soldiers. Miraculously, he was not hit, but galloped back, changed his uniform from white to khaki, quickly returned, and led his men into battle. Untiringly, he rode up and down the line of his soldiers, offering a perfect target to the enemy who knew him very well.

Under the command of Sergeant Müller, a company of the occupation force staged an attack from the military post in Omaruru in order to support the approaching force. The sergeant died a hero's death in the battle; his soldiers fought on with bayonets and the stocks of their rifles. During part of the battle, combatants were separated from one another by no more than three steps. At last, Franke delivered a strong volley at the natives, and with a beastly howl of anger, they fled into the nearby mountains, filled with a superstitious horror of Captain Franke, whom they thought to be the devil incarnate. During the night, the insurrectionists returned to fetch their noble dead, leaving behind their comrades of lesser birth.

Many of the settlers offered to care for our wounded soldiers. Three of the five officers were seriously incapacitated. Lieutenant von Wöllwarth and First Lieutenant Griesbach, both of whom we knew well, suffered terribly. The former died a week after the battle; the latter lived on for several months. First Lieutenant von Nathusius received a flesh wound that quickly healed. As the most experienced officer, Franke took command of Omaruru—which Staff Doctor Kuhn had assumed in his absence. Thanks to the courage and circumspection of Dr. Kuhn, Omaruru was able to hold out for a very long time. Yet who knows how the battle might have ended had he not sortied courageously, thus supporting the arriving force by attacking the Herero from the rear? Eventually, Franke would have entered the fortress, but there might also have been many more casualties, and his forces could possibly have been annihilated completely.

The foreigners still living in their homes on the countryside, as well as Missionary Dannert and his family, had to be moved inside the post as Franke did not wish to be constrained by their interests in his military operations. It may also have been that Franke did not fully trust each of them. But as soon as he was certain that the post was safe, he began searching out the unprotected settlers in his district. From Omburu, he brought in the honorable Missionary Bernsman and his daughter. The poor man had to live down serious disappointments with his Christian Herero converts, for he had been quite unable to prevent the murder of a white settler who had sought refuge in Bernsman's own home. Such was the state of affairs in Omaruru when Themis and his two companions arrived on February 7th. They were enthusiastically received as everyone was astonished to see the men from Okombahe who had been reported dead. Apparently, at the very beginning of the insurrection, the Herero had spread the news of the fall of Okombahe and the murder of all the whites, including wives and children.

Themis was able to report on the steadfastness and loyalty of our courageous Cornelius, as well as the hostile attitude and intentions of his men. He explained further the very considerable efforts we had made in defending Okombahe. Immediately, Captain Franke decided to make a public display of the might of his military force and the "great speaking pipe" (cannons) to the wavering Mountain Damara, and at the same time to bring my child and myself safely to Omaruru. Anxiously, we anticipated the arrival of our troops. Inwardly, I struggled in determining the best course of action—

whether I should leave my husband and take care of my child in safety, or just stay put. I decided to leave. My husband could get on without me, but my child had suffered considerably during the weeks on the post. Life in the tent did not at all agree with him. When it rained, water filled the bottom of our tent. All our belongings, linens, and beds were cold and damp. It was difficult to keep the tent clean, and securing enough good food for the little one was a constant challenge. The cows, that by this point had exhausted the grass just outside the post, gave very little milk. I was adding rye flour and oatmeal to the milk, but soon my stocks of these were depleted as well. The child constantly suffered from an upset stomach and had severe malaria attacks. The poor little thing had become thin and pale; naturally, we were very worried about him. During the final days prior to our departure, I busied myself providing my husband with new clothes as all of his old clothing had been torn to shreds. I was still in possession of some colorful cotton cloth—bright blue with flowers and red climbers on it. And from these lively colors and patterns, I sewed by hand, whether in sunlight or lamplight, and created a stock of new shirts for him.

5. Under the Protection of the Franke Company

One Sunday afternoon we noticed a huge cloud of dust on the other side of the river. A while later we saw that it was our troops approaching our Okombahe fortress. Immediately everyone was up and about: the soldiers fell in place, and the natives came running with their mouths wide open. The building seemed to tremble as the sun-bronzed troopers rode up the small incline to our fortress. We breathed a collective sigh of relief—now, at last, all was well!

After I greeted the officers, I hurried to prepare the evening meal with the help of the sergeant. Soon the table was set with tin plates, rusty knives and forks, tin cups, bread, bacon, and coffee. No matter how much we searched the storage room, nothing else was to be had. Nearly everything had already been consumed. My husband offered his last bottle of Hennessy, some cigarettes, and the last of his stock of imported cigars. A case of beer was opened, and the individual bottles cooled in a water bath. The soldiers were dusty and exhausted, and retired soon after the meal. The rest of us sat up for a long time. A guard was posted at the front of the fortress. It was my last evening in Okombahe, and it would likely be a very long time before I should see it again. The crackling fire burned bright, illuminating the entire area. Saddles, cooking gear, mule carts, pyramids of rifles, and cannon shone in the fire's light; it was a picturesque scene brimming with life. The company camped on the opposite side of the river. We saw their campfires dancing like glimmering dots, sometimes disappearing, sometimes growing larger among the trees and fog along the riverbank.

We arose before daybreak. I still had to pack my suitcases for the journey to Europe. At eight o'clock in the morning, the signal for mounting the horses had sounded. My son, restless with fever, had not slept well during the night. It was clear to me that I had no choice but to leave if I wished to save his life. I was leaving my husband to an uncertain fate in Okombahe. Would we see each other again? And if so, would I be able to return to him with our child? I closed my eyes for a moment as the gravity of what I was doing suddenly overwhelmed me. But I drew a deep breath and pulled myself together. As the cart started forward, those who stayed behind waved good-bye to us.

Reaching the opposite bank of the river, we joined the troopers. Captain Franke took excellent care of me, and I was touched by his concern and attention. His servant, Benjamin, led the mules and followed the orders with such exactness that we passed through the thorn bushes without incident, did not get bogged down in the sand, and not even once did the cart lose its balance. About fifty officers and soldiers took the point; the cannon, the munition wagon, and our cart filled out the midsection, with another fifty soldiers at the rear. The trail was hot, dusty, and in poor condition. My child whimpered each time we passed a rough spot. He appeared to be in pain.

We rested for a time behind the settlement of Kawab whose burned-out pontoks and houses offered but a sad, dismal scene. Franke and I shared our few provisions. Once the air became cooler, our column started to march again. I hardly took note of the natural beauty of the area. I was utterly exhausted and bereft of an inner calm because of my anxieties for the well-being of my son. He refused to eat and whimpered ceaselessly. Nothing I did eased his situation. The horses trotted along the rough trail, the cart often jumping high in the air, other times tilting us dangerously to one side. Thorny bushes on either side of the so-called trail brushed against us, tearing holes in my dress and scratching my hands and face. I had my child on my lap so I could protect him. To make matters worse, it was dreadfully hot and the dust penetrated my lungs, taking away my breath. The radiant yellow sand and bare, glimmering stones blinded me, so that tears flowed from my eyes.

Since the area along the river had yet to be retaken from the Herero, we were unable to travel the most direct way to Omaruru. I watched the setting sun with dullened senses. We camped for the night close to the river. So exhausted was my son that he slept peacefully through the night. Captain Franke and I sat for a long while beside the tall flames of our fire. We had known each other for quite some time now, and he spoke of the trials and tribulations of his campaign. Finally, exhaustion overcame him and he went to sleep. I sat up the entire night watching my ailing child's heavy breathing and trying to shoo away the countless mosquitoes flying about beneath the trees. Except for the distant howls of jackals and the occasional stamping of horses' hooves, the evening was quiet. Stars shone with a festive yet sublime splendor. The Southern Cross sank deeper and deeper on the horizon, and the evening grew very cold. Thick, damp fog settled over the riverbanks. A few stars still shone brightly in the sky when the bugler blew reveille. In no time at all, everyone was ready to march. I quickly downed a cup of hot tea and filled the baby's bottle with the rest. We took our places in the cart and were off.

Day gradually dawned as we drove along. My head ached, and my child suffered from a high fever. I remember that we rested in Otjimpaue at ten o'clock that morning. Franke shared sausages with me and I gave some to my sick little boy, washing them down with watery black coffee. While everyone napped during the heat of the day, I stepped down to the river and took care of myself, then bathed the child and washed all his clothing. It had been a long, arduous ride, and the horses needed extra rest, thus the midday break extended beyond the time originally slotted for it. We resumed, but as the end of our journey was in sight, I felt somewhat less discouraged. Besides this, though my son's eyes were heavily inflamed from all the dust and wind, his health had taken a turn for the better. At last the Omaruru Mountain appeared in the distance. Before us lay the town. We rode down the empty street, passed the church and the mission home, and crossed the river. There we saw the terribly bloated corpse of a Herero man, so bloated that the buttons on his white jacket—except the top one—had burst through the buttonholes. It was a revolting sight. Beyond this, we noticed activity in the fortress' yard: a battalion of marines had just arrived.

6. In the Barracks at Omaruru

Staff Physician, Dr. Kuhn, and his wife cared most tenderly for my child and myself. They both wished to hear our story, but so great was my exhaustion that I battled a nagging fear of falling off my chair. Indeed, I perceived the entire conversation as though I were actually in another room. With all the soldiers needing accommodation, I wondered where I was supposed to be housed—somewhere in the three buildings that comprised the barracks? Settlers were crowded into one of the buildings; soldiers were quartered in the other two as well as on the veranda and in the yard. The remaining refugees had formed a circle with their wagons, as though the carnival had come to town. Behind the wagons stood kraals for the livestock, animals the settlers had rescued and livestock the soldiers had captured.

At this point, where I was to rest my weary head was irrelevant, my only concern was for the well-being of my poor little boy. Again, Captain Franke was my friend in time of need and called his soldiers to prepare a small room reserved for noncommissioned officers. It was far nicer than I had expected, containing a comfortable iron bed with blue and white checked linen, a wash-basin, and a well-worn chair. Herr Kindt, also known affectionately as Fatso, upholstered his bathtub with blankets that I might use it as a bed for my son. A small kerosene lamp, eating utensils, and provisions were also kindly supplied by him. Magically, he somehow procured milk, a fresh egg now and again, and a servant to help me.

Soldiers camped in front of my door and disturbed my child's sleep at night. In order to keep him calm, I had to walk the floor with him, yet in the morning his fever was 40.9 degrees centigrade. Staff Physician, Dr. Hummel, diagnosed malaria. I am truly indebted to his constant care, despite his heavy responsibilities in the hospital, for the eventual recovery of my little one. As my time permitted, I helped care for the wounded in the hospital. There were so many wounded that a nurse would have been most useful. Many of the women who had no children just sat around without ever offering to give us a hand. As much as I would liked to have helped the overwrought hospital attendants, my own child required round-the-clock care. He had to be bathed and rubbed, then given compresses, wrappings, and medicine.

During the second day, I visited the graves of some fallen soldiers. Eight men, most of whom I knew, lay in a single grave. They had given their lives for our colony. In another grave, situated under a large ana tree, rested the body of Lieutenant von Wöllwarth. Two happy days we had spent with him in December. I still see him in front of me as the absolute model of a young officer—lively and cheerful. All of us had loved him. On our journey to Omaruru, a medium-sized dog of dubious breed had caught my attention. I noticed that all the soldiers, without exception, took care of it. If the dog was tired or too hot, they lifted it into a cart. One soldier even emptied the last of his canteen water into the lid of his mess kit so the thirsty dog could drink. I was interested in this animal and enquired after its owner. One of the soldiers replied, "That dog had belonged to Lieutenant von Wöllwarth. We owe it to his master to look after the mutt. We all loved the lieutenant. Why was he the one who had to die?" As he talked, he wiped tears from his eyes.

I met another soldier at the grave who had helped to place a simple cross on it. He was deeply moved when he saw my little bouquet of flowers—gathered from the fields lying next to the grave. We talked. "It's strange," he said, "we all feared for him. Sometimes there's a person one feels, one just instinctively knows, will not survive. He was one of those. Once before, when we were crossing the swift current of the Swakop River, he was in danger of losing his life. Suddenly, his horse lost its footing and was swept away by the waves. By chance, our captain turned around, and before we knew what had happened, he galloped into the water and saved von Wöllwarth. Because strong currents threatened to pull both of them downstream, the captain could not drag him ashore, but had to hold on to him. Corporal Bössow heard him calling and brought the needed help. I sometimes wish the captain had not held onto him in the river. It would have been better had the lieutenant died then and there, and been spared the terrible sufferings that finally brought him to death."

After February 20th, the troops were ordered to move out and join the reinforcements, newly arrived from Germany, and together meet up with the Fourth Field Company coming from Outjo. These three forces were to unite and advance upon the Herero who had amassed near Otjihuenamaperero. Late in the evening, all present assembled in the barracks yard to receive communion. Two huge fires lit the area. Between the fires stood the altar table set with a white cloth. Red flames blazed to the sky, casting light on a scene that those who witnessed it will never forget. The two missionaries administered communion at the front of the table. One by one, the men came forward. First it was the officers, then the rough, weather-beaten warriors—each stepping toward the table, with bared head, to receive the Body of the Lord.

The next morning the troops assembled to take their leave. It was a wondrous scene. The cannons were assembled. Wagons bearing provisions, baggage, and munitions stood in a very long line. Despite the last-minute hustle and bustle, there was exemplary order. Saddles, saddlebags, cooking tins, canteens, and every other needful thing lay in neat rows, ready for the soldiers to collect. Shortly after orders were given, everything stood in readiness. According to rank and file, the soldiers stood in position, muscular and with nerves of steel, their skin tanned copper-brown by the African sun, all of them with mighty beards. These troops were fearsome. Everyone had a bit of the devil in him. Many had lost everything they owned and were out to avenge these losses. Many a face was marked by an expression of savage determination. What they had gained during years of hard labor had been lost. Now they would perform a different kind of work and show no mercy. Then Missionary Bernsmann, with his flowing white beard, stood before the troops, his arms extended upward, and gave them his blessing.

The seemingly endless column of soldiers started to move. I went to Major von Estorff to bid him farewell. My husband was very much indebted to the Major for his many past kindnesses. Fervently, he pleaded with me, "Will you promise me that you'll return to South West? What's to become of this colony if our women give up? You mustn't abandon South West, you must return. Promise me!" I wholeheartedly promised him, firmly convinced that my stay in Germany would be a short one. Our experiences of happiness and sadness had so strongly bonded us to the colony that we could never give it up.

I shook a few more hands and called out to a few more men. Many an eye of those remaining was filled with tears. Everyone knew that the fighting was serious and that many would lose their lives. An uncertain something arose within the mind's eye—battle noise, cannon thunder, rifle fire, twitching bodies, distorted faces, blood. Whom would fate not allow to step upon German soil again? Whom would the Grim Reaper call to enter the gates of eternal silence, to move into a realm where everything is reconciled, where not hatred but love rules? "Dear God, be merciful to them, protect and guard them!"

They crossed the river. At first, we saw only a column of brown figures, then a mighty cloud of dust arose. Finally, they were out of sight. I remained standing there for a long time, even when there was nothing more to be seen. Tears continued streaming down my face though I was hardly conscious of it. Omaruru fell under a deadly silence.

Since Dr. Kuhn had left with the troops, First Lieutenant von Nathusius took command and immediately began the clean-up operations. The entire region was searched, every pontok and every house was rummaged through, and many stolen items were recovered. The furniture from Captain Franke's apartment was found, cut to pieces and painted with the name of the Herero leader, Captain Michael. They also found a large picture of the German Emperor, his signature completely marked out. Written in its place was, "Michael Omuhonga. Michael is Lord." Any dead bodies found were burned. Close to Michael's compound, where the battle had raged most furiously, all the fallen enemies had been burned on a huge funeral pyre. Now thoroughly charred, the grinning skulls and other bones could be seen scattered about the place. All of the pontoks and houses were burned, and like giant torches, the smoke rose to the sky for several days, a warning that the white man metes out terrible punishment when he is offended.

There was much work to be done in the hospital. Wounds were full of puss, and despite every care given, they healed but slowly because the bullets had been poisoned. It was Herero custom to set their bullets, just before firing them, into their own boils and pustules so their enemies would be infected with their own repulsive diseases. But their intentions were never fulfilled as a bullet is heated to such a degree in the barrel that the heat functions as an antiseptic. Still, the intent is nothing short of devilish. Far worse is the poison derived from the many varieties of euphorbia plants, which the Herero use expertly. And after all this, why should such fuss be made in Germany over the reports that soldiers occasionally notch their bullets to make them more effective once the target is struck?

One evening as I sat in my room past ten o'clock, I heard several cracks of a whip with whose sound I was familiar. It sounded to me as if Friedrich, our old driver, were handling the whip. I stepped outside and instantly recognized our cart. Happy as a lark, my husband walked up to me. He had loaded the cart with the remaining crates of duty-free beer so he would have a reason to leave Okombahe and return to active duty in the field. Our cart was immediately requisitioned for troop transportation. Themis could not stay long, for he had telegraphed Major von Estorff and received orders to bring a wagon load of munitions and join von Estorff's staff. Again, the departure bell tolled for us. I was left behind, longing for peace and for my home in Germany. We had patiently waited for news from Karibib that the trail between Omaruru and Karibib could safely be traversed by wagon if accompanied by a strong military detachment when finally it came! Frau Kuhn, whose husband was also in the field, likewise wished to travel to the coast. She was kind enough to take us with her to Karibib.

7. My Return to Germany

On February 28th, around three o'clock in the afternoon, we took our leave of the people of Omaruru. Without delay, we trekked forward along the same trail the Franke Company had taken on their march to Omaruru. The path was marked by broken wagons still hitched to the corpses of fallen oxen, then by dead horses, booty of all kinds, torn-up saddles, and all manner of rags. Without stopping once, we trekked until one o'clock in the morning. The night was wonderfully mild, and the full African moon shone with all her beauty. Wistfully, I noted the charm of the landscape. Though I had not yet left Africa, I was already homesick for it and could hardly wait to return. Had there been no child, I would never have left Africa.

As I was packing and preparing ourselves to leave, the baby's bottle fell and broke. I was desperate for another when a soldier offered me his empty rum bottle as a replacement. It served me well until we reached Swakopmund. Shortly before sunrise, we reached Etiro where we inspected the house of the murdered farmer, Joost. Not one item in the house had been left whole. Frau Joost had planned to seek protection in Omaruru, but was captured by the Herero, robbed, and then delivered to Missionary Dannert. Frau Kuhn and I had become stiff during the long journey and went for an early morning walk. We climbed a rocky hillock to watch the sun rise. Two dead Herero lay there, their bodies facing downhill, each body partially devoured by jackals. One of them still had his leg inside a high riding boot; the other had a silver chain around his arm, an arm now reduced to bare bones.

Under the searing rays of the sun we trekked until ten o'clock, and then again from two o'clock until late in the afternoon. We reached Karibib, but could not enter the town because of a barbed wire barrier. A soldier finally opened the way, and we went from hotel to hotel until we found a room in the Rubien. My son suffered another serious attack of malaria. I felt quite miserable myself. My eyes were seriously inflamed. The following morning Staff Physician Blümchen sent two nurses to our room to care for us. My child had also contracted a typhoid intestinal catarrh, and since it was impossible for him to swallow anything, he was given a quinine injection. He suffered for some months from the effects of this treatment. Our stay in Karibib was not pleasant. Life on the square was boisterous, and wherever one looked there were heaps of rubbish and dirt. I was surprised that typhoid had not yet broken out.

A humorous anecdote is told about the inhabitants of Karibib. One evening, they thought the Herero had surrounded them again. They saw something dark sneaking along Store Street. In the excitement that followed, they all assumed a spy had been searching for a weak spot in the barbed wire fence that would allow him to breach it despite the sixty soldiers posted there as guards. After an intense council meeting, it appeared that the "dark thing" was creeping ever closer, hence, the command for rapid fire was given. Many shots were fired in the direction of the danger, and the shots appeared to have met their target. The "dark thing" stopped and groaned. At once, several courageous men declared themselves ready to brave the danger and investigate the angry, and now impaired, enemy. In the end, they decided against this course of action. It was night, after all, and "one could never be quite sure." The next morning there would be daylight and the "thing" could be safely inspected. In the morning, the "something" was still there and still moving. It was a sow with her little piglets. They had somehow escaped from their stable. The sow was not seriously hurt, and her little ones lay cheerily beneath her!

The Herero never did advance on Karibib. There were far too many whites in the town, and they could not have caused much damage. Besides, the whites were constantly practicing and fully prepared for any attackers.

My child was not improving, and the staff physician advised me to remove to the coast as quickly as possible. On March 2nd, a freight train departed for the coast, and we were taken along. I still had a bandage across my eyes, but I was able to see well enough when I pushed it back. In Jakalswater, we slept in the living room of the kind and hospitable Frau Heyn. The next morning, off went the train, and we arrived in Swakopmund during the night.

Coming from the interior of the colony, I noticed at once how cold it was along the coast. The hour was very late, and a damp fog covered the town. I had no idea where to find lodging. I had met the local administrative officer in Karibib—the staff physician had expressly asked him to look after me—but it was up to me to find a place to stay. The train station is situated quite a distance from town, and I ended up carrying my hungry, frozen, crying child through the deep sand of the dunes, which made walking very difficult indeed. It was quite a chore to find any kind of hotel at that hour. There was a bed left in a boarded-up room of the Central Hotel. I set my child in a wooden crate and, after massaging the little one, he calmed down and even drank a bit of canned milk, then fell asleep. The cold fog penetrated the wooden boards, and all our blankets felt damp. Indeed, the lamp itself was surrounded by a veil of dampness.

The next day, First Lieutenant von Bülow and several other gentlemen arrived and offered their assistance. Von Bülow was the same officer who courageously brought relief to the disheartened force in Okahandja. He deserves the highest admiration for his bravery. Dr. Zacklehner was touching. He took my child and gave me all kinds of nice things for the journey at sea. Frau Funck, the hostess, was also very attentive to my needs. They both offered me money and anything I desired, but I graciously declined and thanked them—there were other women whose fate was far worse then mine.

On March 7th, Frau Kuhn, Herr Wecke, Dr. Zachlehner, and I boarded the *Lucie Woermann*. Kindly, Dr. Zachlehner wished to help me with the child. The *Pionier* took us from the mole to the *Lucie*, which was anchored farther out at sea. The sea was completely calm, and we passengers were hoisted aboard in a wicker chair lowered by a chain from the *Lucie*. Time passed quickly. The band played a march, then with a mighty jerk the anchors were hoisted. The siren wailed and the large ship slowly began to move. Gradually, Swakopmund disappeared in the distance. We could still make out the houses, then only the coastline with its white breakers; eventually these faded into oblivion leaving only a bank of fog. At last, there was nothing.

I stood at the railing with a heart so heavy it felt ready to burst. You, South West, the land of our ideals, the land of our dreams, how shamefully you have deceived us! We came to you filled with love and hope, but you gave us disappointment after disappointment. You withheld your gifts from us yet took everything we had. I leave you with empty hands and an aching heart. Nevertheless, I love you more than any other land! A deep longing for you has already awakened in my heart. May God grant me that before long I may see you again, your defiant rocks with their deep blue shadows where baboons and rock dassies live. How I wish to listen again to the natives' monotonous songs as they sit round their fires and the howls of the jackals during the night. How I desire to look out over the endless yellow Namib Desert during the heat of the noonday sun, or when the silvery moon shines through the branches of camelthorn trees. I would like . . . , I would like . . .

The dinner bell rang and tore me from my deep melancholy, from my longing. My young child, who had fallen asleep, needed me now.

As though caught in a bad dream, we have been aboard ship for three dreadful weeks now. From day to day, my child's illness only strengthens, and I have prepared myself for the worst. The doctor offered no hope at all. The little one only vegetates, and I have barely kept him alive with egg whites fed to him with a teaspoon. It was difficult for me to witness this little creature's sufferings—it was worse than anything I had experienced up to this point in time. The doctor advised me to wait for him to die in Madeira so that I would at least have the consolation that he would not be buried at sea. As desperate as I felt, I decided to remain on board—despite every reason for not doing so. I arrived in Hamburg. My mother met me at the docks for what was a sad reunion.

In Berlin, our doctor determined that my son had had typhoid fever and was now suffering from acute pneumonia. For almost three months I vacillated between the hope of keeping him alive with my round-the-clock care and total despondency thinking that all my efforts would be in vain. I also received unfortunate news from my husband. He had been infected with typhoid fever at Otjehaenamaperero. He was seriously ill and had lain unconscious for weeks. Again, I prepared for the worst. Gradually, however, he recovered, and in August 1904, returned home. Because of his long residence in South West Africa and the unbelievable stresses he had endured, he had contracted a heart condition that would not permit him to return to South West.

We moved to Weimar where my husband attended the art academy. Thanks to a talent he inherited from his father, he now sends his works to art exhibitions. He prefers painting African scenes from his memory. Both of us will probably never be entirely happy in Europe again. We love Africa too much, particularly the country we conquered suffering a thousand pains to do so. Once a person has lived there and then leaves, he longs to return and cannot find peace anywhere else, yet because of the unrest in South West we were advised not to return. Despite his poor health, my husband joined the German East Africa Company in September of 1906. My husband is near the settlement of Lindi, and I regret very much that I was not able to accompany him to German East Africa. However, my children need my care, and because of the unhealthy climate there we cannot take them along. Thus we moved to Braunschweig.

It will always remain my sincerest desire to return to Africa, to make our fortune there, a desire that, to this point, has eluded us. I can only hope that my wish will be fulfilled in the near future. We fought for Africa, we suffered there, and our two sons, and we belong there.

Frau von Eckenbrecher with her son on the veranda of their home in Okombahe.

Herr von Eckenbrecher at the military post holding his son on the latter's second birthday.

Our home in Okombahe.

Breakfast on the veranda.

My husband holding orphaned jackel pups.

At the militay post in Okombahe with other settler families.

Our son with his nursemaid.

Our son with his playmates.

My husband in the study.

My desk and writing space.

Book II

A NOTE TO MY READERS

My book on Africa, first published in 1907, went through several editions and thus met the objectives I had set out for that book. In it, I recounted my experiences as a pioneering farmer's wife in German South West Africa beginning in 1902 and ending with the Herero Insurrection. I wished to describe our life in the colony and to encourage women endowed with strength, fortitude, and a cheery disposition to become pioneers in this venture of spreading our Germanity to Africa, side by side with our vigorous men. I find great satisfaction and am delighted that to this day many a courageous pioneer woman says to me, "Frau von Eckenbrecher, your book helped me in deciding to come here—and I do not regret it."

For several years now I have had repeated requests to continue telling about my life out here during the Great War and now under the Mandate Rule. I have finally consented to do so. Twenty-two years of personal experience form the basis of my current writing, and I have carefully checked and cross-checked my material to prevent errors from creeping into my account. Certain chapters I discussed with close friends; I am also indebted to my editors. It is my intention to present things in a straightforward manner, though by doing so I hope no one takes umbrage with me.

This book should not be taken as a piece of propaganda. I have not written it to persuade or compel my fellow Germans to immigrate here, only to relate to those in our homeland how we Germans lived and live "out here" before, during, and after the war, how we suffered and endured, how our colony's destiny blossomed and then changed. I am unable to report on the impressive personal experiences of the well known and well heeled in this colony during these past twenty-two years, but only on my own rather insignificant fate that is submerged and intermingled with the tremendous contestations of our great era. Behind me rests a difficult life of struggle and toil, of twenty-one long years of dedicated work with our German children here in South West. My work has helped me endure those periods of deprivation and the heavy bitternesses of my life so that my burdens did not consume and overwhelm me.

It has become lonely, very lonely, for me, though my loneliness is not without its deep satisfactions. From early on in my youth I have had a difficult life; still, it has been rich and rewarding, and I feel that I have been favored to be at the forefront of defending our culture and our Germanity for the past twenty-one years. My beloved South West has revealed all its grandeur, beauty, and harshness to me. It has given me much, but it has also taken much from me and required much of me. I have become part of South West, and it has become part of me.

Thus it is no easy task for me to send this book on its way. It is meant as a greeting to all those settlers who came to South West but have returned to the homeland. It is to remind them of the time they lived and toiled out here in the land of the sun—and I know it will reawaken in some of them those ardent yearnings to return. But this book is primarily intended to acquaint those who have never lived here with this piece of our Fatherland—a piece we have lost—and to arouse in them a deep love for it, along with the assurance that we in South West are not a lost people. We are loyal to our old homeland, loyal to the Reich and its Führer, and we are proud to be German pioneers.

Margarethe von Eckenbrecher, née Hopfer
Windhuk, South West Africa

THE NEW SOUTH WEST

1. My Return to South West Africa

It was May 1914, a full ten years since my journey home, and it was the final night before our landing in Swakopmund. The ship steadily forged ahead through the rolling waves. During the long voyage I had become accustomed to the ship's gentle rocking and pounding, as well as to the sound and feel of the waves as they splashed incessantly against the sides of the ship. On deck, we celebrated our landing with laughter, joking, and the popping of corks—much the same as on my first voyage. All of us were in a cheerful mood, or at least we pretended to be that we might fool the others and participate in the merrymaking. Each of us was filled with hope and excitement, busily making plans for the future.

I had already seen much of the world. Africa was no longer the land of promise it had been twelve years ago. I was, without question, looking forward to being back, but it was a melancholy joy, and my worries about the future weighed heavily on my mind. I had no firm plans as to what I wished to do with my life and the lives of my children. The last year in Europe had been terribly difficult, and conditions at home had become increasingly unbearable for me. For four years I had labored at the Higher Pedagogical Trade School for Women in Braunschweig teaching French, English, Economics, and Geography. I had been in charge of courses in female social work, had given lectures, and discussed books. Despite all these responsibilities and time commitments, my anxieties continued day and night. Deep into the evening I lay awake brooding over my situation.

On one such difficult night, an incident from my first stay in Africa came to mind. Looming before me were the blue, elongated shadows falling from the back of the Okombahe Mountain. It was just before sunset, and the crimson rays from the setting sun filled the sky above a sulphur-yellow line on the western horizon. I stood there gazing at this scene, rejoicing at the golden splendor set within infinite space. Someone walked up to me from behind. It was none other than subcaptain Joshua, who asked, "Madam, why do you stand there gazing at the setting sun? You have not been in Africa very long, yet every evening I observe you standing there and watching the sun go down. This I can tell you: anyone new to this country who does this day after day is deeply rooted in Africa and must forever live in this land. Sometimes this is good; sometimes it is not. But if such a person leaves, Africa will beckon him back—and he will return whether he wants to or not. You will see."

I looked into that setting sun exactly as I had during the first days after my arrival in Africa. I knew that old-time settlers were aware of the call of Africa; I also knew that I would follow that call. What was there to keep me in Germany? I was divorced. And though I was awarded sole custody of my two sons, I was lonely. I had to find a new start somewhere, and Africa was beckoning me. I decided to immigrate again to South West Africa, the land of my dreams. But I did not do so blindly trusting my luck. Instead I carefully consulted old friends of mine, asking if it would be wise for my children and me to settle in Africa, whether or not I would be able to find employment there, if the children would receive proper schooling, and if it would be possible to rear them there in such a way that their future prospects would be good. I received favorable answers to these enquiries, especially (and gratefully) from my friend, Rudolf Kindt. At that time he, along with his sister, was owner and editor-in-chief of the South West newspaper. They consistently proved to be my most loyal and selfless friends in every imaginable way. They were Friesian, and it was said that Fräulein Kindt could actually glimpse into the future and see things that did eventually come to pass. At the time of this writing they have both lain several years beneath the lawns of the Darmstadt Cemetery. They were among the first to be evicted by the British (South Africans). Every day I walk past the house where they lived; in my thoughts they continue to live.

With heavy heart, I shook, for the second time, the dust of old Europe from my feet. It was difficult to leave my beloved homeland—the beautiful Harz region with its dark spruce trees. But along with my eight- and ten-year-old sons, and my loyal Anna, I was aboard the *General* anchored just off the beaches of Swakopmund, before the country of my new hopes and my new future.

At daybreak I went on deck. Fog hung like a dense, gray curtain about the ship. I could see nothing, but heard the incessant howl of the sirens. It was cold and damp. It was Africa. Sometime during the morning the fog finally lifted. A launch came up alongside the ship, a tugboat appeared, and passengers were ushered into the waiting boat by being set in a chairlike contraption attached to a steel cable that was then lowered by winch to the boat below. The now familiar process of disembarkation, of both passengers and luggage, proceeded slowly, though this time we were landed much more comfortably on a pier. We quickly cleared customs and trudged our way through deep sand to the Prince Bismarck Hotel. The hotel had changed greatly, and for the better, as had the whole of Swakopmund; indeed, there was scant resemblance to the town I had left in 1904.

The next morning we boarded a train that took us into the interior of the colony. Gone was the romance of the narrow gage railway. The new one no longer stopped as the old one did—whenever the engineer or the passengers were hungry and wished to cook a meal! Now we sat in comfortable leather seats and looked out at the dreary, monotonous Namib Desert through heavily tinted windows that blocked much of the blinding sunlight. Yet the sun, that dear African sun, was there. At last, far in the distance I could see the peaks of Spitzkoppies bathed in glowing light. They were my old friends; this was my fairytale land. On went the train, past Karibib and then toward Brakwater, the final station before Windhuk. As a fair was scheduled in Windhuk and every hotel and private home had been fully booked, the Kindts had arranged temporary quarters for us on a model farm belonging to Frau von F. The next day I left my children in the care of my faithful helper, Anna, and drove to Windhuk to visit the Kindts. The fair was on, and I admired all the new and beautiful things created since the insurrection. With a feeling of pride and joy, I noted progress and growth everywhere. During the past several years, South West had become a model colony—thanks to the diligence and perseverance of the settlers.

I met many of my acquaintances from before, and everyone encouraged me. They had no doubt I would find something good, but I needed to be patient before making final decisions and guard against acting rashly. Many people kindly invited my children over so I could think about the situation in peace, yet I felt uneasy about accepting their hospitality since I was not sure I could reciprocate their kindnesses. My financial situation was tight.

2. Our Idyllic Life in Klein-Windhuk

Windhuk appeared to be the most suitable place for us to wait, and I began looking for an apartment. As a place of residence, I did not favor Windhuk proper. Its houses were crowded together and its streets were busy. This was not Africa! I wanted to provide as much room as possible for my boys to play and be active, and for this reason I rented a pleasant, fully furnished little house with a garden in the suburb of Klein-Windhuk. The house belonged to the area secretary in the government (who was leaving on vacation) and was located on Jan-Jonker Weg, at the foot of the Wasserberg. The house boasted four furnished rooms and two large verandas beautifully draped in green vines. Most of our days were spent on the verandas, and it was always the place we gathered to spend the evening. A private windmill pumped water to the house, and a large metal basin close to the road became a favorite swimming pool for the children. More than once did a thirsty ox jump the fence and land in the water, and it was only with valiant effort, and the help of our neighbors, that we succeeded in pulling the beast from the water. Sheep and goats drowned in our pool again and again. Behind the house was a vegetable garden. Citrus trees had been planted there and were now bearing fruit. I had rented the house for six months with no inkling that a full eight years would pass before we moved again.

My next objective was to situate my children in school so they would not fall behind in their education. My older boy entered the sixth grade of the Imperial Middle School; the younger attended preschool in the government primary school. The schools were quite far away—a good thirty-minute walk—but the children were healthy and enjoyed their freedom and the unrestricted space for play.

I had always maintained that children should have pets, that the responsibility of caring for animals instills love in them, affords them opportunity for reflection, and imposes a necessity to undertake maintenance tasks that cannot be delayed. Thus, all kinds of animals found their way to our home. The first creature to arrive was Fips, a baby baboon. He was only a few weeks old, so we fed him sugared rice, dried fruit, and warm milk. During the day one or the other of my sons kept Fips beneath his shirt and against his body, and Fips was thus taken everywhere. At night we tethered Fips with a soft belt to the inside of his wooden box and carried it to the catch-all storage room where we carefully covered him so he would stay warm through the night.

As Fips grew, we had to confine him to a gnarled tree trunk in the yard. From the top of the trunk he had a splendid view of his surroundings. We had replaced his soft belt with a long iron chain that he skillfully looped. He then set leftover bits of food in the loops, perched himself a good distance from the loops, squinted into the sunlight, and waited for a chicken to recklessly venture into a loop. When one did, he immediately seized the chain, pulled on it, captured the chicken, snapped off its head, and began to pluck its feathers. All this was done with such lightning speed that we were never able to save the poor chickens.

Fips now spent his evenings in an old wooden cement barrel into which air holes had been bored. Two old cloth sacks served as his blankets that he aired and dried in his tree every morning. It was strange that he had premonitions about the weather: if the day promised to be beautiful, he set his barrel in an upright position; if it threatened to rain, he laid it flat on the ground, tore the sacks into pieces, and stuffed rags into the little holes. Fips carefully checked the direction of the wind and aligned the barrel accordingly. If the wind suddenly changed direction, he let out a piercing scream and quickly moved the barrel, then wrapped himself in the one sack and stuffed the other into the largest opening. This was also his course of action when a thunderstorm threatened. He was terribly afraid of the lightning and screamed and covered his eyes whenever it struck. For many years our Fips was the children's amiable playmate, but eventually he became too large and too obstinate and we placed him in the Windhuk Zoological Gardens.

The actual guardian of our house and yard was Nero, a strong Airedale. Nero was Fips's best friend as he always successfully defended Fips against other dogs, but he was also the most notorious roughneck in Klein-Windhuk. Every dog avoided him. Unfortunately for us, his very favorite food was chickens' eggs, and Nero had dug his own secret tunnel into the chicken coop. We eventually devised a punishment for him and placed an egg—filled with mustard and pepper—into the nest. Hiding, waiting, and watching, when he bit into the egg with such great anticipation, we burst out laughing. The poor thing was very ashamed, and from that moment forward we were never able to persuade him to visit the chicken coop with us. He was a most loyal dog and never accepted even the most delicious scrap of meat if it came from General Botha; on the other hand, he eagerly devoured the orange peel Colonel Franke handed him. Sadly, one night as he was on his way to court a lady friend of his, he ate a piece of poisoned meat some boys had set along the road and died.

Hans Gombo (Herero for goat) was a strong, white, long-haired, beautifully horned Merino ram that prevailed against all of his enemies. With commendable patience, Hans pulled a cart the children had built—even when Nero took the lead and Fips sat upon his back. His favorite spot was a quiet corner hidden far in the back of our garden, and so much did Hans regard this place as his that it came to blows when any other animal tried to pass a few moments in his beloved spot.

My favorite pet was an eagle we named Polly. She was a strong, beautiful animal with a wingspan of at least one and a half meters, and lived on a tree trunk opposite Fips, to which she was tethered by a chain linked to a ring around her foot. Natives had taken Polly from her nest sometime before she came to us, and she had never known freedom. At feeding time I let her off the chain and she fell in line with the dogs and cats—all of which held her in high regard. Like the dogs, she snapped at the bits of meat and skillfully snatched them up. Her food was expensive, so the boys captured rats and mice, sometimes even a meerkat, with their slingshots to supply her with fresh meat. Polly also loved the chickens Fips caught and prepared! Around noon, she flew to the very top of the tree and watched for the children to come home. As soon as she caught sight of them, she began with her loud screeches, and once the dogs and monkeys recognized her call, they too joined in announcing the approach of the children. Polly also announced other visitors, though with different sounds. When natives approached, she called hoarsely and ruffled her feathers. Somehow she knew she was my favorite and was quite jealous of others. Inside the house, she enjoyed sitting at my feet and pecking at the children if they dared come too close to me. When we left her untethered and unsupervised, she would chase after our chickens until she caught one of them. In 1922, I took my holiday in Germany, and before departing, I brought Polly to the Zoological Garden in Windhuk, where she lived for many years. Each time I visited her, I approached her cage and called for her. She always came close and snuggled her head on the fencing so that I could stroke it.

These were the most interesting of the many animals we kept as pets. We also had dachshunds, terriers, cats, and a very tame hen named Julchen that attempted to lay eggs in the strangest places: on a lamp cylinder, in the corner of the sofa, and even on the inkwell atop our desk. We had meerkats, rock badgers, fourteen young ostriches, owls, eagle owls, and crows. It did take a lot of time to care for all of them, but I cannot imagine what it would have been like to have lived without these animals in and around our house.

My children adjusted well and were happy, but had no idea of the worries their mother had and how she struggled to give them this life. Nonetheless, our idyllic life in Klein-Windhuk was delightful. We led a simple, frugal, and contented existence in our home. The friendly, stimulating atmosphere in the governor's house was the high point of our social life. I was uncertain what the future had in store for me, but for the time being I took care of my garden and raised chickens to reduce household expenses. Still, I anxiously noted that my savings were gradually diminishing and realized that soon I had to look for a way of generating additional income. I contemplated starting a kind of trade school and teaching accounting, courses in economics, languages, and business correspondence. Such courses were not offered in town then, and to this day I wonder if such an educational undertaking would not be perfectly suitable for Windhuk. But it was not to be.

HOW I EXPERIENCED THE FIRST WORLD WAR

1. The Impending Storm

It was a day like all the rest. The sky was a brilliant blue; now and again a burst of wind swept through the calm afternoon streets and swirled dust, leaves, scraps of paper—and the occasional tin can—into the air. For a moment, a giant whirlwind stood tall and erect, then the singing and humming of the whirling increased until the once-proud windy structure collapsed, covering houses and people with a dry, prickling dust.

We had been invited by our good friends, the Kindts, for an afternoon visit. Coming from Klein-Windhuk, the children and I had strolled down Sperlingsluster Pad and now found ourselves sitting in a cool, airy room in the house, enjoying afternoon cake and coffee. Rolls with jam and honey, apple cake, and all kinds of other delicacies were heaped before the two boys. After the long walk on the hot and dusty road, the boys were in good spirits. A parrot chatted in the corner of the room; two dogs and three cats circled the table, several eventually settling beneath it. The windows stood wide open. Outside in the garden, a few small green parrots with brilliant red heads screeched their happy songs. We heard the laughter and chatter of the natives. It was a peaceful scene, and we enjoyed the African afternoon very much. My elder son, Bitz, sighed and said, "How beautiful it is here." And we, the three adults who had experienced the struggles and deprivations of life, concluded, "Yes, how beautiful this land really is—and may it never change." Suddenly, the dogs started barking and a native came running toward us, sweaty and out of breath, holding a freshly printed piece of paper in his hand. It was the page proof of the first special edition published by the *Südwest Newspaper*. The news had been forwarded by radio signal via Kamina, Togo, and contained the Austrian ultimatum to Serbia. I will never forget the impression it made on me. Fiery red beams from the setting sun fell through the open door. A deep blue darkness had settled at the foot of the Kaiser Wilhelm Mountain. For a long time we sat in complete silence and fear as the world held its breath, as the die of fate, now cast, rolled across old Europe.

I immediately thought of my old beloved homeland, the Harz Mountains with its highest peak, the Brocken, and the mixture of dark spruce trees and the light green of birches and larches. In my mind's eye, I saw the old Oker House with the large sliding windows that forever clattered in a wind. I saw the massive brass knocker on the heavy door, the rather shrill doorbell, the brick floor that always smelled musty, the old-fashioned rooms with dark pictures on the wall. This old house had been loved and revered by members of our family as our home for more than a hundred years. Poplar trees that had once been hit by lightning creaked in the spacious, park-like garden, and beside them, a fountain trickled and trout jumped in the pond. Beneath the canopy of old linden trees, all my loved ones, among them many children and youth, were seated round the coffee table—uncles, brothers, cousins—nearly all of them active army officers. The old folks were probably anxious, but the youngsters cheered enthusiastically. I saw each one of them before me, all of them together, yet here sat I completely alone with my two sons in far-off Africa. "Do not become discouraged! Grit your teeth!" I thought. "Germany! Germany above everything else!"

The first news was received with mixed emotions. People simply reacted in different ways, but whether the news evoked indignation, rage, or enthusiasm, everyone had a burning desire to be part of it. No one wished to be left behind this time; each desired to make a contribution as Germany's power and honor were at stake. The steamer, *Gertrud Woermann*, lay anchored off the coast of Swakopmund. How fortunate, I thought, were those persons able to return home to participate in it. To be left behind is a terribly difficult thing to bear. The large *Ostafrikaner* had already sailed on to Cape Town.

Serious discussions commenced regarding the role the German colonies—including South West Africa—would play now that war was imminent. Many felt our fate would be decided on the battlefields of Europe without us being involved. Culturally, no European power would accept responsibility for involving colored people in this conflict—the consequences would be unthinkable. We did, however, have second thoughts about the role of South West. The discovery of significant quantities of diamonds, as well as the positive development of farms and other mining operations, might cause England to enter the war. Some time ago, Louis Botha, who later became supreme commander of the South African troops in the field, had warned that war would come. Most people had not believed him, yet it was rumored that South African scouting parties had already collected highly reliable information about us. Excellently detailed English maps existed in which water holes, roads, and even narrow pathways had been accurately recorded. It is said that this information was gathered by prospectors, some of whom worked for the British (South Africans) and were able to inconspicuously collect such information under the guise of "prospecting activities."

Since geographically South West Africa borders on the Union of South Africa, we wondered what South Africa's reaction would be. At the time there were two political parties in that country: the first was the Botha Party, which, one assumed, counted among its ranks many wealthy, influential South Africans who were completely under English influence; the second was the Herzog Party that strove for freedom and independence. Many Germans felt the Boers were well disposed toward us and were simply waiting for the chance to rise against English domination. Those who were better informed knew that only a tiny minority openly regarded themselves to be of German descent, yet still deeply etched in their memories were the recent deaths of thousands of their women and children who, during a futile struggle for freedom (the Anglo-Boer War), were penned behind barbed wire in English concentration camps. Now they appeared content to be left in peace on their small plots of land.

The governor of South West sent trusted persons to the Union to discover what the people were thinking. I heard him say repeatedly that he regarded an attack by our side on the Union as pointless, that he merely wanted to defend our borders as best he could. His position contrasted with those who adamantly contended that an attack would be our best defense.

I stayed as informed as possible about the state of our brave troops. Germany had seriously curtailed financial support of our colony, despite the many requests made by responsible parties in South West. The possibility that some day our colonial troops might have to go to war against another white nation was certainly never seriously entertained. Our men were too poorly equipped for that possibility, and our army was just shy of two thousand soldiers. Including the police force, if a general mobilization of able-bodied men were called for, the number would only increase to a mere five thousand. Besides, their equipment and weapons were completely outdated. Our force was barely strong enough to subdue a black enemy, but against a white enemy equipped with all the latest weapons of modern technology it would not stand a chance. Even before the outbreak of the war, not only were their machine guns in need of perpetual repair and their cannons outdated, but there was a constant shortage of ammunition. The army possessed two old aeroplanes and a total of five cars! Without support from Germany, how could our soldiers possibly prevail against a modern enemy in an area one and a half times the size of Germany with poor roads, scarce water, and never enough pasture for the draught animals? How could the army be supplied with the necessary provisions for both men and animals, let alone with clothing, shoes, leather goods, tools, and other materials? How could we supply them with just the bare essentials? The horses were in good shape, but there were certainly not enough horses and mules to sustain a war effort. Furthermore, the coastline could not be defended as Swakopmund and Lüderitz Bay were wide open and fully unprotected.

2. The Future Is Foreshadowed in Everyday Events

When I went into Windhuk to shop, I saw that general excitement prevailed. As usual, I went to replenish my supplies and asked for a hundred pounds of flour and sugar, but could only purchase small amounts of them. I was also informed that goods had become more expensive. What was I to do? Instantly, I thought about other things I might need—soap, candles, kerosene, canned foods—and purchased as many items as I could. The customary credit system in general use here was cancelled. Everyone had to pay in cash. Under the circumstances, this was understandable, but immediately daredevil entrepreneurs attempted to take advantage of the new situation. Legally or otherwise, they began speculating on essential goods. Thankfully, the governor intervened and forbade price gouging. Nobody, however, could have anticipated how those who had hoarded and stockpiled provisions by the roomful later profited by selling their half-rotten goods at unbelievably high prices.

The government acted justly and responsibly: "Remaining calm is now the citizen's foremost duty!" We had a competent and reliable governor who was backed by a staff of proven officials. We had loyal soldiers led by good and seasoned officers who had gained victory and honor ten years earlier. We were German men and women united with our homeland beyond the ocean blue. Our confidence, trust, and enthusiasm were boundless. The black, white, and red colors waved in the wind; patriotic songs filled the air.

3. We Colonists Prepare Ourselves

Preparations for war were undertaken in our colony with special attention given to radio installations. If the use of the English cable were lost, radio communication via Lomè and Kamina could be maintained. Our hope rested primarily on the radio tower in Windhuk—the only transmitter between us and our distant homeland. The tower was surrounded by a barrier of thorn bushes, a high tension line, and trenches. We knew that thorns merely provided illusionary protection, but it was done to impress the natives. In an emergency, one match would suffice to burn down the entire wall. In the beginning two soldiers protected the tower, then a machine gun was added, as well as more soldiers, until an entire company guarded the tower.

As much military equipment and provisions as could be spared were moved from Swakopmund and Lüderitz Bay on the coast to the interior. In this, the railroad was the primary means of transport. Duties on all goods were reduced. The ships *Gertrud* and *Frieda Woermann* had disappeared, but had not sailed on to Germany. Eventually we learned that the officers and crew of the *Frieda* had lacked the courage to unload her four thousand bags of flour and that both she and the *Gertrud*, along with all their passengers, had steamed to South America under the protection of the German gunboat, *Panther*.

I learned of this on August 1st, 1914, and immediately went to the post office as my most recent letters to Europe were on board the *Gertrud*. I wondered if they would ever reach their destination. Since this would likely be the last chance to do so for a good long time, I wished to contact my elderly parents at home and express my love and assurances. I entered the post office and read the notice on the board:

1. Telegrams at reduced rate. From now on, weekend telegrams are no longer permitted.
2. Telegrams are only permitted in English, French, or Dutch, and they must be signed.

This notice was supposedly mandated by the postmaster general in Cape Town because all telegrams were routed through Cape Town via the English cable line in Walvis Bay. I was hardly alone at the counter, and just as I handed my telegram to the official, a second official entered and said, "Telegrams are no longer being accepted." At this very moment another notice was pinned to the board: "According to paragraph X, letters and packages will only be taken if the sender accepts complete responsibility for them." In the same vein, the following announcement was attached to the Woermann Shipping Line's notice board: "Due to the insecure political situation, the steamer, *Kronprinz*, scheduled to sail for Germany on August 13th, will most likely not leave port."

It seemed that all lines of communications were now severed. I was cut off from my homeland, alone in the colony with my two children; I had little money, no income, no protection, no other family at all. As I stood in a corner of the post office, the realization of my predicament washed over me. I did not want anyone to see how I felt, though I believe our feelings were quite similar. Now none of us here would have any contact with friends and family in Germany. The horrors of war had already begun; all we could do was wait and remain calm amid all the uncertainty.

The natives had to be watched very carefully. Several times, British agents had tried to incite them. The Herero and Hottentots were certainly not to be trusted as it was possible they might take advantage of the political situation and start another insurrection. But we had learned from the events of ten years ago that their homesteads were inconspicuously guarded. No native was permitted to be outside of his homestead after eight o'clock in the evening. A native without a permit always had to give exact information about where he was going and where he had just been. Even so, one observed many suspicious activities. The importance natives placed on learning the most current news from the people for whom they worked was obvious. They did their best not to be left out of conversations we whites had among ourselves on the latest developments. They listened at the door, consulted with other servants in the household, and spelled out the words of dispatches. Those who had masters who allowed them to ride horses raced through the streets with such speed that it was almost life threatening. Imitating the warrior customs of the local Oorlog tribe, they set their hats and caps askew, decorated them with ostrich feathers, fastened the chinstraps under their chins, and rode their horses at full speed—always surrounded by a cloud of dust. As they did so, both older and younger native girls squealed with delight and feverishly pinched one another. But the police resolutely put an end to this thrilling spectacle by forbidding anyone to ride at full speed through the streets.

The danger of a native insurrection in Windhuk was quite insignificant despite the many rumors and stories circulating out of control. The danger was far greater in the countryside where the Baas—the white farmer—governed with a strong sense of justice and with a strong hand ready to mete out corporal punishment. Unfortunately, some of our farmers dismissed a portion of their labor force upon hearing of the impending shortages and price increases of foodstuffs. These dismissed workers roamed about the bush hungry, with nothing constructive to do, and in a mind-set favoring rebellion. Fortunately, the government acted calmly and with specific ends in mind. The commissioners, who were generally trusted by the natives, worked furiously as they rode through the country to explain, to make promises, and to calm people. They were successful in their efforts: the excitement diminished and work on the farms continued. Maize and potato crops were actually increased in the hope of a good rainy season. Slaughtering calves was strictly forbidden, and unemployed natives were given food and sometimes employment.

The malicious Bondelswart Hottentots lived near the English (South African) border, yet worse than they, the already rebellious Bastars lived at the center of the colony. They were a bad lot, being neither black nor white but of mixed race. In addition to them, there was another group of no-good folks, people without roots or homeland, who had been attracted by the promise of resources. They had neither conscience nor fatherland, and stood ready to serve the person who paid the most, but not with loyalty—for should things take a turn for the worse, off they would run. North of Windhuk lived the mostly reliable natives of Okombahe. The Ovambos were trustworthy, but their numerous armed servants were anything but trustworthy.

We anxiously worried about the immediate future. The offices of the two newspapers, *Südwest* and *Südwest Bote*, were so inundated with news seekers that they agreed to inform the public when a new dispatch arrived by use of an external sign. Flags were posted on the roofs of the two newspaper buildings—a yellow flag for the *Südwest*, and a green one for the *Bote*. In addition to the flags, three long whistles were sounded, one immediately after the other, from the nearby Felsenkeller Brewery when fresh news had been received. During the course of the next few months, we anxiously awaited the latest news from the outside world.

On August 2nd, a telegram was received from the Imperial Colonial Office informing us that the colonies were not in danger of war. At first the news was kept secret since no one actually believed it. Gradually, however, the message spread abroad, and on August 5th, via cable from Kamina, we learned that the British had declared war on Germany. On the following day, August 6th, bright red posters calling for volunteers to join the colonial forces were posted on our houses. Those who volunteered would serve as mounted soldiers without rank. It was also stressed that independent farmers and administrators who were unable to find suitable white persons to take over their affairs should not volunteer. The same held true for persons who employed large numbers of natives.

The governor and the military commander, together with their respective staffs, worked tirelessly and intensely. The army was already holding maneuvers in the area where the Bastars lived, thus communication between the government and the military was well established. On August 8th, a general mobilization was ordered.

4. Mobilization

This order was immediately followed by a ban on the export of weapons, cattle, and food from the Protectorate. Our normally calm Windhuk was turned into a war zone. Vehicles and riders raced along the streets, and people gathered in front of the Alte Feste fortress. There seemed to be a continual coming and going as horses galloped and mules stamped their hooves. Ox carts and donkey carts, accompanied by the usual shouts of their drivers, creaked from the strain of their unusually heavy loads. Signals blown on the trumpet, dogs tangled in the throng, uniforms, and hoards of people in front of the bookstore and the newspaper offices only added to the confusion. Natives of every hue and tribe pushed their way through the crowds. Uniformed officers' servants, aware of their importance, found themselves surrounded by admirers. Glittering clouds of dust whirled in the air, and our beloved German colors waved in the silky-blue sky, contrasting marvelously with the dark green of trees growing in the governmental and army gardens. The entire scene was bathed in gold by the African sun.

Overcrowded as they were, the trains arrived late. Everyone answered the call of the Emperor; nobody wished to stay home. Calming the anxieties of their husbands, many a German woman now took up her post and, with brave heart, loyally fulfilled her duty on the lonesome farm. Because the hotels had no more room, private homes opened their doors to the influx. Again and again, additional housing was sought and, surprisingly, there was always another spot where someone else could be quartered. All the rooms of my house, even the veranda, were filled with guests, and my dear Anna labored busily to feed so many people.

It was a wise move to induct the Bastars and Bersheba Hottentots into the military. The officers assumed they had been well trained and would prove useful, though they were not entirely trustworthy. They were supervised as inconspicuously as possible so as not to insult or offend them, yet this strategy allowed for them to be rapidly disarmed if the situation called for it. The Bondelswarts, unreliable as they were, were shipped to the north.

Later, a South African Free Corps was formed under the leadership of the seasoned Boer fighter, Andries de Wet. Not only did Boers enlist, but also German settlers who, for one reason or another, could not be used by the regular army. To give this paramilitary corps legitimacy, the area councilman from Grootfontein, Mr. v. Z., was assigned the most difficult task of functioning as the corps's auxiliary commandant. This he did, and fulfilled his commission both skillfully and successfully. The corps was a colorful group, to say the least. Despite the lack of discipline, its members knew various languages and were tenacious, resourceful, daring, and adventurous; they managed to get along without most necessities—I imagine the mercenaries of old were just like them. The Free Corps accomplished a great deal, but such a fighting unit could only be usefully deployed in a place like Africa.

Natives also volunteered to work for the army, and many were accepted. To help create a supply line, the army bought horses from every region of the colony. In the south, the Bastars delivered very good horses, but the selection of horses from the north was limited due to equine illnesses. The colonial government undertook the responsibility of rationing supplies for the army, the white population, and the natives. In this, the government was entirely successful, and we never went hungry. Of course, after many months of being in storage the flour became musty and worm infested, but we always had bread—though it was of lower quality than before the war. During this time no one baked rolls or cakes. The governor's wife set a good example for us. She also encouraged attendance at medical courses given by physicians and accredited nurses. These lectures were filled with practical advice, and most ladies did attend as everyone was ready to help.

During these interesting times, my children spent more time on the streets of Windhuk than at home. One evening they came home dirty and sweaty, and I sat them down for a talk. I explained how serious our situation was, how we in the colony, as well as our Fatherland, were under pressure and had to make sacrifices. I told them from now on our slices of bread would be thinner—even though corn meal would be mixed with our regular bread flour. I told them there would be no more cakes, desserts, or fruit, and that they would have to take better care of their trousers and shoes. Enthusiastically, they suggested they would go barefoot from now on, which they did, gleefully at first, but soon their feet became painful yet they would not hear of wearing shoes. My sons adjusted very well. On the way home from school at noon, they learned not to walk on the sidewalk because the paving stones were so hot. In the deep dust of the roads in Klein-Windhuk, they skipped along for a while, then rested in the shade of a thorn-tree, shaking the dust from one leg and then the other. My boys never complained.

My unsettledness and concern about the future grew, though I never once regretted my decision to return to South West with my children. I was fully able to work and perfectly willing to accept any kind of employment. However, because of the war, I had to postpone my original plan of opening a trade school. I needed to find some kind of work that would allow me to look after my children. I was sad to think that I might have to take a job that would separate me from my boys. I kept searching for employment, and my savings continued to decrease. We lived on corn meal and the produce from our kitchen garden, we sat in the dark, and often my supper consisted of a glass of water. But I kept on looking and asking, asking and looking. Finally, out of the blue, an opportunity presented itself. Only male teachers had been employed at the Imperial Secondary School, all of whom, with one exception, had been conscripted, though for the time being they remained in Windhuk and taught in the morning. Their afternoons were spent in military duty, but one of them had been transferred at the very beginning of the mobilization. Immediately, I gathered together my transcripts and asked the governor if I could be employed as a substitute teacher at the school. How fortunate! I was needed. I had a physical examination and was found fit, so I was able to begin teaching on September 1, 1914—my youngest son's ninth birthday.

So thrilled was I to receive the confirmation of my application that I had to take a deep breath to avoid fainting. It was simply too good to be true, but true it was. At least for the next while I would have the means to ensure that my children would not go hungry! Not only was there no reason to be separated from them now, but I would be teaching at the school they attended and would be able to walk to and from school with them. I had always been their mother, their friend, and their comrade, now I would be their teacher—and even get paid for it. I was deeply grateful for my good fortune.

I felt a bit apprehensive on my first day of school. As the only female teacher in the school, it first appeared that I would not be very welcome in the male quorum. However, I was soon a part of the group, secure in the curriculum, and found it a new and delightful kind of work. Cooperation within the quorum was soon established, and we got on splendidly. I ended up teaching there for eight-and-a-half years. Only in June 1923, after a six-month vacation in Germany, did I transfer to the German Governmental School where I taught without interruption until 1935. In the course of these twenty-one years I had contact with many, many children in South West.

In the Secondary School, I was originally given charge of the first grade. German, English, French, geography, and history were my principle subjects. In mathematics and singing I was hopeless—but I am very grateful to my four directors during those twenty-one years who kindly tolerated my inability in these two subjects. During the eight-and-a-half years at the Secondary School, I taught in all classes up to the fifth grade, and I taught my own children until they graduated.

With my new position, I felt at ease about our future but anxiously anticipated the coming developments.

5. South Africa Begins Hostilities

Our government anticipated hostile action from our neighbor to the south, and for this reason attempted to protect the extensive southern border as best as possible. Securing the border was accomplished by clearing a fifty-kilometer-wide swath of land stripped of its human population—whites and blacks—of all cattle and provisions, and by the destruction of all water holes. The latter was a measure not without controversy. Carrying a passport became compulsory. The southern part of the colony had been settled mostly by Boers whose intentions and loyalties could not be taken for granted. Indeed, despite the restrictions on movement within the fifty-kilometer-wide zone, some Boers attempted to drive their cattle across to the British side but were spotted inside the German border by a military patrol. As soon as the Boers saw the patrol, they began to shoot and killed two German troopers—our very first casualties of the war. The action was falsely reported as having taken place on British territory, ridiculously blown out of proportion as an attack on the Union of South Africa, and finally proffered as the long-hoped-for justification to begin hostilities.

Another ridiculous trumped-up charge was the alleged appearance of a German scouting patrol on English territory. We later learned that these fabricated incidents grew out of several firm agreements between the Union of South Africa and the United Kingdom on the subject of how to annex South West Africa. As it turned out, the rather late start of hostilities was principally due to differences of opinion between the British and South African governments. South Africa wished to prevent the conquest of South West by the British, as it already had its own expansion plans—most likely to gain control of South West's diamond mines. It was for this reason that the South African prime minister, Louis Botha, decided to attack the Protectorate with Union forces.

In mid-September, our police post at Ramansdrift (on the Orange River) was attacked, after which the free town of Lüderitzbucht, formerly Angra Pequena, was seized and occupied by eight thousand men. Unfortunately, the enemy was able to capture vast quantities of supplies because the local companies had given false information about their inventories to the government but had then failed to move their holdings out of sight. It seemed strange that one of the leaders of the assault was the English Consul Mueller, a man who had enjoyed cordial relations with many inhabitants of the town. It displeased me greatly; indeed, I found it distasteful that in the latter stages of the war it was primarily enemy officials of German descent who treated us worse than anyone else. In particular, I remember a prison warden with a German surname who treated our prisoners in a cruel, despicable manner.

Lüderitzbucht was not defended. Women and children stood under the protection of a local militia because the men of the town were serving in the field. The local population was entirely peaceful, and the South African occupation forces acted in like manner. Although Lüderitzbucht was cut off from all traffic and communication, the remaining population was transported in cattle boats to the Union of South Africa. Initially they were sent to concentration camps; later, however, the prisoners were quartered with German immigrants who had become naturalized citizens of South Africa.

The first English incursion came from the south. After a few successful skirmishes, our commander, Lieutenant Colonel von Heydebreck, began a counteroffensive. On September 26th, he soundly defeated Union forces at Sandfontein. He took two hundred and fifty prisoners and captured large quantities of ammunition. However, the South Africans countered this version of events with the story that South African troops had surrendered due to a lack of ammunition. Von Heydebreck's victory was announced in church on Sunday, and we rejoiced over our first success. But the victory was not without sadness—our troops had suffered losses, especially the loss of the well-liked Major Rappard.

After the victory at Sandfontein, things along the southern border were quiet for a time. Botha now realized that the conquest of South West would not be the leisurely walk he had boasted it would be. In the end, it required sixty-five thousand Union troops to confront our seven thousand in the field. The Sandfontein prisoners, along with their chief commander, Colonel Grant, were marched north by way of Windhuk until they reached Outjo, where the officers were quartered. The two captured cannons were placed in front of the Troopers Monument in Windhuk.

While this was going on, a substantial Union force had landed in Lüderitzbucht, and light skirmishes were constantly being fought by various patrols. The Union troops worked on repairing the destroyed railroad line and gradually advanced. Our soldiers, on the other hand, worked tenaciously to destroy the repaired line again, yet the South Africans, with equal vigor, mended it over and over despite the constant interruptions and disturbances meted out by our men. One of our detachments was successful in saving most of the provisions and food supplies from the diamond fields near Lüderitzbucht. For months on end, the goods were transported from the area by ox cart and later by camel caravans. It was only the treasonous activities of the natives that finally brought this difficult and heroic undertaking in the lonely, desolate sand dunes to an end.

After the battle of Sandfontein, the enemy departed from the German side of the Orange River. This German success encouraged the Boer leader, Maritz, to gather men loyal to him and join with the Free Corps in an assault upon the town of Upington in South Africa. In return, Germany gave certain guarantees to the Boers amid their continuing struggle for independence from Britain. The guarantees were granted, but because of indecisiveness, disunity, and mistrust of the Germans, the plan was eventually shelved. When the Boers finally did attack, they found their opportunity was gone, for in the meantime the South Africans had reinforced their troops so strongly that after a small, courageous encounter, the insurgents had no choice but to return to the German side. A few weeks later, the Boers attempted another insurrection that caused substantial South African casualties, yet because the Boer leaders were hesitant and untrusting of one another, their success was minimal. The insurgents made another attempt to conquer Upington but failed because of shameful internal betrayals from the very beginning. On January 31, 1915, the insurgents surrendered.

6. The Treachery of Naulila

We were upset when news of the failed Boer insurrection reached us, but from that point onward, things only grew worse. From the very beginning of the war, the governor of South West had made arrangements to receive provisions through Angola so that we colonists would not suffer unduly from want. Angola, a Portuguese colony at our northern border, was, like its mother country, neutral. Our government, however, did harbor some doubts about their neutrality, yet Angola was the only neutral territory bordering us from which our government could import provisions for men and animals. In early negotiations, the Portuguese agreed to the deliveries, yet the transports were postponed. In order to determine the state of affairs, the head government official at Outjo, a Dr. Schulze-Jena, as well as First Lieutenant Lösch, a volunteer by the name of Röder, a Danish translator with the surname of Jensen, three troopers, and a handful of native policemen were sent to the Portuguese border.

In fourteen days, the group rode five hundred kilometers to the Kunene River that separates South West from Angola. Commandant Sereno—may his wretched name be forever cursed—and the Portuguese stronghold, Fort Naulila, were just ten kilometers away from the German camp. Sereno cordially invited the German delegation to enter his fortress. Shortly after the declaration of war in Europe, the Portuguese government published its own declaration that crossing the border between South West and Angola was prohibited, and that the life and property of transgressors would not be guaranteed. Dr. Schulze was aware of the declaration but accepted the commandant's invitation without hesitation. Lieutenant Lösch, on the other hand, was reluctant and mentioned the Portuguese decree, but for reasons of comradery he hesitantly accompanied Dr. Schulze, Herr Röder, and Jensen. The police sergeant and the troopers remained at the encampment. Thus, accompanied by their Portuguese host, the Germans crossed the border and entered the fortress while servants held the horses' reins in the courtyard.

I suppose no one knows the precise series of events that transpired, but according to the survivors' accounts, the ancient tragedy of Siegfried was reenacted. It was said an unfriendly discussion ensued in that room and that the German delegation quickly repaired to their horses standing in the yard and rode off. In the process they were shot in the back. Their loyal servants were also killed; one was even fed to the crocodiles. The wounded interpreter, Jensen, was taken prisoner but succeeded in handing a note to one of the servants who pretended to be dead. It read, "Beware, they have set a trap!" The native was able to escape and brought word of the treacherous murder to the men who had stayed behind at the camp. They were already worried and had prepared themselves to come after their comrades because of a note from the Portuguese commandant, deceptively delivered by the hand of a native. In this note they were told that some natives had unfortunately mistaken the actual identity of the Germans and killed them. They were invited to the fort for a complete briefing. The Portuguese commandant most likely wanted to eliminate all German witnesses. The German troopers, however, hurried back and delivered their report. Both the governor and chief military commander of South West agreed at once that this cowardly murder required the most stringent of punishments. Straightway, they decided to send a punitive expedition to Angola under the command of Captain Franke, notwithstanding the fact that his detachment was urgently needed in the south.

At the beginning of November 1914, a British auxiliary cruiser, the Kinfouns Castle, appeared in the waters off Swakopmund and fired on the open city. Damage was minimal. Because of what had happened in Lüderitzbucht, we anticipated the possibility of enemy troop landings in Swakopmund and evacuated the city. Refugees were transported inland, primarily to Windhuk. As the evacuation order had been given so quickly, most of the evacuees had too little time to gather such necessities as linens, clothing, or even food. For the time being, a coast guard detachment remained in the empty port, and natives who attempted to plunder were arrested. However, a Jewish man named I. began a lucrative trade in stolen goods and furniture, but was soon incarcerated. A woman who had been inducted into service but whose husband had a leading position in another town was able, with the help of natives, to appropriate great bales of cloth, linens, and clothing. Later, when these items were scarce, she became rich by selling them. Even though many of these stolen goods were recognized, nothing could be done about the situation. Rumors circulated that she was a British spy. Happily, England thanked her by deporting her—and she was among the first of those sent away.

To this point in time, we who lived in Windhuk had fared quite well and were not seriously affected by the war. Food was certainly rationed, but supplies were sufficient and we actually did not suffer at all. The worst we endured was the ongoing uncertainty and our complete isolation from home. Everyone went to work quietly, though not without a heavy heart, still hopeful that the war would conclude favorably.

The refugees who had come to Windhuk nearly doubled our school enrollment, but our teaching continued with few interruptions. Maps and a large atlas were available in the teachers' common room, and we eagerly followed the events back home in Europe—at least as often as news arrived via telegraph. As soon as we heard the familiar three whistles from the Felsenkeller Brewery, the director dispatched a pupil to the bookstore. Panting for want of breath, he returned with the special edition newspaper, the ink not yet fully dry. Almost immediately, everyone assembled in the hall where the news was read to staff and pupils alike.

As time wore on, however, the situation in the Protectorate became increasingly tenuous. We had no chance of receiving any kind of help as our fierce enemies threatened us from all sides. We had hoped for the appearance of the German East Asia Fleet, and Major Franke had even offered one hundred German marks to the first person who was able to announce, via telegraph, the arrival of the fleet. Even the natives of the colony were interested in the matter. One day I was visited by the village elder from Okombahe, Zedekias, who had recently arrived in Klein-Windhuk. He came and asked me to inform him of the arrival of the German Kaiser's battleships as soon as they came into port—he and his entire village wished to help suppress the British/South Africans.

7. Our Commander Dies

It was during this time that our beloved commander, von Heydebreck, who had gained everyone's confidence and trust, was killed by a locally produced gun grenade during training exercises. The grenade entered his lower body, and he expired during the ensuing operation. His soldiers built him a simple, unadorned casket. His body was taken to Windhuk and laid in state before the altar of Christus Kirche (the beautiful Lutheran Church). After the service, a huge crowd participated in the procession from the church to his gravesite. My children and I stood on a hill overlooking the cemetery and watched as the procession arrived, saw the many wreathes and bouquets of flowers, and heard the honor salute at his grave. He was laid to rest beside his wife, whom he had married just a short time before, and whose life had ended prematurely because of typhus. I saw that my children were deeply touched. They held their sailor hats in hand and stood quietly. Finally, my ten-year-old said, "Momma, there are so many things I do not understand." I finally replied, "Neither do I, my son."

8. A Punitive Expedition to Naulila

The death of our proven and dependable leader was deeply mourned. His successor, Major Franke, was far away, marching through a desolate wasteland, over sand dunes and through fever-infested swamps toward the Kunene River in order to avenge the cowardly murders of his German brothers. Would our misfortunes ever end?

Shortly after Christmas, we accepted the invitation of the radio tower commander to visit him, and while on our way, a rider raced toward us bearing the news of a successful attack by Major Franke's detachment on the Portuguese fort and its complete destruction. The successful completion of the mission by Franke and his regiment was a monumental accomplishment. Men and animals had succeeded despite a lack of water and pasture, roads that had yet to be graded, insufficient food, both typhus and dysentery among the soldiers, as well as sickness among the horses. Advancing through lands controlled by unfriendly and sometimes hostile natives, they had endured attacks from wild animals and unending difficulties with the supply line; indeed, their already meager rations were cut again and again. The entire expeditionary force consisted of some three hundred and fifty men, including both commissioned and noncommissioned officers.

On December 18, 1914, an exceptionally brave attack was launched against a superior force of some 1,200 to 1,300 well-entrenched enemy soldiers. Our ranks were decimated from the very beginning by native sharp shooters who had camouflaged themselves among the treetops. The commander was severely wounded by one of these sharp shooters, and Captain Trainer had to assume command. Their situation was bleak; as a last-ditch option, a direct frontal attack was considered: the attack would be all or nothing. A barrage of artillery fire set the fort ablaze, and with unparalleled bravery, Captain Trainer and Captain Weiss, together with twenty-eight men (twenty-two from artillery, six from the infantry), stormed the fortress. Utterly bewildered, the enemy fled, and from far and wide our victory flag could be seen waving above the smoldering rubble. Twelve noncommissioned officers and soldiers were lost in the battle; along with the commander, an additional six officers and nineteen noncommissioned officers were severely wounded. In consequence of a miscalculation of the distance and locality of the fortress, the second company, which was supposed to attack from the rear, did not take part in the battle.

Many Portuguese, including civilians and officers, were taken prisoner, and among spoils we captured was an ambulance, together with all the accompanying equipment, in perfect condition. It was most appreciated. Besides Naulila, several other fortresses along the Kunene River were destroyed by small units of the German police. Marauding Ovambo tribesmen supposedly killed many of the fleeing Portuguese, and to this day, December 18th is celebrated as a day of mourning in Portuguese Angola.

The effect of this punitive expedition against Fort Naulila was tremendous. The Portuguese never again dared to cross into German territory for the duration of the war. The Ovambo tribes along the Kunene River rose against the Portuguese, creating a general insurrection among the natives of southern Angola. Hence, we had nothing to fear from the Portuguese; Germany had avenged her murdered sons. Captain Mandume of the Ovombo destroyed three Portuguese border posts on his own, and sent three prisoners to our governor to prove his loyalty to the German cause. When the German Protectorate Forces eventually surrendered, Mandume was unable to understand it. He had even established a messenger service to our officers' camp at Okawayo and offered the use of his warriors to drive out the South Africans. He himself refused to surrender to the South Africans; his loyalty to us sealed his fate. A few years later, he was killed while leading his men against a second South African attack meant to force his surrender. Now, close to Railroad Square in Windhoek, in the middle of a small, cultivated garden, stands a simple obelisk known as the Mandume Column. On it are engraved the names of all the Union troops who perished during the attacks on Mandume's tribe, thus the name of our loyal Mandume will never be forgotten by our descendants!

We received the good news of the attack and fall of Fort Naulila on Christmas Eve. I had decorated a thornbush in a corner of our veranda for my children and placed under it a single gift—a small-caliber rifle. Though we were in the midst of a war, my boys were still able to receive a present. I was touched by their joy. The rifle was the last item I could buy (along with the last available bullets) from the regional office.

On New Year's Day, the sixty-five-thousand-man strong South African Force, under the supreme command of General Botha, began their general advance on South West. While forces landed in Lüderitzbucht and Swakopmund under the umbrella of British warships, Union troops again crossed the Orange River. Other troops advanced from a fourth direction across South West's eastern border, which our forces had held for the previous five months. Gradually, the south was overrun. For a number of weeks, our troops in the hamlet of Aus had anticipated an attack. Their health declined as an outbreak of typhus spread. Pasturelands around Aus were utterly exhausted, and only the horses used for patroling were fed from the stores of oats and other dried fodder.

A substantial enemy force of some six thousand men penetrated our land from the south and the east, but marched passed Aus, rather than taking it, in the direction of Keetmanshoop, threatening our ability to offer support from the rear. With seven thousand men and numerous guns, Botha advanced from Swakopmund. Our scouting patrols reported that strong cavalry reinforcements had landed. On March 24th, a retreat of our troops from Aus to the north was ordered. Any materials potentially useful to our enemy had to be taken along or destroyed; hence, springs and water holes had to be rendered useless (though not by poisoning), and the railroad lines, with all of the accompanying installations, had to be completely disabled. The remaining whites and natives were also transported north.

The rainy season was off to a late start, but when it came, rainfall was nothing less than furious—as the Prussian king, Frederick the Great, had said: "God always helps the strongest battalions," now the Lord was helping our strongest battalions. All water holes, swamps, streamlets, rivers, and flood-planes overflowed. Beautiful green grass covered the land, and everywhere water was plentiful for both man and beast. In the Namib, as well as along the lower Swakop River, where not a drop of rain had fallen in years, life-bringing waters flowed everywhere. Unfortunately, despite all the rainfall, the enemy advanced quickly and easily—outfitted, as they were, with the latest vehicles and military equipment.

The Free Corps was disbanded, and those who no longer wished to fight were sent home. The remaining men were unconditionally incorporated into the Protective Forces. The Southern Border Guard was also disbanded and its men integrated into various companies. The southern part of the colony had to be cleared, as there was no way to protect it, and our handful of soldiers retreated before an overwhelming enemy force. The forces on the eastern border of our colony successfully engaged the enemy several times and temporarily slowed their inevitable advance, though they were eventually asked to give up and join the main force retreating to the north.

Botha's and Smut's combined divisions of sixty-five thousand troops chased our small retreating forces with heavy guns, armored vehicles, and the most technologically modern equipment, rapidly passing through vast stretches of sand without suffering thirst, for the rains had left in their wake abundant water and verdant pasturelands. By contrast, our small, weak German units made their retreat with both men and animals suffering greatly. Whenever possible, though, they attempted to slow the enemy advance by sudden attacks or defensive maneuvers. They held as much equipment as they could in reserve, destroyed the railroad lines, and by using Creolin, they made the springs and water holes unusable. Moving beyond Windhuk and Okahandja, they steadily drove northward, always northward, on haggard, worn-down horses. In March of 1915, when the seat of government was evacuated from Windhuk to Grootfontein in the north, only the Third Reserve Company remained in Windhuk to protect the German colonists.

Up to this point academic instruction continued undisturbed in the three local schools: the Secondary School, the Government School, and the Advanced Mission School for Catholic Women. Then, without warning, a diphtheria epidemic quickly broke out among the children. Unfortunately, our vaccine was in short supply; some of it was also out of date. To prevent the further spread of the disease, schools were closed, and all the male teachers reported to their various military units. Friends gave me various items they were unable to take with them: weapons, letters, files, medical instruments, etc., for safekeeping. I promised to care for them as best I could, which I did—though I had no conception of the trouble these things would cause me.

The natives saw, heard, and watched everything, biding their time and remaining quiet. The officers' servants loyally accompanied their masters to the north.

9. The Baster War

Rumors of all sorts concerning the Rehoboth-Basters circulated furiously, and while they could not be verified, they were nonetheless frightening. Basters are a group of people of mixed origin, arising from the offspring of Boer men and Hottentot women. Afrikaans and Namaqua are the languages they speak, and sometime near the middle of the nineteenth century, they migrated from the Cape Colony and settled in an area sixty kilometers south of Windhuk known as Rehoboth. Many is the time they have proven themselves unreliable and unworthy of any trust. They are herdsmen, as well as good hunters and marksmen, yet seem to have no appreciation for physical labor. Tobacco and alcohol they certainly love, and they are known to engage in questionable business affairs. Because they are neither white nor black, both races distrust them. In 1895, they concluded a treaty of friendship with the German Empire that placed them under an obligation to serve in the military. In 1896 and 1903 they dutifully observed this obligation, but in 1904 their conduct fell into question. Our government had granted them a number of special privileges and treated them with great consideration. Their tribal area had excellent grassland, pastures, and water, and was regarded as a reserve. Only a very few white settlers were allowed to settle in that area after having purchased land from the Basters.

In the beginning of our colonial endeavors, it sometimes happened that white settlers and some missionaries actually entered into legal marriages with Baster women because white women were simply not available. In those days, entirely false and baseless ideas regarding racial questions prevailed; indeed, marriages between whites and Baster women were regarded favorably by both the government and the Mission. At that time, Basters were still regarded as desirable settlers, especially once they had intermingled with Germans. Children and grandchildren came from these marriages. Yet today, in a time of racial awakening, these descendants carry the heavy burden of their parents' sin, so thoughtlessly committed by couples who most likely never considered the future complexities of their action. Among these children and grandchildren are loyal Germans, honest and hard-working men. In my twenty-one years at school I taught many a dear and gifted child from these mixed marriages. Their fate is hard and undeserved, and innocently they must pay for the fact that their parents and ancestors sinned against blood and race. Of course, German blood does not flow in all of these children. Some completely favor their mother in looks, behavior, and business activities. I am firmly convinced that more than any other people it is we Germans, by virtue of the new racial views, who are destined to be bearers of culture and civilization in Africa.

Among South Africans, one often sees “coffee with cream”—as the descendants of mixed marriages are called. These mixed-race people are not at all regarded as equals to white South Africans, and there, such marriages are also forbidden. In fact, a new law threatens a white man with both corporeal punishment as well as hard labor in prison for having sexual relations with a native woman.

As early as August 1914, the commander of the Protective Forces drafted all men who were able to carry arms and organized them into a cavalry unit. His decree, however, met with stubborn resistance. The Basters insisted that European matters were no concern of theirs, that their treaty only called for their participation in a local war. Besides this, they maintained the treaty had long since expired; furthermore, only about fifty of their old soldiers were still able to carry arms, and the younger generation had not received any military training whatsoever. In consequence of this, the commander promised they would not have to fight against a white enemy but solely against native insurgents. After lengthy discussions, a company of one hundred fifty men was drafted and equipped with model 71 rifles. They then received their basic training, but in comparison with the Basters’ previous performance, their current horse riding and shooting skills were nothing less than disappointing.

The Baster company undertook relay services, guard and patrol duty, and caught thieving natives who plundered the evacuated houses of Swakopmund. Later on, it was left to them to clear the Baster sand dune area of the gangs of Hottentot and Bushmen robbers who roamed about that place. For a while, all went well—but it was an uncalculated mistake when they were ordered to guard a British prisoner-of-war camp. Colored men should never have been placed over white men, especially not those who were related to these prisoners by blood and language. This measure was only undertaken due to a deficit of white soldiers. However, when we bear in mind what Britain, and above all France, did in the World War, perhaps this act was not so bad after all—though it created an unusual excitement among the Basters. In mid-February, 1915, all their ammunition was confiscated except for three rounds, yet by then it was discovered that the Basters had a goodly supply of mostly carved and hollowed-out bullets in their possession.

On April 17, 1915, the commander of the Defense Forces personally traveled to Rehoboth and requested that the Basters surrender their arms—unless they preferred to march north with him. The council asked for three days to fully consider the request—it was, of course, only a strategy to gain time. The following night, a group of Baster soldiers charged with guarding the prisoner of war camp deserted their posts, while in another camp, two Basters refused to give up their arms, ran away, and were subsequently shot. This only served as a signal for them to attack small guard posts, various police officers, and farmers. A white woman and her seven-month-old baby were murdered, while other white women, clothed only in their nightgowns, were abducted from their homes and taken prisoner. When our soldiers finally came, the rebellious Basters used the women as a human shield to protect themselves. The last news we received in Windhuk was that the governor's request for them to lay down their arms was met with a Baster plea for extra time. This was now a grave concern, as the South Africans were advancing on every front and inciting a vicious enemy against us from the very center of our country. Our commander was not to be envied.

I wish now to recount an incident in such detail that it will help those in Germany (who have no conception or understanding of our difficulties and problems) understand why, in the end, our brave soldiers were forced to surrender. Between these simple lines lies much quiet heroism and courageous endurance to the end.

One night I was awakened by the long, sustained whistling of the brewery siren. I listened: five long whistles at short intervals, then silence. Again, five whistles shrilled through the night—sinister and frightful! A moment of silence was followed by five blasts of the whistle. Finally, I understood: gar-ri-son-a-larm! My heart stood still, I wondered what had happened. Soon I heard sounds and voices on the road—running, shouting, screaming, as well as the pounding of horses' hooves. I dashed outside and listened attentively in the darkness. At first I thought it was a South African attack, but I heard nothing suspicious, and no shots were fired. Perhaps it was a native insurrection, but all was quiet in both of their villages.

It was morning before I learned those vicious Basters had mutinied. That was their gratitude for all the special considerations and care they had received from us! All the salt shipments they had fetched in their ox carts from the British-held Walvis Bay had been shipments of arms and ammunition provided by the British!

The alarm had been sounded for the reserve company stationed in Windhuk. A train stood ready at the station, and the reservists, with their weapons, boarded for Rehoboth, without coats, blankets, provisions, and other necessities. They were needed to reinforce our troops and to subdue the Basters. This latter task was most difficult because the Basters had scattered singly or in small groups across the rocky terrain. Several small skirmishes took place, several thousand cattle and sheep were confiscated, and many Baster ox carts burned. Near the water hole of Tsamkubis, we attacked their main force, and they suffered considerable losses. It was unfortunate that time did not permit our soldiers to take full advantage of the victory, yet due to the larger concerns of the war the entire company was ordered to march on to Rehoboth, then further north that they might evade the Union forces advancing toward Windhuk from the south and the west.

The Basters paid a heavy price for their disloyalty, for the Mandate Government canceled some of their special privileges and failed to keep the new promises made to them. They nearly started another war, and a South African company of soldiers was dispatched to Rehoboth as hurriedly as our German soldiers had been before. Many arrived in athletic shoes, most of them without so much as a blanket. At the last minute, the Basters finally gave in—what a pity!

Now they are poor and in debt, and their complaint to the League of Nations went unheeded. They have repeatedly stated their desires to sell their farms in the border areas (preferably to Germans), but are forbidden to do so. A good many gold claims have been staked near Rehoboth, yet the large mining companies have failed there while a few small, individual enterprises still hold on. Investments from the Union of South Africa do not come, and the Basters do not have sufficient capital to mine the gold themselves. They despise the British (i.e., South African) police and cry, "Give us just one German police officer for the entire region. We'd much rather obey him than the many Boer (South African) policemen—policemen we don't want." But by then it was far too late.

10. The Germans Must Be Cleared Out!

As the last soldiers marched through Windhuk, an indescribable tumult and agitation fell across the crowd at the train station. It was difficult to bid the men farewell, but at long last the endless succession of railroad carriages rolled their way north. In the last open car, an old soldier stood erect, tightly gripping a large German flag that fluttered in the breeze. I had to hold back my tears. It is terribly hard to remain behind; nonetheless, heavy duties awaited us here in Windhuk.

Across from our house, on the Wasserberg, a heliograph station had been erected and was our very last connection with the outside world. Only a few people knew about the station, and the operators who stayed on top of the mountain were never seen by the passersby below. Secretly, they came to our house to fetch water. The nights in May were quite cold, and many a pot of hot coffee and plates of warm food were carefully—and secretly—taken up to them by my two boys. Yet because my children often played on the mountain, nobody suspected anything.

One gray morning long before sunup, our dogs started barking and I heard the soft clopping of horse's hooves. I went outside and met three German troopers, one of whom was Count B. They were the last patrol from the south and had ridden through the mountainous region along the Auas pass. Along the way, they had dynamited the very last stretch of the southern railroad, but now the South Africans were on their heels. I quickly served them breakfast while the boys fed and brushed down the horses. The men rested briefly, then mounted their horses and quickly disappeared into the gray dawn. The morning star was still shining brightly in the eastern sky as we carefully raked the entrance and the grounds of our property to cover the horses' tracks. All the while the natives slept; they had seen nothing.

A few days earlier, just after the soldiers left Windhuk for the north, all the stored military supplies they were unable to take with them had to be dealt with. To destroy these remaining supplies would have been shameful, so it was decided that they be sold very cheaply to private parties before they fell into the hands of the enemy. We paid with "Seitzscheinen," a local currency Governor Seitz had authorized to be printed when the supply of German banknotes came to an end. All kinds of goods were for sale—carts, wagons, tools, leather, saddles, leather cords, bridles, entire bales of khaki cord, and blue uniform cloth. Some of the rich farmer's wives purchased several ox-drawn wagons full of these goods.

With the soldiers now gone, we decided to form a militia to protect our city. Every night, all the militiamen of Klein-Windhuk presented themselves for duty at the guard's office. Many were old and unable to serve; others simply shirked their duty. Many women and children were left alone at night when they would have preferred to have had their husbands and fathers at home with them. The Windhuk City Council called for volunteers to join the militia and mandated the registration of all males and females above the age of fourteen—these registration lists were to be handed over to the anticipated occupation forces.

On May 12th, Union troops occupied Okahandja, and the news, via telephone, that the enemy was treating the population quite fairly had a calming effect. On May 8th, the mayor had made the following announcement: "So that peace and order may be maintained, the white population of Windhuk is requested to stay inside when the occupation troops arrive; they are also requested to keep their servants off the streets."

Fearfully, we awaited the events of the next few days. It hurt to see white flags waving from both church towers and city hall. The Military Hospital, the Catholic Hospital, and the Elisabeth House Maternity Hospital had mounted red-cross flags. Critically ill and wounded soldiers, as well as the seriously wounded Portugese prisoners who had been left behind, were under the care of Dr. K., a private physician. Since the fall of Kamina, no news had been received, and our beloved radio tower stood silent. How strange that this radio tower was supposed to have been the principle reason why the British government had demanded an invasion be launched against South West.

High up in the clear air, the antennas hummed as the radio operators had much to discuss: "Someone has come and taken a very small, but critically important part of the radio that has passed from one hand to another and now is lost. Rumor has it that the enemy had promised a fortune to the builder of the tower, but he knew nothing. He was threatened, sent to a concentration camp in South Africa, but he knew absolutely nothing. Highly intelligent people were brought in to repair the tower—ridiculous, we have to get this thing functioning again—but it did not work, they established a connection with South Africa, but that was all." In reply, the bright antennas hummed, "We know that a tiny part is hidden beneath someone's threshold in Klein-Windhuk, but it has disappeared from its hiding place and where it now sits gathering dust, no one knows!"

11. General Botha Occupies Windhuk

It was said that our mayor drove out to meet the South Africans on May 11th, to hand over the key to city hall and thus in reality to the city itself. On May 12th they made their entrance proper. Surrounded by mighty clouds of dust, they arrived by car, on horseback, in carts, and on foot. All we could see was the dust settling over the Brakwater Valley. At midday, the cavalry rode down the main thoroughfare. We did not move. My two little boys hid themselves in a corner like frightened birds. Thursday, May 16th, was Ascension Day, and Pastor Kriele had chosen "let everyone be subject to the authorities who have power over them" as the text for his sermon. He did his best to admonish his flock to be sensible and peaceful—but we did not think much of his sermon.

The next morning notices were placed everywhere, passed around to everyone, even handed out at the bookstore in Swakopmund. The notice contained a proclamation by General, THE RIGHT HONOURABLE LOUIS BOTHA, Supreme Commander of the Union Forces in the Field: "Pursuant to the occupation of the capital city, of all important cities, both seaports and all railroad lines, with the exception of the narrow gauge line to Ongati, martial law is now in effect so that peace and order can be maintained." This was followed by a reproach of our civil and military authorities who had elected to continue fighting despite their losses, and who now bore responsibility for the misfortune that would befall the civilian population because of their decisions. At the same time Botha committed to the populous the protection of his officers and soldiers. Civilians were allowed to stay in their homes and to go about their business on condition that they would not aid or abet the "enemy," or cause harm of any sort to his forces. If these guidelines were not respected, all or many of the civilians would be deported. The notice continued, that despite these warnings the commander of the German troops continued to poison the water, hence, Botha threatened a strong, vengeful response that accorded with what he deemed appropriate. And finally he informed the population that from this moment forward everyone had to align himself with official time—which meant that everyone had to advance his clock by fifty minutes!

All of this did not strike one as particularly dangerous, though I naturally had my own idea about the meaning of not supporting the enemy! At first, advancing the time by fifty minutes was truly bothersome. We simply used the movements of the sun as our timepiece. The sun rises at six o'clock, sets at six o'clock, and reaches its apogee at noon. Many farmers continued using sun-time, but we who lived in the urban areas had to grow accustomed to "South African" time.

By all appearances, the Union troops received the meagerist of rations, for even on the first afternoon, troopers came by and begged us for bread, milk, and butter. Of course we gave them some food, but more and more of them appeared. When we had nothing left to give them, they opened our cool larders themselves! One soldier, after smelling the bread baking in the oven, opened the oven door, retrieved the half-baked loaf, cut it in half, then threw a shilling on the table and disappeared with half a steaming loaf. All these were but small inconveniences, and I cannot remember even one occasion when the occupation forces acted impertinently, and I never heard of a crime being committed.

The time had finally come for me to set certain items entrusted to my care in very secure places. However, this was easier said than done. Whenever I thought of a clever hiding place, I convinced myself that others would be just as clever in discovering it. Many items wanting concealment had accumulated in my house—items whose discovery would endanger me: rifles, pistols, files, documents, and many valuable and useful medical instruments. I decided to deal with the larger items first, so I carefully greased them with glycerin, lined a large tropical suitcase with oilpaper, and set the items inside. But now what should I do with the suitcase? I had a brilliant idea. My children had a play tent, and as I had done many times before, I asked them to set it up in our yard. Inside the tent, we dug a deep hole in the earth, and on a dark night, when the moon was not shining, set the suitcase inside the hole and covered it with earth. My boys felt very important, and one pair of blue eyes, one pair of brown, glistened as I told them, "You may help, but you must keep quiet." The most loyal assistance was provided by my children's best friend, J.v.D., our boarder at the time. The next morning we arranged for a joyful native dance, accompanied by the proper incantations! In order to compact the earth on top of the hole, my children played in the tent for several more days, then shifted the tent a few feet from the original site. We planted a thorn bush into the soil above the buried suitcase as it was still a bit on the soft side.

Several years later, when the proper time had come, we unearthed the items. My elder son was in Germany receiving his training, and Büdi and I were unable to entirely clear the way for the suitcase to be pulled out of the hole, so we decided to remove the items while leaving the suitcase in the ground, and to cover it as best we could. All the stored items are now back in Germany with their rightful owners.

I had also buried papers—but so well that I had to hunt for quite some time before I was able to recover them. Though I had wrapped them in oil-cloth, they had been partially devoured by termites. I realize now I should have put them in tin containers, but I did not dare set them in that suitcase because they were so very important—I simply wanted them to be in a more secure hiding place.

I still had weapons to conceal, and since we had a small shed in back—which I used as an incubator—we dug up the clay floor in one corner, buried the pistols, then made a nest for the brooding hens above the small cache. I also dismantled the rifles and, for the time being, put them in a buffet that belonged not to me but to the owner of the house and was otherwise standing empty. Finally, I tied a roll of important papers into the folds of my living room drapes where they would be ready to be collected in a moment's notice should a certain person call for them. Despite hiding the other weapons, I dutifully registered my own Browning and my boys' small-caliber gun at the district office.

On Friday, May 14th, a notice was posted that all citizens were required to deliver their weapons to the old district office next to the Christus Church. I had to give up my little Browning after all, but it was an especially difficult task for my children to lose their beloved gun. They kept it in peak condition, always ensuring that it was cleaned, polished, and properly oiled. It was also constantly held by one of the boys. But it had to be done, for in order to safeguard the many items I had hidden for my acquaintances, it was necessary that I give up my own property. Before I left on this errand, there was quite a scene—the children hung onto me, faithful Anna had tears in her eyes, and black Anna, our washer woman, clung to me from behind. It seemed the walk to the now hostile district office in Windhuk was a walk to my execution. All of them imagined the horrible dangers I was about to face. To his disappointment, I declined my eleven-year-old's gallant offer to accompany me to the office.

I noticed the white flag had disappeared from the church tower and that above the city hall the Union Jack flapped in the breeze. The Union Jack also waved above the red-cross flags on the military and Catholic hospitals. What we had long feared, but had put out of our minds again and again, became reality—a foreign flag waving proudly over our capital. The morning sun was shining as I arrived at the district office. Various groups of people, mostly women, stood restlessly in front of the office, and for the first time I saw the sand-colored, well-fitted uniforms of the South African officers. The Union soldiers certainly had no need to dip their rain- and wind-bleached cord uniforms into coffee or tea, as our soldiers had to do, to make them more camouflage. The two connecting rooms of the district office were in disarray, with shelves half empty, and partially strewn and disorganized files, crumpled-up papers, and broken crates languishing about the place. Disconnected telephone wires hung from the ceiling. In front of the doors, as well as in the rooms, British (South African) soldiers stood guard, their rifles resting against their shoulders. Behind a table standing in the middle of the connecting rooms sat two officers.

The office was hot and dusty, and smelled of sweat. Our carefully registered weapons were placed on the empty shelves. Conversation between us was difficult because hardly anyone in the occupation forces spoke German, and only a handful of Germans could speak English. I, however, was fluent in English, and as I overheard some of the misunderstandings between the South Africans and the Germans, I feared it could lead to trouble for some of our people and thus stepped forward to the table and offered to act as interpreter. My offer was immediately and courteously accepted. From that time onward, I was able to assist in both speaking and writing. I mediated such problems as women, who lived far outside of Windhuk proper, requesting that their weapons be returned to them, milk deliverers wishing to enter the city at seven o'clock in the morning, and a midwife seeking permission to leave her house at night when she needed to attend to a birth.

The next afternoon I left my boys at home and ventured into town to shop and to scout up some news. I also visited Fräulein Kindt (whose brothers had accompanied the troops in the north as journalists). What I learned that day was quite distressing. Since the morning that the Union troops had entered Windhuk, certain German women began openly seeking contact in order to make friends with the enemy. Some of these women were members of society—one had discovered a distant cousin, another a common friend, a third someone she had met on the steamship, etc. Proudly they rode through town in enemy cars while their own men were on the frontline.

They even invited the enemies to their homes and in turn accepted invitations from them. In my opinion, this kind of fraternizing with the enemy began all too quickly. These women could surely have waited for a decent amount of time to have passed; certainly they must have known they would be found out. Immediately after the occupation began, a deserter from the militia provided the enemy with a list of obliging women. Furthermore, he created a list of prominent German men and proposed their evacuation to South Africa. It was whispered about that this proposal was accepted; the enemy, however, failed to show much gratitude.

I was often asked to translate, but I refused to accept payment. Translation also placed me in an awkward position from time to time. Once a woman came to lodge a complaint, and I had to translate the following, "When the soldiers arrived, enemy officers entered her house, located outside of town, and demanded tea and breakfast. She freely offered what she had, and laid the table very nicely. She was horrified when she noticed some of the gentlemen hiding pieces of her silverware in their leggings. She has come to lodge a complaint because she wants her silver returned. At the time this occurred she was too afraid to confront the gentlemen." I then had to ask her why she had set the table with her best silver. She answered frankly, "Until now, I have only known German officers and was hence unaware that officers would steal." The provost marshal looked first at me, then at her, and said: "I did not understand that correctly. Will you please translate what the woman said again." I repeated what I had said, but as to whether or not the silver was ever returned to that woman, I do not know.

One day the provost marshal entreated me again to accept payment for my services. Though I would not accept payment for the help I provided my people, I did ask a favor. Fräulein Kindt, my elderly friend who now lived alone, was afraid and felt she needed protection. I also was alone with my two young sons in Klein-Windhuk. I asked that our weapons be returned to us. I received a certificate that would allow us to fetch the weapons from the arsenal. Fräulein Kindt and I went together. When I showed my certificate, the duty officer contemptuously looked us up and down, and after making us wait a while, he disappeared altogether. The waiting was unbearable. Finally, I had waited long enough, and when he failed to open the door after I knocked, I pushed it open, and there in the adjoining office stood the man, beside a window, smoking. He had done nothing, so I told him that war was no excuse to be impolite to ladies who had come with an officially signed and stamped certificate simply wanting to have their property returned to them. I added that I would complain to the provost marshal. He squinted, then went to use the telephone. When he returned, he was a changed man. Several times he apologized for his discourteous behavior and admitted to having made a "mistake."

We also learned that we were not the first to have had their weapons returned. Several German "ladies" had come with such permits, but a later check revealed that these had been forgeries. Several South African officers had indeed signed the certificates, but as a return service . . . ! He even showed us a list of their names. Tears of anger and shame filled my eyes. Fräulein Kindt received her revolver, but my Browning was gone. Magnanimously, I received another in its place, though my boys' rifle remained impounded.

There was no news of our soldiers. The South Africans published nothing, and the rumors that circulated about the war were vague though nonetheless frightening. A battle was to have taken place at the Waterberg during which the major and the governor were taken prisoner. What mail we did receive from the north was scrupulously censored; most of the time it was not delivered at all.

We had requests for help from the military hospital since the English nurses attached to the occupation forces suffered terribly from war psychosis. They even refused to give minimal care to our badly wounded soldiers who had been left behind. Consequently, the wounded soldiers were not bathed, smelled awful, and lay in a filthy hall. A few German women risked going inside to clean, to nurture, and to help. They all came prepared with brooms and buckets.

Several days after we had received our guns, an endless train of cavalry, cannons, munitions, and supply columns marched down the road past our house toward the east. We knew they were on their way to cut off our Fourth Reserve Company, which was marching from the south via Seeis in an attempt to join our forces in the north. We learned a while later that a German turncoat in Seeis was largely responsible for their capture. It was but a few days later that our captured soldiers returned, and like cattle in a kraal, incarcerated in the old uniform depot. The front yard was lower than the street and separated from the road by a high iron fence. There they sat as the guards patrolled in front of them. Though we were not allowed to converse with them, the guards kindly allowed us to walk up to the fence and hand the prisoners underwear, food, and tobacco. The captured soldiers had suffered unbelievable strains and difficulties—hunger, thirst, scratches from thorns, and cold. Now there they sat, overtaxed, hopeless, and indifferent. I was granted permission to look after some of our acquaintances who were ill, but there was very little we could do for them.

Germans who had not been drafted were required to assemble every morning at nine o'clock sharp outside the city hall. Latecomers were fined five pounds (one hundred German marks). This order even included missionaries and priests, who, when asked to be granted exception from this, were given no reply. One morning some of the assembled men were allowed to leave, though a number of them, including some of the most prominent citizens of Windhuk, were detained and then led to the railroad station. At the station they were unceremoniously loaded into cars and shipped to the Union of South Africa. Thanks to the list provided by a cooperating traitor, these men had to leave without being permitted to take anything along with them. News of their departure quickly spread, and caring relatives rushed to the train to bring their men clothing, coats, blankets, and food. Even though the station was cordoned off, they broke through and delivered the needed supplies. The train was delayed for hours before the signal for departure was finally given. In the Union, the men were incarcerated at various camps.

We German women were afraid we would be treated in like manner. The boys and I had already buried our silverware, while Anna and I came up with a method of inconspicuously transporting gold coins. In those days women still wore corsets, so we sewed small, firm pockets on the side of the corset touching our body and set the coins inside. We had no idea if we would be transported as hastily as our men, but we knew there would be no one to bring us supplies at the station. Durable clothing for the children was packed and ready, as well as three small suitcases, one of which was filled with food items. Day and night we nervously awaited our fate—with cold, solid gold coins pressing against us.

All prisoners were shipped to South Africa; only those who were ill remained behind. A secondary school teacher, who had been taken prisoner as a sergeant, was also allowed to stay because his wife was expecting their first child in just a few days. Sometimes the enemy acted in a rather humane fashion. Women were usually treated with courtesy, offered a chair, and taken care of before the men. Quite a few women were granted requests which, when sought by men, fell on deaf ears.

12. An Honorable Surrender

Like a wildfire sweeping across the veld, on May 20th, news spread throughout the city that an armistice had been concluded between our forces and the occupation army. We hoped peace would come quickly, but alas, the negotiations between Governor Seitz and Major Franke on our side, and General Botha, on the other, held at the Giftkuppe (located between Karibib and Omaruru) ended in stalemate. Botha did not accept Governor Seitz's proposals because his political decision, in case war broke out in the first place, had already been made in London. So Botha rejected the governor's proposals and made demands that impugned the military honor of our forces. For this reason, negotiations were broken off. The armistice ended at noon on May 22nd, and our forces recommenced their exhausting bush warfare.

Small radio stations were erected in Namutoni, Tsumeb, and near Sesfontein. Over and over, rumors circulated about the fall of Warsaw and Calais and of a German invasion of England. This news was disseminated in such detail that it was even printed by our small German newspaper, the *Buschlaus*. My friend, Herr Kindt, who had gone along to Tsumeb, printed these rumors with the help of a handheld printer. Besides intermittent news items, the paper contained reports about the movement of troops. The four-page paper was the size of a notebook, yet filled with anticipation, we ripped them out of one another's hands. The *Buschlaus* was printed in fifteen editions and is still a rare collector's item today.

On account of positive news from Germany, our forces hoped to somehow carry on until a quick decision was made in Europe. They planned a retreat via Namutoni into Ovambo territory. They had nearly two months' worth of provisions, and the commander set up a supply line along the track to Ovamboland.

It sometimes appeared that the Union troops were ill provisioned, and that their disappointing progress in the campaign against our troops seemed to have affected them, so their strategy changed. A direct consequence of this was the plundering of many farms. Robbed of all their possessions and animals, farmers and their wives were forced to leave. The farms owned by men who had been drafted suffered the most. Certificates of requisition were rarely handed out; instead, the enemy simply took anything he could use without thought of compensation.

The natives, of course, dreamt of a golden future and already regarded themselves as the new owners of German farms and cattle. Bribed with money and cattle, as well as promises of even larger rewards, they were an easy target. Native military servants received extra pay, and every possible means was used to encourage them to cooperate. Failure to do so was deemed treasonous, and other despicable threats were made. Even before the enemy arrived, a group of native and white spies, the latter mostly of Boer descent and who knew the land area well, rendered invaluable services to the South Africans while causing us great harm. It was therefore no surprise that the fortunes of the Germans gradually waned in a country they had settled with their own blood and sweat. In the course of all these years, I have often asked myself whether or not the enemy's policy of playing the natives against whites was truly necessary or prudent. In time, they will regret this tactical approach, and I fear, unless all signs prove me wrong, that time is not too far off. Then woe to today's victors!

Our troops marched farther north. The railroad lines were blown up, water holes destroyed, the main roads mined, but all was in vain. The enemy pressed hard against our forces; they were well rested; had fresh, well-fed horses, vehicles, carts, and mules; and received a continual supply of provisions.

The British/South Africans officially denied the reports of the capture of Calais and Warsaw, and posted a reward of twenty-five pounds for the capture of the person who had spread those rumors. In turn, Provost Marshal de Waal promised to publish the latest news from Reuters, though he failed to do so even once. Due to the continuing lack of information, we could only hope that Germany was fairing well. Every day, the men who had stayed behind had to report with the occupying forces. On one particular day, the Italians failed to appear for registration; many of them no longer greeted us, while others proudly walked about in South African uniforms. Those who knew the country well served as couriers, and it became obvious to us that the Italians had defected. Now they received perks, were well paid, and knew the latest Reuter news—which some of them secretly passed along to us.

Women whose husbands had been drafted were able to draw a stipend from the local government, collectable at the city hall. It was there that I received the last installment of my schoolteacher's salary.

On July 9, 1915, the fate of our valiant forces was sealed, yet the conditions of surrender are ample proof of their accomplishments.

The Surrender Conditions of the Colonial Forces of the German-South-West-African Protectorate, signed on July 9, 1915, on Kilometer 500 of the Railroad Line between Otavi and Khorab.

1. The Colonial Forces of the Protectorate, under their Commander in the field, surrender immediately to General Botha, Commander of the Union Forces. Brigadier General H. T. Luckin C. M. G. D. S. O., deputy of General Botha, is commissioned to supervise the details and the execution of the surrender.
2. Active officers of the surrendering Colonial Forces may keep their weapons and, on their word of honor, may choose their own domicile. Should the Union Government be unable to grant the wishes of one of the officers concerning his domicile, he must choose another place.

3–10.

9 July, 1915

Signed by Seitz
Imperial Governor of German South West Africa

Signed by Franke
Lieutenant Colonel and Commaner of the Colonial Forces
of German South West Africa

Signed by Louis Botha
General, Commander of the Union Field Forces

I, the undersigned, give my word of honor not to participate in any of the present military hostilities between Great Britain and Germany.

As my final comment on this chapter, I wish to quote verbatim the concluding remarks made by Major von Ölhafen in his work on the German general staff entitled, "Der Feldzug in Südwest" (The Campaign in South West).

"On July 9, 1915, the Colonial Forces of South West Africa have ceased to be, for the duration of the current war, a sharp-edged cooperative instrument maintaining the honor and glory of the Fatherland. After having spent an entire year resisting a superior enemy under the most unfavorable conditions, they were forced to surrender. Thanks to their tenacity and loyal service, they succeeded not only in protecting the country against the hostile invasion and the atrocities of natives, but also required the enemy to make enormous efforts and huge expenditures in order to subdue them. During this year, they prevented the enemy from committing considerable armed forces to European and other theaters of war. Finally, the end of this desperate struggle has created unique conditions in this war, and the return of our reservists to their places of work within the community certainly guarantees that German influence in this country will remain constant despite the influence of the British and the Boers."

THE TERMS OF SURRENDER

The terms of surrender of the military forces of the Protectorate of German South-West Africa, as agreed to by the Government of the Union of South Africa, and accepted by his Excellency Dr. Seitz, the Imperial Governor of the Protectorate of German South-West Africa, the commander of the military forces, which was signed on the 9th of July, 1915, are that—

(1) The military forces of the Protectorate of German South-West Africa (hereinafter referred to as the Protectorate) remaining in the field under arms and at the disposal and the command of the commander of the said Protectorate forces, are hereby surrendered to General the Right Hon. Louis Botha, Commander-in-Chief of the Forces of the Union of South Africa in the field. Brigadier- General H. T. Lukin, C.M.G., D.S.O., acting on behalf of General Botha, shall be the officer in charge with arranging details of the surrender and giving effect to it.

(2) The active troops of the said forces of the said Protectorate surrendered in terms of paragraph (1) shall, in the case of officers, retain their arms and may

give parole, being allowed to live each under that parole at such places as he may select. If for any reason the Government of the Union is unable to meet the wish of any officer as regards choice of abode, the officer concerned will choose some place in respect of which no difficulty exists. In the case of other ranks of the active troops of the said forces of the Protectorate, such other ranks shall be interned under proper guard at such place in the Protectorate as the Union Government shall decide upon.

(3) Each non-commissioned officer and man of the ranks last referred to shall be allowed to retain their rifles, but no ammunition. One officer shall be permitted to be interned with the other ranks of artillery, and one with the other ranks of the remainder of the active troops, and one with the other ranks of the police.

(4) All reservists (Landwehr) of all ranks of the said forces of the Protectorate now remaining under arms in the field shall, except to the extent as is provided for in paragraph (6) below, give up their arms upon being surrendered, in such formations as may be found most convenient, and after signing the annexed form of parole shall be allowed to return to their homes and resume civil occupation.

(5) All reservists (Landwehr and Landsturm) of all ranks of the said forces of the Protectorate who are now held by the Union Government as prisoners of war taken from the forces of the Protectorate, upon signing the form of parole above mentioned in paragraph (4), shall be allowed to resume civil occupation in the Protectorate.

(6) Officers of the Reserve (Landwehr and Landsturm) of the said forces of the Protectorate who surrender in terms of paragraph (1) above shall be allowed to retain their arms, provided they sign the parole above mentioned in paragraph (4).

(7) All the officers of the said forces of the Protectorate who sign the form of parole above mentioned in paragraph (4) shall be allowed to retain their horses, which are nominally allotted to them in the military establishment.

(8) The Police of the Protectorate shall be treated, as far as have been mobilised, as active troops. Those members of the Police who are on duty on distant stations shall remain at their posts until relieved by the Union troops, in order that the lives and property of non-combatants may be protected.

(9) Civil officials in the employment of the German Government of the Protectorate shall be allowed to remain in their homes provided they sign the parole above mentioned in paragraph (4). Nothing, however, in this statement is to be construed as entitling any such official to exercise the functions of the appointment which he holds in the service of either of the Governments aforesaid, or to claim from the Union Government the emoluments of such appointment.

(10) With the exception of the arms retained by the officers of the Protectorate forces and by other ranks of the active troops, as provided in paragraph (2), all war material (including all field guns, mountain guns, small arms and guns, and small arm ammunition), and the whole of the property of the Government of the Protectorate, shall be placed at the disposal of the Union Government.

(11) His Excellency the Imperial Governor shall appoint a civil official of the Protectorate Service who shall hand over and keep a record of all Government property of the Civil Departments, including records which are handed over to the Union Government in terms of paragraph (10), and the Commander of the said forces of the Protectorate shall appoint military officers, who shall hand over and keep a similar record of all Government Property of the Military Department of the Protectorate.
Given under our hand this 19th day of July 1915.

(Signed) Louis BOTHA,
General Commanding-in-Chief of the Union Forces in the Field.

SEITZ,
Imperial Governor of German South West Africa.

FRANKE,
Lieut.-Colonel, Commander of the Protectorate Forces of German South West Africa.

13. Living under Martial Law

Beginning on July 9, 1915, the date of "formal surrender," martial law was proclaimed within the entire Protectorate and was to remain in force until rescinded by proper authority. Seventeen different decrees were listed on August 13, 1915, the first day of publication, the day when the occupation of the Protectorate and martial law were announced. Persons found in violation of these decrees were to be incarcerated and heavily fined.

Since they had been issued to support military operations against the forces of the South African Union, the "Seitz" banknotes were declared null and void. Governmental property that had not been purchased, or property bought for less than its proper value, now had to be purchased at full market value. Goods imported from South Africa were duty free. Until further notice, German banknotes could remain in circulation but were discounted by 25 percent. Concessions for mining operations, as well as trade generated since the beginning of the occupation, were now null and void. Persons prospecting, mining, trading, purchasing, or selling raw or uncut diamonds were fined one thousand pounds, or condemned to a prison term or forced labor for up to five years. Immigration and emigration without a written permit were forbidden, and a fine of one hundred pounds or six months in prison was imposed. Persons who owned, supervised, or kept cattle, vehicles, or other property belonging to the former German government had to surrender these things or face prosecution for theft. The sale of alcoholic beverages after nine o'clock in the evening was prohibited; Sunday sales could only be made to hotel guests. Persons who spread false news were to be prosecuted by a military tribunal. Persons pointing out or finding hidden ammunition would be rewarded. Persons able to prove that their private cattle had been commandeered by occupation forces could reclaim them.

One afternoon, Fräulein H., a secretary employed by the railroad, arrived at our home utterly exhausted. Distressed and weeping, she collapsed into one of the chairs on the veranda. Gradually she began to talk. Her boss, Station Manager B., had been arrested by the enemy because he "was unable" to produce important documents regarding financial transactions and troop and supply movements. He had noticed the arrival of the authorities from his office window, and since he knew exactly what they wanted, he quickly pushed these under the door of the adjoining room where Fräulein H. was working. She understood what this meant, gathered up the documents, and quickly set them under her seat cushion. So frightened was she that she could not even bring herself to rise from her chair. Trembling, she remained seated while the authorities searched the room, then rolled the documents in legal paper. Being the daughter of a pastor herself, her first thought was to flee to the local pastor and ask him to keep these documents—the same ones she herself had hidden for a time among the flowers on her veranda. The enemy threatened to deport her to South Africa if she did not produce the documents. She was willing to be deported so long as the documents were safely hidden away. The pastor vigorously refused to accept them; now she was sitting before me in a state of desperation. Worst of all was the fact that she was constantly followed wherever she went. And since she knew a bit of English, she had been summoned to city hall where she sat at a long table together with English people so that her work could be properly supervised.

Clearly, she was in need of assistance—and I had an idea. I thought it best to temporarily store the documents in direct sight of the enemy. With complete astonishment, her mouth and eyes opened widely when I advised her to take the roll of documents with her to the city hall and set them directly in front of her. She would only risk it for one day. When the office closed at five o'clock in the afternoon, my good Anna—who seldom left the house and thus was not known in town—would dress in white and call on her as a friend. A few minutes before closing time, Anna would help her get her things ready. As they tidied the office, Anna would "innocently" exchange a similar looking roll of papers for the important ones. Because of my translation services, I was well known and therefore not someone to be seen; thus, it was left to Anna to do it. She agreed without hesitation. The plan was successfully executed the following afternoon. Both women hid their anxiety well, though afterward, poor Anna hardly knew how she had found her way home—but the documents were now safely in our possession.

When Fräulein H. arrived at her apartment that afternoon, guards were posted in front of it. For weeks she had been under constant surveillance at work; now another thorough search of her apartment had been ordered—was this ever going to end? The stationmaster was to be escorted to a prison camp, yet despite all the bustle and difficulty involved in his departure, his wife was able to let him know that the papers were safely hidden away. It was during this time that general house searches began, not wide-ranging, systematic searches, but we did hear of them occurring here and there. It was often the case that longtime neighbors and "loyal" friends fomented these searches. It was difficult for me to understand why these local residents would do anything other than keep quiet, yet they did not and seemed to be very busy informing the enemy of our neighborhood gossip—gossip the enemy found most helpful. Many houses were searched, and a number of men from Windhuk were actually deported to Union prison camps. Most of these searches were occasioned by the denunciation of a so-called friend.

One afternoon around three o'clock I heard some unusual noises on our property. I opened the door to the living room and was confronted by a South African officer. Guards had been posted at the three gates as well as beside each exit. They had come to search the house, and though the officer was well mannered and apologized for having to perform this unpleasant task, he stated that "an order is an order." I nodded. He asked me to accompany him on the search. My children were scared; Anna looked green. I asked him about the object of his search, telling him I had handed over my weapons, and that they were returned to me. To prove this I showed him my documents, but he was fully aware of the matter. The real object of his visit was to ascertain whether or not I had any idea of what had become of the governor's silverware.

"You are acquainted with it, aren't you?" he asked.

"Yes," I replied, "it has been my pleasure to dine with the governor on many occasions."

So that was it—someone had denounced me, someone who had been jealous of the many kindnesses extended to me by the governor and his wife!

"But no," I continued, "I have no idea where the silver has gone; the governor did not give me any of his property for safekeeping."

With a single gesture from the officer, everything on the veranda was turned upside down, every flowerpot lifted up, wires searchingly poked into the ground. This was getting ugly. They entered the living room, and I open every drawer in the desk. Each drawer's contents were emptied onto the ground. The desk itself was moved so the men could tap all over it in their search for secret compartments. Naturally, none was found. The sofa and the armchairs were turned upside down, in case something had been hidden among the springs. But nothing could be found! Next, they began to ruffle the curtains within whose uppermost folds I had tied the roll of documents. Now I began to panic; fortunately, though, the documents were not discovered. Between the armchairs stood a small table with an exquisitely carved lace pattern in almost transparent sandalwood, upon which lay a removable plate. Beneath this table was a very precious microscope, wrapped inside a light brown leather etui. A doctor from the colonial forces had given it to me—at the very last moment—to keep safe for him. I had completely forgotten it until this moment. Now the yellow leather shone through the wooden lacework; they had to notice it. It was an object whose registry was on the military hospital's inventory, and as it had not been located, several enquiries had been made concerning the missing microscope. There it stood, the leather gleaming through the arabesques of wood. The South Africans continued their search, turning over every stick of furniture, and this was a vulnerable item—all one had to do was press against the table and the cover would move. I planted myself beside it. Reacting to the stifling heat of the room, the officer removed his helmet and wiped the perspiration from his brow. I seized the chance and took his helmet from him, saying, "It is terribly hot, but even without the helmet you are on official duty." I set the helmet on the table. That did it! The rug beneath it was lifted on each corner, but the little table was untouched.

When they searched the fireplace, clouds of soot fell on them. They removed the books from the cases and thumbed through them. The floor was thoroughly examined, and they pounded against the planks on the ceiling. The search of the living room left them without any leads, and when they turned their attention to the two bedrooms, I unconcernedly relaxed.

The dining room was next, and I opened the buffet for them—it received only a passing glance. One of them crept under the dining room table; nothing of interest there. Now they turned to the linen closet whose contents belonged to the owner of the house (who resided in Germany). I asked the soldiers not to open it since it did not belong to me and since I had no key for it. To the best of my knowledge, I told them, it contained only linens. With a polite "alright," they made their way to the kitchen, where they poked into and under the stove, then opened the flap to the chimney. A bag of potatoes was emptied onto the floor of the pantry; they rummaged through the maize and the flour, then decided they were finished with the kitchen.

Now I opened the last room for them to search. In it stood my large linen closet containing all kinds of important papers, as well as a buffet belonging to someone else. The buffet held the owner's guns and ammunition whose care he had entrusted to me. I opened the wardrobe for examination and leaned myself against the buffet because my legs were weakening. The wardrobe looked nice and orderly, and the sweaty soldier's fists actually seemed hesitant in touching the brilliant whites. Carefully, one of them reached into the back when something clanked! At once I jumped forward and removed the front stack of linens, among which my papers were stored. In utter desperation, I piled up the linens in front of the buffet door. The clanking was caused by some of my silverware. The gentlemanly officer became suspicious again and had my silverware spread out along the table.

"The imperial silverware!" barked the officer. "Look at the crown! Madam, I'm very sorry!"

"Why?" I asked. "This belongs to me. You ought to recognize the German eagle with the imperial crown. Look, here is an E with the crown, and here is a TE, and here an HE, and here is my seal. This has nothing at all to do with the governor or the government; if you will look here you'll see the same insignia on my tablecloth and bed linens!"

I encouraged him to continue his search of the wardrobe, which he did, setting one stack after another in front of the buffet. I also reclined my body against the buffet.

My mouth was dry, yet I forced myself to ask, "Is there anything else you wish to see?"

"Thank you," came the eventual reply, "but there is nothing else in the house we want to see."

Outside they were more than welcome to rummage through the chicken coup and the lumber storage area. The thornbush above the buried suitcase had bright green leaves. They searched the outhouse and even removed the bucket. A soldier pushed open the side space in the hen house where the revolvers lay hidden beneath a nest. "Gwaaaaak," sounded an honest hen as she ruffled her feathers.

All of us returned to the house, and I asked permission to offer them a cup of tea. After their strenuous activities, conducted in the heat of the day, the soldiers settled down on the veranda and drank theirs while the officer sat with Anna, the children, and me in the dining room. On his saucer lay a spoon with an E and the crown. Thoughtfully, he played with it.

After the clap of the horses' hoofs had subsided, the children performed an Indian dance round the thornbush. Anna and I felt quite miserable—to have a pained conscience is awful. Yet since the British consoled themselves with the maxim "My country right or wrong," I felt I could do the same.

In all, our home was searched three times. The first one was the worst; with the others, we felt more relaxed, and just to be on the safe side, we buried all the suspect items from the house in the yard. The soldiers never found anything, and after the war was over, all items were returned safe and sound to their rightful owners.

I still remember a most unpleasant train trip I took with the children to Swakopmund in December 1916. At that time the governor and his wife were living in Swakopmund, and I had been enjoined to take with me one of only three copies of the naval code present in South West. One copy was aboard the ship, *Panther*; a second was hidden somewhere and could not be reached, now the third one had to be transported to Swakopmund. Smuggling items such as this was risky as suspect travelers and their luggage were often searched. But I could not refuse, so I placed the code book in an open basket filled with the first grapes of the season from Klein-Windhuk. The trip lasted forever, and all of us were thirsty. The conductor came along and checked our compartment and our bags.

When he saw the grapes, he asked, "Nothing but grapes in there?"

"Yes," I replied, "only grapes. Will you have some, please?" And without waiting for his answer, I handed him several beautiful bunches. The danger had passed.

In Karibib, I got off the train to fetch additional papers from the commander so I might deliver them to the governor as well. With my hat in my hand, the papers were inconspicuously placed within it. When I returned the hat to my head, however, it no longer fit me very well. The papers inside proved so heavy that it began to slide. When finally we pulled into Swakopmund, unchecked and very happy, I felt able to breathe a huge sigh of relief.

During all these weeks and months the schools remained closed; in fact, soldiers, as well as Bastars with their families, were quartered in our beautiful new school. One day I happened to be strolling past the school and noticed that the soldiers had abandoned one wing of the school. I was determined to maintain the school and help the schoolchildren who were now bored to tears from the prolonged vacation break. I placed a notice on the board at the bookstore stating that school would begin again on the following Monday, entreated the German parents to send their children to school, and signed my name.

I was a bit anxious for the following Monday, but to my delight children big and small were milling about the school waiting for me. Some had blond hair, others brown, some had short hair, others came in braids. Instantly, I loved them, and with what enthusiasm did we greet one another. At first everyone had something to say, and once each had had his say, we began cleaning. Soon the desks and benches had been set in proper order in three different classrooms. With the help of the older ones, the children were divided and sent to the proper classrooms. We combined several of the age groups and began working hard. The older children supervised the younger ones in reading, arithmetic, and dictation while I instructed one of the more advanced classes. At midday, the younger ones were sent home, and I continued teaching the more advanced students (of course, mathematics and physics were not offered!). I explained to all of the children that what was most important was that they act in a disciplined manner and do nothing to upset the South Africans. In every respect, their behavior was completely above reproach.

As I anticipated, a few days after school began, I was summoned to appear before the Provost Marshal. He interrogated me and asked who had authorized me to open the school.

"No one," I answered.

"And just who is paying you to teach?"

"No one," I replied.

"Do you expect the Crown to pay you?"

"Certainly not."

"Then why do you conduct these classes?"

"Because I simply cannot stand by and watch these German children idle away their days."

Satisfied, he dismissed me.

The doors to the school's boarding houses were soon opened, allowing the farmers to send their children to school in Windhuk. Peace and quiet were maintained, and classes were conducted with strict discipline. At the end of the year, the governmental commissioner presided over the final examinations.

About three weeks after the school reopened our soldiers surrendered in Khorab. Four or so weeks later, our first teacher returned, Professor G.—the mathematics and science specialist. By and by, the other gentlemen returned as well; the last to return was our director, Herr Z., who had been sent to a prison camp in the Union.

Not long after the surrender was formally signed, a regular rail service to the Union was inaugurated. Soon provisions arrived and several Union-based companies opened their doors in Windhuk. Naturally, after a season of privation, the stores were mobbed. Initially, we were allowed to pay in Seitz banknotes, though at a 33 percent loss; later, however, the notes were no longer accepted and even German silver coins were barred as legal tender. All of us had clearly felt the lack of sugar, but now jams, fruit salads, chocolate, and sweets were available to those who had money: all our German money went to the English businesses. Later on, our local German provisioners also began ordering goods from the Union, which allowed us to have plenty of everything.

Our troops from the north finally arrived back home on an unending freight train that transported the professional soldiers to the settlement of Aus in the south, where they were interned behind barbed wire. Several officers had volunteered to guard these men and were joined for a short while by the governor and his wife until they were able to take up residence in Neu Heusis. The commander and his staff went to Okawayo, near Karibib. Officers were released on their word of honor and then allowed to move about freely within the country; those who did not give their word were held in an officer's camp. The released officers had to find their own lodging as their former apartments and quarters had been taken over by the occupying troops. Many were granted hospitable lodging on farms where they made themselves useful.

I made room in my house for three officers who did not, unfortunately, come alone but with their own servants and two horses apiece. They had been allowed to keep their mounts, and now lively riding exhibitions were held on our property. The boys were constantly involved in the care of the horses: they rode, cleaned, watered, and fed them, as well as maintained the saddles and helped to clean the stables. No wonder they smelled of ammonia—an unpleasant stench that hovered over them like a cloud! Eventually, all our rooms and furnishings smelled of horses—especially the beds. I was not able to host my guests on my own income, so they paid their share; they also paid for alcoholic beverages. One of them was a passionate hunter, and every now and then he disappeared for a few days, then returned with game both large or small. How well I still remember the springbok and the kudu bull hanging in the shade of a thorn tree.

Provost Marshal de Waal (Botha's son-in-law) had been replaced by Captain Lewis, a fine, courteous, well-bred English gentleman, who displayed a great deal of empathy for our difficult situation. From South West, he was sent on to German East Africa, and when the war ended there, he became commandant of the camp in Aus. Captain Lewis is now a retired general and lives in Durban. Any German who had contact with him remembers him well; he was actually the one who granted me the return of my weapons and other items.

We continued to receive only the kinds of news the government considered good for us. South African newspapers were allowed, translated sections of which were published in our new little paper *Der Weltkrieg*. The following Reuter telegram, printed in edition fifty-eight of the *Weltkrieg* on May 17, 1916, caused great indignation:

"Lord Robert Cecil, quoted speaking in the lower house of parliament in answer to a British government question concerning the policy of 'No annexations' and 'No compensation,' has said: 'We did not participate in the war without consideration to enlarging the empire. We did not conquer the German colonies in order to free the natives.' He continued that he shuddered at the thought of returning responsibility for the natives to Germany—the natives who had been liberated from German rule." The lie of colonial culpability!

Mail was strictly censored. Overseas correspondence was possible via Switzerland and Sweden. Often, however, after one had looked forward to finally receiving a bit of news from home, and after anxiously opening the letter, it was empty or simply contained a note that the letter had been destroyed. For a while, a Jewish-born German, Herr H., was the main censor. He took pleasure in stopping someone in the street to tell him that he had destroyed his letter but would later offer to disclose the contents of the destroyed letter for a certain amount of compensation. For example, he told the local physician, Dr. F., that he had news from his mother, then asked him how much he was willing to pay. In the end, the doctor put out a fair sum only to learn that his mother had passed away.

Some of us Germans were also spies, and I could not help thinking how fitting it was that these people who considered themselves safe were among the first to be deported. Additionally, the Department of Criminal Investigation—also known as the secret police—carried out its work in the harshest of manners under the supervision of a German-Swiss. As a citizen of Switzerland, he spoke perfect German and pretended to be a friend of the Germans, creating a false sense of security that induced many persons to be rather free about what they said. For example, this man was able to tap into the telephone connection of our military commissar, Major T., who was initially embarrassed at not having recognized the tap until his spies discovered what was going on and reported it to him. After that, the Swiss supervisor heard only kindnesses and news especially fabricated for him.

I myself had to justify my actions before this Swiss officer for reportedly dealing with the "enemy." The retired owner of my house lived in Germany, and as I was unable to send the rent from here, I had asked my stepfather in Berlin to pay him from my savings account in Germany. The censor got a hold of a receipt for a payment and I was subsequently accused of dealing with the enemy; but because the money had not come from the colonies, nothing happened to me. This Swiss devil was the most hated man of all—loathed by friend and enemy alike—and died in 1918 during the influenza season. No one shed a tear.

It is unbelievable what one can become accustomed to simply because one has no other choice. Even though we did not like it, we actually got used to having the enemy in our country! In the beginning we were consoled by the thought that it was only a period of transition. But during that time, long trains loaded with governmental property rolled into the Union. Entire trains loaded with horses and cattle were removed from the North, and not even private property could be vouched safe. For me, however, it was simple proof of the excellence of German products! They even pulled up the bridge poles of the Swakopmund jetty and shipped them out. They tried to dismantle our cavalry monument. I later saw the Cape Cross (a stone cross erected on a stretch of beach north of Swakopmund by sixteenth-century Portuguese navigators who explored the coasts of Africa) in a Cape Town museum. Only when one considers that the author of the mandate was convinced from the beginning that he would eventually be awarded South West Africa as a mandated region to be run as his own fiefdom does it become clear why he confiscated as his own items of private property that had considerable value but which actually belonged to the people of South West. General Smuts probably knew exactly what his intentions were when the mandate system was put in place. "Mandate means annexation," he declared, though the League of Nations, the people of South West, and the South African Nationalists protested against this statement while they still had power.

Work on the farms began again; maize and potato fields were prepared; cattle herds increased. Even though there was not much of a demand for beef, the prices were still high. Once the war was over we expected our farm goods to sell at a handsome profit. Business was booming, and many South African shops were opened. Many people immigrated to South West, particularly Jewish businessmen, and for a time Yiddish was permitted as a language for official notices. Strange names, business signs, and goods were everywhere to be seen. Actual news of the war was rare, and when it did come, it had a distinctively British flavor; thus, much of what we read we simply did not believe.

A cinema opened in the large hall of the Stadt Hotel in Windhuk. Entrance fees were low, and some of the soldiers (brought to Windhuk on a work detail from the prison camp in Aus) were admitted without charge. We occasionally attended the cinema where people were seated at separate tables and watched these old, kitschy films, most of those present completely unable to understand the English dialogue. When the picture appeared on the screen, the voice of a German translator called from the back of the room. The hall was warm and muggy, and the lamps high upon the ceiling whizzed and rustled as countless swarms of flying nocturnal insects buzzed about them. Thick, blue smoke rose to the ceiling, and at the end of each performance "God save the King" was played—a custom in South Africa as well. Everyone stood up when it began, though we Germans rose immediately after the film and left the hall before the hymn was played. Strangely, after a while they ceased playing the hymn. Occasionally, weekly newsreels of the war were shown, but these were insulting to us, and several times, in protest, we Germans left the hall en masse. Sometimes fights broke out, and after that we Germans no longer patronized the cinema. The government eventually formed a commission of South African and German officers to censor the films before they were shown, and in due course, many films were deemed unsuitable for public screening.

The look, the feel, and the general sense of things on the street had changed in Windhuk. White people strolling down the street had different facial expressions, wore different clothing and hairstyles, and spoke other languages. New and different flags fluttered atop all the public buildings. It was all so very strange. Even the natives began to act differently—and continue to do so. During German colonization they did not walk on the sidewalk but along the side of the street and courteously gave way. Now groups of them shuffled along on the sidewalk acting silly, impudent, and loud; it was we who had to give way to them. Our "dear black brother" has equal rights! The dirty native children romp, loll about, and scream on the street corners while we had no choice but to walk around them.

SOUTH WEST BECOMES A LEAGUE OF NATIONS MANDATE

1. Dissolution?

One evening in November, Major T., the military commissar—and legal guardian of my children should something happen to me—invited us along to the cinema. As we made our way by dogcart to Windhuk proper, we noticed an unusual commotion; drunkards stumbled from the pubs amid screaming, laughter, and howling. Even before we came to a halt, friends called to us that an armistice had been signed, that a revolution raged in Germany, that we local Germans were in danger. People were assembling in the street, and it looked to us like a massive fight. Immediately, the major turned us around, and we made our way back home without exchanging a word. Wondering if what we had heard was actually true, we drove on to see Dr. S., a lawyer and friend of ours. He confirmed the truth of what we had heard, after which we sat facing one another in complete silence for a very long time. Quietly, we drove through the moonlight into the Klein-Windhuk valley. Was this the end? I swallowed hard. Major T. ground his teeth. It was the end—but how could this be possible?

Soon enough we learned that German South West Africa was to become a mandate of the Union of South Africa. At that point we were far from certain what a mandate was; now we know exactly what it is! The mandate was an invention of General Smuts, and much has been written and spoken about it. A mandate country is an independent state over which a guardian country has been given custodianship by the League of Nations. The guardian has the duty to promote the material and moral welfare of the entire population, and for us it is the Union of South Africa that is responsible to the League of Nations and must render account of its guardianship over us. South Africa is not allowed to derive any kind of advantage for itself from this trusteeship, neither is it allowed to annex the country in any manner, nor attach it to its own state.

In the beginning, the Union was not the country administering the mandate and did not become such until December 1920. In the meantime, a substantial ethnic cleansing began with the desire to remove as many Germans from South West as possible; indeed, for a time it seemed that all the Germans would be deported. I found it strange that here, in our corner of the world, the Union officers were content to sit idly about while their British comrades in other parts of the world bled and died. This was an unacceptable form of peace; somehow, there had to be progress. Surely the time would come when they could do something to help. The brave enlisted men, as well as the policemen, were hoping for an improvement.

There was nothing for our German government officials to do either. The South African administration was eager to send all of them home as quickly as possible—and the German officials were happy to leave. After all, what was their position now? Many had been displaced from their homes and depended on the hospitality of friends. Our German government had taken generous care of the officers and officials who had been transferred here, providing them with apartments furnished with every imaginable amenity—even with tables, beds, and bath linens. The furniture was of the finest German craftsmanship (called Prächtel Möbel) and was designed to be self-assembled, durable, plain, practical, made of solid oak to withstand the rigor of the tropics, and had solid brass locks and hinges. Even now, twenty years later, this furniture is greatly valued and fetches high prices whenever it appears at auctions. The wives of officers and bureaucrats decorated their apartments with special items that gave them a more personal feel. The South Africans were only too happy to confiscate the government furniture and to force the German officials to sell everything else they had at bargain prices since little time was allotted them to prepare for their removal. Only true necessities were allowed to be taken along. Naturally, opportunists took advantage of the situation, and sharp bargaining and haggling commenced.

Botha had published a proclamation guaranteeing a free voyage to anyone who wished to return home to Germany. Many took advantage of this offer, some motivated by homesickness, others by the desire not to live under foreign rule, with a remaining few who simply wanted a cheap trip home paid for by His Britannic Majesty. Later though, some were very sorry to have made such a hasty decision—especially when they faced the stresses of living under such difficult circumstances in postwar Germany. Gladly would they have returned, but returning was made nearly impossible, for they were required to obtain a passport, an immigration permit, and to deposit sixty pounds sterling for a first class return ticket on an English steamship with poor food and accommodation. Of course, all these immigration expenses had to be paid from their own pockets.

Many, many households were dissolved and valuables given away for next to nothing. The South Africans were delighted, especially the ladies who considered it good sport to go from house to house and greedily gather together, at minimal expense, as many of our solid German goods as possible. In groups of five or six they appeared, brazenly touching and examining each object before tussling with one another over the goods. Then began the quibbling and bargaining. Most highly desired was our valuable needlework, after that, our jewelry. It was curious and amusing that most of them were completely unable to distinguish between crystal and pressed glass. Even a picture painted by an artist and a simple colored print seemed indistinguishable to many. Thick golden frames struck a deep chord with them.

In the beginning it seemed that all German teachers would be "repatriated." My own future in South West was unclear. Would I have to leave since I was a teacher at the secondary school, or would I be permitted to stay since I had come at my own expense and had, during the war, helped out on an auxiliary basis? The German government commissar had no answer for me. Shrugging his shoulders he advised me, "Be prepared by selling whatever it is possible to sell." Well, all right then. I owned no government property, only some very beautiful personal items that were quickly sold. My genuine Persian carpet from the dining room went for three pounds; my polished crystal chalices brought an astonishing one pound each; my beautiful lamps, chandeliers, and many other things went at rock-bottom prices. Strangers came and sniffled over everything, handled everything, turned it around and around and said, "I'll take this thing and that thing and give you so much for this and so much for that."

British coins and banknotes lay atop my table as people left with my possessions. I let them take whatever they wanted and accepted whatever price they wanted to pay. I no longer cared. I felt dull and dizzy and wished the end would come sooner rather than later. After a few hours they returned, these human hyenas, searching for further bargains. At some point I lost my patience and felt as though I were awakening from a dead sleep when I heard myself say, "No, I am selling nothing more! Now please leave immediately and never come back!" I wondered what on earth they must have thought.

I was upset and angry, and decided I would not sell another thing at such ridiculously low prices. Everything I owned, I suddenly thought, I had brought with me under most difficult circumstances; every item had its own history and memories. I considered it to be my duty to maintain a proper home for my sons. I won't leave, I thought, but if my children and I are forced to, then we'll beat everything to pieces and burn those things that can be burned! No one will be able to brag triumphantly about how cheaply he obtained goods from the Germans who were thrown out. I locked my doors and let the carts, wagons, and riders sit outside. Let them knock on my door and shout. I refused to stir—enough was enough!

My children were terribly upset by all the bargaining. Close to the road in a thick camelthorn tree they had built a tree house, and when well-fed horses pulling a cart came trotting by, a well-aimed little stone, fired from their catapults, hit the horses which then galloped off like lightning. Even unsuspecting troopers on patrol were frightened off! Occasionally the boys trotted along the road generating a thick cloud of dust with their little homemade cart. My sons, along with Jochen v. D., had fastened a rattling piece of tin on one of the wheels while behind them they dragged a very strong chain. All this apparatus kicked up so much dust that those following them would choke. Given our circumstances, it was often very difficult for me to punish their naughty behavior.

After the surrender, many officials and businessmen, along with their families, came from the Union to settle here. Spats that often developed into fights with these foreign boys could not be avoided. The foreign boys acted like victors and insulted our children by calling them "Huns," "Bloody Germans," or even worse. No amount of admonition helped; they simply had to endure it. One day I was ordered to appear before the Provost Marshal who asked me if I were the mother of the three boys on Jan-Jonker Weg. "Yes," I replied, "two are mine, the third is my foster son." Some parents had complained that their children had been roughed up by mine, and he asked if I would put an end to it. I then presented my complaint, my side of the story, and asked for understanding, promising at the same time to do everything in my power to influence my boys. I then asked that the parents of the other boys please do the same. He smiled, and we separated understanding each other very well.

I was nonetheless angry with my children and called all three of them to me. I told them how ashamed I was to have listened to complaints about their poor behavior from a foreigner. Well-bred boys would not be the cause of such things, I told them, especially under the current circumstances that necessitated us being particularly disciplined.

Like poor little sinners, they stood before me with defiant expressions on their faces.

"Confess now," I demanded, "tell me what you did."

"Well, you see, Mama, there are three kinds of boys who annoy us, and we won't let them do that."

"Three kinds?"

"Yes, three kinds. The first are the Tommies, who are the most decent of them because if we disagree with them, we fight hard—but only with those who are bigger than we are. We're certain they didn't complain."

"And the second?"

"The Boers, they curse and throw rocks and shoot slingshots at us—but only when we have our backs turned."

"And you?"

"We throw rocks in return."

"And the third?"

"They're the louts who pretend not to understand German; they're the worst."

"And what do you do to them?"

Silence reigned.

"Jochen, you tell her."

"Well, we don't fight them or shoot at them."

Again there was silence.

"Büdi, you tell her."

Finally, the youngest blurted out, "We press horse-apples into their mouths."

Incidentally, I was hardly the only one to appear before the Provost Marshal on account of squabbles among children. Even the director of the German boarding school had his difficulties in this respect. My children's guardian decided a change of scenery would do them some good and took them with him to a farm located to the east of Windhuk, a farm he and several friends of his were planning to buy.

On the Wednesday before Easter, a mounted policeman appeared at our door and handed me an order requiring my sons and me to leave the country on the following Saturday. The moment had finally come! At daybreak, in one of the few rental cars available in the country, I drove the eighty-kilometer distance to the farm with the major's expulsion order in hand. The drive to the east is beautiful as the road leads up and down hills amid the strangely shaped peaks and folds of the Auas Mountains. Eventually, the road approaches more level scenery, then winds through scrub forests, hackies, and riverbeds. I must admit that I was too preoccupied with my own dire circumstances to notice much of it. We hurriedly returned and were back in Windhuk that night. Arriving at our house, we found another note affixed to our door: our expulsion had been postponed until further notice. Because the major had just received a new appointment as transport leader, these matters were placed on hold. We had gained time, a good deal of time, and this would allow me to get my affairs in order. The uncertainty, however, was dreadful; day and night I waited and worried.

South West, you harsh land, you dear, dear land, again you abandon me! You beckoned me to return, and I did—with everything I owned—to perform a cultural work. My sons were to become Africans, though in harmony with German character. Alas, now you are pushing us away—just as you did before. Oh, South West, I have given and given, and you have but taken and taken!

I became very bitter. The children were also beside themselves. "Mama," they would say, "let's stay here. We don't want to leave. Africa is our home. Our friends and our animals are here."

On the Saturday before Easter, the major and I drove to the railroad station to bid farewell to the people who were leaving. A very long train stood on the tracks. An ancient, care-worn locomotive belched thick, black, dirty smoke into the air (the coal currently comes from Natal and is filthy). All the German locomotives were now in poor condition because the South African stokers did not know how to care for them. Punctuality had long since become some kind of vague concept. But there stood the train with its freight of German settlers who were no longer wanted. We had almost been on that train ourselves.

While still at the station I was informed as to why I had not been deported: the government commissar told me that I must not leave because I was urgently needed here to continue teaching in the German school. Since I had immigrated as a private citizen, my voyage had not been paid by the government, and since I had been employed here as an auxiliary teacher, my case did not fall under the expulsion order for government officials; hence, I was allowed to stay. The commissar had negotiated my case by arguing along these lines with the new government, and during those few minutes our fate had been decided. We were allowed to stay, and I wanted to stay. I now had work and could provide for my children and myself! What a good thing it was that I had maintained the home for my little ones. The stresses and strains of the past instantly lifted from me; Africa not only wanted to keep me, she needed me! And how willing I was to faithfully fulfill my duty for my distant Fatherland. I was chosen to help, to be a support for Germanity! I had a lump in my throat and quickly turned away lest the commissar see me in this state.

Slowly, the long train puffed and groaned its way out of the station. People both laughed and cried; they screamed, sang, sobbed, and waved farewell. In my emotional state, I witnessed the happenings as though through a veil.

I could not sleep that night. It was unbearably muggy, so I quietly arose and went outside. Unusual for this season of the year, I saw black clouds gathering behind the mountain with lightning striking against the tropical night sky. It was beautiful. At the moment when the lightning illuminated the trees, they stood tall against a blue light, their black branches reaching sharply toward the light. Then everything disappeared—like a Fata Morgana behind the yellow sand dunes. It was not long before the first thunder rolled across the mountain, echoing amid the old trees that stood along the rivercourse. Swaying a bit, a fragrant rain of flowers fell from the trees to the ground. A breath of wind cooled the pale grass. Suddenly, a bolt of lightning struck brilliantly across the top of the mountain. Treetops appeared entirely white while the echo of the undulating thunder broke against the rocks. Then, total silence. Again a bolt of lightning struck! The motionless branches magically glowed for a moment or two before growing dark. Another pale blue flash struck. Then darkness, complete silence, and warm, soft air. Black clouds now covered the sky overhead, yet not a drop of rain fell. This was a thunderstorm in the tropics—and I was allowed to stand outside there and experience it. No longer would I have to wander homeless with my children back to Germany. I was permitted to work for Germany, and South West would remain our home!

Against his will, Captain T. remained military commissar until the beginning of July 1919. As he was the guardian of my children, as well as someone who advised me and helped me with my work, I was delighted that he had to stay on. He lived in the house at the foot of the mountain above the Klein-Windhuk valley that I had owned now for several years. On weekends, the children were able to visit him. He owned beautiful horses and was an excellent riding instructor. He gave much to my children: besides riding and shooting, he also taught them to play chess and a card game called skat. The mandate government watched him very closely. In his possession were all sorts of documents that had to be kept from the enemy, and with these he was most careful. All of us, actually, were constantly on guard, and as soon as he was warned that something was going to happen, his servant Pupke, a crafty Hottentot, came charging over carrying a note with a red exclamation mark on it. Immediately then, my two boys walked along the valley and collected "suspicious" documents from him and brought them to a place where they would be safe.

2. Expulsions

By now the new government had ordered most of the soldiers, policemen, and officials to leave the country. Some surveyors and teachers were allowed to stay, while others left voluntarily. However, some people's names appeared on a list under the new governmental designation of "undesirable." Since it was impossible to obtain money from Germany, many were in dire financial straits. Some had no work; others had no income due to a lack of trade and poor business conditions. There were also those who took advantage of the situation and simply loafed. Whatever the reason, such people had become dependent on the Mandate Government to provide the basic necessities for themselves and their families—mainly foodstuffs. Anyone who applied for support was labeled "undesirable" and eventually sent away. In addition to poor Germans, there were plenty of impoverished South Africans now living in South West.

Some Germans voiced strong opinions against the new government; others acted against it as well. These persons were also sent away—even when it meant that they had to leave behind their homes and farms. Those whose names appeared on the list of drunkards and loudmouths were labeled undesirable and deported. Those who had broken the law during the German period (unfortunately, many nearly forgotten documents were still extent) were sent away. Barmaids and other immoral women, as well as those who had simply been accused or denounced by former friends and neighbors, were deported. Families whose members were caught in perpetual rows were likewise sent away.

Men and women who had served the enemy (the South Africans) as spies were also branded as "undesirables." For once we agreed with them! Many were actually delighted to be expelled, thinking, as they did, that they had nothing to lose, while nurturing the hope of a fresh start in Germany. Others were beside themselves at the injustice being done to them. A lucky few were allowed to stay at the last moment because they had been able to justify themselves or had had someone speak in their favor. The rest were not given the time of day.

It was not only men who were punished but also entire families with children. We Germans provided such people with clothes and blankets. Our two German women's organizations—the Overseas Red Cross and the German Women's Colonial Society—combined their efforts and opened a collection point for clothing. Everyone gave, though no one really had enough for himself, and the ladies of these organizations repaired, changed over, and sewed clothes. It was amazing how much was done and how many of the so-called undesirables looked neat and proper when they left. It should go without saying that everyone was given a large package of provisions for the journey. Mothers with small children received extra rations of evaporated milk, and it was only later that we learned how inadequate the food supplies on board were.

Detectives from the secret police spied on everyone, and the main spy, dubbed the Silver Jackal, was known by everyone. Completely fluent in German, he was responsible for the expulsion of certain people, while to others he granted permission to stay. He went to Walfisch Bay, where the deportees boarded ship. Passengers who were under suspicion because of having been denounced by someone else were carefully checked. To invalidate a denunciation, officers needed only to give their word of honor, enlisted men their "word as a soldier." This was enough. Higher officials who had become distinctly unpopular were in trouble. One such passenger had the heels of his boots torn off, another's pocket watch was demolished, while yet a third man had the buttons on his coat torn off because he was suspected of hiding gold pieces beneath them. There was one woman whose tins of melted butter were confiscated and found to contain gold pieces. She was suspected because, unfortunately for her, a piece of gold fell from its rather intimate hiding place! As a consequence, the rest of the passengers were all the more carefully checked.

Each passenger was allowed to take forty kilograms of luggage. One farmer arrived with a very heavy metal suitcase at customs.

"Open it at once!" he was commanded.

It contained nothing but sand and earth.

"What's this all about?"

"Just a remembrance of my farm."

"Nonsense! You've hidden gold or raw diamonds in there!"

"I wish that were true, but this is all you swine have left me."

"Empty it then and let it run through a sieve."

It really was just sand and dirt from his farm—a final bit of Africa he wished to take with him to Germany.

It was painful to see the faces of the expelled. When they had first arrived in South West they were filled with courage, hope, and energy. Many had worked very hard in this harsh country, but now, under enemy occupation, they were chased off their property like stray dogs and branded as "undesirable." The Union Parliament even debated the expulsions of so many people and the execution of this policy in such a rigorous manner. In June 1920, an investigation was called for, but by that point troop transports and expulsions had affected nearly half of the German population—an exercise intended to weaken the German influence before the end of the war. It was not until 1921 that the state of war was officially ended and the expulsion of Germans brought to an end.

3. The Influenza Epidemic of 1918

During 1918, a scourge of diseases raged through South West and caused much suffering among us: Malta fever, malaria, and typhus caused death; throat infections, diphtheria, scarlet fever, and measles broke out among the children. In August and September we read for the first time that certain towns in the Union were battling outbreaks of Spanish influenza. We read about it but thought nothing of it and really had no idea what it was or how such an illness could be treated. Gradually, however, it dawned on us that Spanish influenza could well find its way to us. Now and then, our German paper published brief notices about the disease, but suddenly we began reading about a widespread epidemic and actual cases of death in the Union of South Africa.

At the beginning of October, a circus visited us in South West for the first time. The first evening performance was completely sold out, and some of Windhuk's blacks were permitted to sit in the gallery. Even during this first performance the director announced that some routines were cancelled due to illness. Captain T. and I had agreed we should not attend despite the fervent pleading of the children. After all, the circus came from the Union and we had no way of knowing whether or not these cases of illness were influenza related. We certainly did not want to expose the children to the massive crowds of people there.

The very next day a train from the Union arrived on time, though its conductor was dead and a critically ill waiter died shortly after being admitted to hospital. The evening circus performance was cancelled as almost the entire troupe had fallen ill. At the native village in Windhuk, widespread illness and several deaths were reported. The next morning as I walked to school, a dead native lay in the road leading to Windhuk proper. At the corner of Kaiser Strasse, natives were helping a collapsed woman.

Among the whites, the first death was a young Jewish woman who lived at Ausspann Platz. Several hours later, a young girl died. From then on illness stalked almost every household. All of us, whether German, English, Boer, or native, began to panic. Schools were closed, offices were emptied, many retail businesses shut their doors, trains ceased running, and no goods were brought into Windhuk at all. Milk and butter were no longer available and bread was scarce, but meat could still be had. The epidemic had struck with such force and had spread so rapidly that it had been impossible to send the children from the boarding house home to their parents on the farms. Now public transportation was no longer available.

In Klein-Windhuk, we started a neighborhood "check and help" brigade. Everyone took a turn checking the inhabitants of each house for the disease. This turned out to be a necessity as many bachelors, German officers, and officials lived by themselves and were in urgent need of help. Of our two local doctors, one had immediately fallen ill, the other, our beloved Dr. K., made superhuman efforts to care for those afflicted. Night and day, he went up and down each street in both Klein-Windhuk and Windhuk proper, mounted on his large horse, his white doctor's coat flowing behind him, a tropical helmet on his head, visiting, helping, counseling, and consoling. The former missionary, Herr T., stood at his side. The city and military hospitals could no longer accept patients, so the empty school building was commandeered to serve as an auxiliary hospital. Farmers and farm families from the countryside came to Windhuk unaware of the severity of the disease and quickly found themselves stricken; many such persons simply collapsed to the ground, and all over Windhuk people lay on the ground because they were unable to do anything else.

The two pharmacies were practically stormed and emptied of medications. Many people hung little bags filled with camphor about their necks, which, of course, had no medicinal value at all. Medicine bottles were soon finished, and whoever wanted medicine had to bring his own bottle.

It seems that no house was without a sick person, and as everyone was busy caring for those in his own household, it did happen that some of the more isolated houses remained unvisited. Suddenly, it dawned on us that in some of them there were no signs of life. Upon entering such a home we found the occupant dead.

The hearse was in constant motion, making trip after trip to the cemetery; still, it alone was not sufficient to handle all of the dead and an oxcart was loaded with several caskets at a time and driven, creaking and groaning, along the dusty Kirchhofspfad to the cemetery. When the gravedigger went missing, there was no one to dig the graves. In the early morning, from that point onward, a number of men from Windhuk, joined by soldiers of the occupation forces, could be seen digging graves. Among the Union population, illness and death were even more severe than among us. In fact, I heard that there were twice as many cases.

The death of so many young people was tragic. Before they finally succumbed, they scratched their fingernails on the walls, tore the bed linens, or bit into them. Their poor young bodies fought valiantly to the very end; then there was peace. The dead child's face was covered with a piece of cloth until the body could be carried away. Where were they to be taken? Next to the military hospital stood a small funeral parlor that was nearly always filled with dead South Africans, caskets already standing beside the entrance. It was the same at the Catholic Hospital. In our school, the bathroom walls had been removed so we had a place for the dead. Corpses laid out on boards (so numerous that they appeared to be stacked on top of one another) were set in this space where they remained until a casket became available for them to move on to the next step. This process took a long time—and it did not help matters that our few carpenters were sick as well. The lumber supply was finally exhausted. It was October and the weather was hot and humid; soon, it began to rain.

Many in our immediate neighborhood had fallen ill. Luckily, Anna and I remained healthy, for every day our help was requested and the fire in our stove never ceased burning, our large soup kettle was never empty.

Early one evening, my children's guardian came and asked for help. He was living with three other officers in the house I currently owned. Two were absent, one was very sick. Since the illness had come on so suddenly, I worried about my boys. They were still in good health, and I did not want to leave them alone at night since Anna was also afraid, so we switched places: he stayed with the children and I went to his house. Two of his servants had already left and the other two were to stay and help me. I entered the house and found only one servant; the second lay unconscious in the kitchen of an outbuilding. I looked in on my patient. He had a very high fever and initially failed to recognize me. As he was quite blue and bloated and had difficulty breathing, I decided to sit him up. Hardly had I got him onto my arm when he vomited all over my face and dress. He needed a compress, and I fashioned one from packing paper and towels. After I applied it, he began to calm down.

I had told the healthy servant to stay close. I soon heard him drop: my helper, my pearl Pupke, had fallen ill. Dragging him toward the other one in the kitchen, I gave both of them cold tea and aspirin, covered them, and returned to my patient. It was a terrible night, and I did not think I would survive. No one answered the telephone, and I could neither leave nor reach the doctor. I refreshed the compress, supported the sick person when he was shaken by terrible coughing and choking spells, and stayed awake all night long. Towards morning he fell asleep. Later on I took him to my house where I nursed him back to health. He survived and now has a farm in Angola.

Captain T. had promised to relieve me early the next morning, but he did not appear, and I began to worry about the boys. Finally, around ten o'clock, a horse cart appeared driven by my elder son. Nearly falling, he climbed off the cart and groaned, "Mama." I caught him and put him in bed. He demanded sparkling water, lots and lots of sparkling water. The others also were very thirsty, so I left them and went to Windhuk to buy some water. Besides that, I needed to notify the emergency station for natives and ask them to come fetch the two stricken servants.

Bitz and I drove into town and found it empty of people. Kaiser Strasse, the main thoroughfare, was deserted, and in the bright sunlight the hoofs of my two horses clanked loudly through the silent streets. Many stores were closed; others had their doors wide open but no salesperson was in sight. We drove into a side street where sparkling water was produced. Everything stood open, though not a soul was about. We called. A weak voice from one of the side rooms asked what was wanted. We presented our request, which was promptly denied, yet I could not return empty handed, so I asked again. We got permission to see if any was left. We found a case of it, I thanked him through the door, wrote the number of bottles on a piece of paper, and set it on the counter. We went into a butcher shop and bought beef to make soup. At the end of this street was a large grocery store where I used to shop back in those days. It was open, and we went inside. I called out, but no one was to be seen. Again, we called. Again, there was no reply. Finally, someone did answer us and asked what we needed. "I can't get up and serve you, but take what you need and write it on the counter." I went over to the poor man and asked what I could do for him. Nothing, was his reply. Should I contact the hospital, I wanted to know. "For heaven's sake no," came the answer. Three days later he was dead.

We drove on in the direction of the large native village and were met by two oxcarts.

"Look," said Bitz, "a farmer is bringing in sacks of cabbage."

As the first cart passed, the driver pointed backward with the handle of his whip. "Look, Missie," he called out, "everyone in that village died last night."

The fear of illness drove many natives mad. Senselessly, many ran into the bush where they took ill and died miserable deaths. All along the main roads, in the mountains, and beside water holes lay their corpses.

The day after the captain fell ill, my oldest went to bed; the following day my youngest was unwell. Anna and I remained healthy. My youngest became very, very ill, and in my distress I went to see Dr. K.

"Please," I pleaded, "please come by."

He shook his head, saying, "Child, I simply can't. Don't you see how many are dying under my care? Please, you'll have to bring him here to me."

No, that I would not do. It was far better that he should stay with me—we would try to save him. He required many days and nights of nursing, but he did recover.

As soon as our boys felt better, Anna and I took turns visiting other stricken persons in the neighborhood. No longer did we have any help, for our servants and washer woman were all ill, and under the current circumstances it was impossible to find new help. The illness had raged for about ten days before I found a moment to visit our school's boarding house for the first time. I wanted to find out how the children there were fairing. The scene was absolutely heartbreaking. The director and his wife were down, their assistant had been taken to the hospital deathly ill, and all forty children were sick except for two girls who had recovered enough to be able to prepare some things in the kitchen.

First I went to the boys' communal bedroom. An unmarried teacher from our secondary school, Herr Sch., tried to help. But what a scene, what a stench, what filth! The boys, who formerly had been so active and always ready for any kind of fun and games, looked at me with hollow, feverish eyes. Many lay without linen on terribly soiled mattresses. All the supplies were spent. The washhouse was piled high with soiled clothes and linen, some of which had been soaking and now stunk, but no one had come to do the washing. Beneath the beds bits of bread, bowls, paper, dirty clothes, and much, much worse were to be seen.

The girls' rooms looked a bit better, but how the poor girls themselves looked! One could hardly recognize them as the same girls who had once been so happy. Some were on their way to recovery; others lay critically ill. Shortly before the influenza epidemic, measles had broken out, and while many children were down with measles, they were struck with the flu as well. Now their little bodies were weakened to the point that they had no resistance left in them.

I returned to the director's wife, who was crying bitterly. A messenger from the community had just come. Leaning against the door, he asked, "I was sent to enquire about who precisely will pay for the caskets for your helper and the child. They were both delivered to us last night and died."

A few days later we lost another dear little thing, only twelve years old. We dressed her in a white lacy dress and placed a wreath on her head. The casket was an ugly brown and unfinished; I helped nail it shut. We then draped garlands over it and, together with her two sisters who had just recovered, took her to the cemetery. Two teachers and several of the larger boys carried the casket and filled in the grave. Missionary T. gave a single sermon for the several deceased persons. This dear little child's parents lived far away in the south and could not even be present in thought as their darling was turned over to the earth.

The cemetery could no longer accommodate all the dead, so its borders were extended towards the north, and one little mound of soil after another could be seen laid out in a neat row. Volunteers now dug the graves and covered them over. Herr v. B., a baron from Saxony and former dragoon from Ludwigslust, was one of them. Nobody could have imagined that he would give of his time to do this kind of charitable work. On a particular afternoon, he came to me and asked for a cup of coffee. "I don't feel so well," he said, wearing a dinner jacket with a torn-off sleeve—which made a rather strange impression to say the least. I bade him to sit and remove his jacket. I then took the jacket and quietly sewed on the sleeve again (we Africans can be a bit eccentric!). He said nothing, and I, sewing diligently, asked him nothing. Sitting in complete silence for quite some time, he finally turned toward me and said, "I like to wear black clothing anyway, but when I dig graves I always wear black." He was such a dear person. Later though, he took his own life with a single gunshot.

Every day new graves appeared in that terrible cemetery. Many a poor soul was quickly buried without a single mourner looking on. Every day, horses drawing the hearse trotted down the road; every day, an oxcart ground through the glimmering dust carrying South African soldiers, each casket draped in a Union Jack. One particular coffin had been so poorly made that its back came off while being driven along the rough road. The driver took no notice of this and drove on while the head of the dead man slipped out of the coffin. Fortunately, one of the few chauffeurs in town was driving his car behind that cart, noticed the mishap, stopped the driver, took out his tools, carefully pushed the head back into the coffin, and nailed the panel back in place again. Immediately, the cart continued its journey toward the graveyard as so many bodies awaited burial. Such a gruesome experience had little effect on those who had already witnessed the horrors of the war. But after having seen this one myself, I was bothered for many, many days. Images of that same horrid oxcart bearing caskets that often dripped with blood have never quite left my mind.

October 18th was the epidemic's highpoint, but from then on, it began to subside in Windhuk, though illness spread rapidly to the farms and other more remote areas of the country. Cart after cart loaded with the ill who hoped to be cured by a doctor's skill came to the city: most of them closed their eyes for the last time here. The Union sent serum, but it came very late. Salvarsan is supposed to have helped many, and on the advise of Dr. K. I was also vaccinated but developed a rash that itched terribly. Several of our acquaintances, including the tribal chief, were vaccinated to avoid a recurrence of the illness. Dr. K. had strictly forbidden the drinking of alcoholic beverages, but the gentlemen did not comply as they maintained that the serum's effectiveness would be increased if they had at least two inches of alcohol in their stomach!

At last the epidemic was over. It had taken so many persons from our midst, and though the pressure and worry was lifted from our minds, sorrow for our dead remained. Every Sunday people who had lost loved ones journeyed to the cemetery, now greatly enlarged though not yet properly tended.

For months on end, dead baboons were found at lonely water holes in the veld—they had also been affected by the influenza. The natives continued to remember the time when influenza raged uncontrolled, for just a few years ago when the circus came to town again, it hardly turned a profit because so few natives attended the performances. When asked about this, they replied they would perish if they went because years ago the circus had brought death.

4. Changes

The children's guardian had been ordered to accompany one of the last transports to Germany, scheduled to sail from Walfisch Bay at the beginning of July. At his request, I had decided to send my sixteen-year-old son with him, hoping he would obtain a better education in Germany. Here in South West, we could not yet predict how the educational system would develop. It was very difficult for me to let him go, but I knew it was for the best. My younger son remained with me. To cheer him in his loneliness, after all the recent sorrows and excitement, I promised to take him to the cinema on the following evening. At increased prices, and for the first time, the famous South African film, *Die Voortrekker*, was shown. The film is a depiction of the large-scale Boer trek of 1837 and their battles with the Zulu. The hall was filled to capacity, and the South African audience played along enthusiastically. When twenty great trek wagons forded the Orange River, the Boers in the audience cheered every single team. "Trek, trek; daar de kant; alemachtag; man, pas op!" (Pull, pull; mind the edge; almighty; hey, watch out!), they shouted, stamping their feet, howling, and whistling. The heat in the hall was beyond endurance, and together with the foulest of odors, clouds of dust, and filthy tobacco smoke, at quarter past eight the film began, finishing at half-past midnight. It was an experience, to say the least.

During the performance my mind generated all kinds of thoughts. While filming, thousands of Zulus, a very warlike Bantu tribe, had been engaged to demonstrate with historical accuracy the brutality of the battles against the Voortrekkers. Blacks fought against whites and initially won. Our enemy, the Union of South Africa, had sent hundreds of thousands of colonial blacks to war, even to fight in Europe, and it was there that they had learned warfare. An account of this, written by a native soldier and translated by a French professor, titled, *Assim Affenga, Die schwarze Welle*, published by Wabbel & Raumann, Regensburg and Leipzig, is a truly worthwhile read.

The Negro author, René Maran, published a genuine Negro novel under the title, *Batouala*, that won the 1921 Goncourt Prize in Paris, which is comparable to the German Schiller Prize. Thousands of copies have been distributed in France, Holland, and England, and have caused considerable excitement. The author is a Negro from Martinique who later became a French colonial official in the Congo. The content is despicable and begins with an introduction that does nothing less than incite the Negroes to declare war against the whites and stands as yet another proof of the awakening of the black race. "Revenge is not a meal to be eaten hot," and "was it not you, curious little Eve, who asked if a white woman may marry a Negro?" By having incited these uncultured people to fight the whites, and by having made them lords over the white man, these colonizers have committed a disgraceful outrage against the white race. Such material shall reap nothing but vengeance.

We Germans have been robbed of our colonies. And because we Germans are "incapable" of governing colonies, one certainly owes it to these poor natives, for reasons of "humanity," never to leave them at Germany's mercy again. Throughout his book, the French native calls his white colleagues torturers, cowards, tormentors, drunkards, bloodsuckers, and murderers of the black race. And the Académie Goncourt awards its prize to this *cultural document*!

In German East Africa, Cameroon, and Togo—our former African colonies—the natives witnessed two great European powers warring against each other. One was defeated. Up to that point the defeated nation had been regarded as the perfect example of power and strength, yet because of that defeat the African natives no longer accept the white man taking their land by force, lording it over them, or compelling them to do drudge work. I fear the time will come when the old, ardent cry, "Africa for the Africans," will be raised from the Cape of Good Hope to the Mediterranean Sea. Then heaven help the white man in Africa! Where hatred mediates peace, peace will not last—not in Africa, not anywhere in the world.

The deportations had not yet ended. A large and final group of people was to be taken aboard the *Guildford Castle*, but influenza broke out again and caused considerable delay. Even in the heart of the country a reoccurrence of the illness was expected. The government had learned its lesson from the first outbreak and now made careful preparations. A combined medical commission was formed, and nurses, cooks, and washer women were called up. Our school was to be outfitted with one hundred forty beds. Fortunately, the plague passed quickly except for a prison for natives; it was heavily affected and, consequently, all street and road construction ground to a halt. A few people in the native village also died.

At the beginning of September 1919, my good Anna married a farmer. For nearly thirteen years she had stood loyally by my side. I assumed responsibility for her wedding and let her go—though not without a heavy heart. She found little happiness in this late marriage, for where there once had been affluence, now worry, distress, and illness prevailed. The second farm had to be sold; someone cheated her husband out of the mortgage he owed on a piece of real estate in Windhuk; large and small animals died; the grown children from the first marriage became ill, probably from an inherited disease. Two children had died before their father's remarriage, and three additional adults passed on, one after the other. After ten years of marriage Anna buried her husband next to his deceased children, turned over the farm to a stepson, and at the age of sixty-two courageously earned a hard living by sewing, cooking, and helping out on various farms where the wife was absent. Truly, she was everyone's auntie, and everyone needed and loved her.

Rent in Windhuk proper rose out of all proportion. The South Africans lived in all the beautiful apartments while opportunists purchased houses at give-away prices and sold them for enormous profit. The house where I lived—whose owner resided in Germany and was forbidden to return—was to be sold as quickly as possible, and I, under these difficult conditions, had to search for another place to stay. My younger son, Büdi, and I spent a whole afternoon looking and finally succeeded in locating a place. The house on the mountain above Klein-Windhuk valley was vacant, the house where Captain T. had lived. For nine pounds a months I rented it, even though it had no water source. Two little donkeys had to pull a water cart up the mountain each day. Because of this, it was impossible to have a garden or even flower beds.

Every morning our horse, Hans, pulled us to school in our dogcart. The school was quite a ways from our house, a fifty-minute journey by foot. Horse and cart were both given to my son as a going-away present. The horse, like all of our other animals, was unbelievably tame; never did we have a problem with him. And such an ambitious creature was he that every car going in our direction had to be passed. As he galloped, our light little dogcart bumped up and down—it was pure delight though we entirely failed in teaching him to mind us.

5. New Unrest

The new government had made many promises to the natives and kept but very few of them. Consequently, the natives were bitterly disappointed, and as early as 1916, we noted a certain unrest among them. Itinerant preachers of the Ethiopian Movement, "Africa for the black race," had arrived from the Union and began holding meetings on the reserves as well as on farms. Our old nemesis of 1904, Samuel Maharero, sent his son Friedrich from Lake Ngami (in Botswana) to South West. In the native village near Windhuk, a podium had been erected from which he would deliver his father's message to the Herero. He also visited smaller settlements, even single farms, giving his inflammatory speeches. Interestingly, his remarks were directed less against the Germans and far more against the South Africans, who had given him permission to enter the Protectorate.

The Union administration punished less with imprisonment or beatings and more with fines. Only when a convicted native was truly unable to pay his fine was he incarcerated. Any minor infraction could be settled with a fine. If the fine were high, relatives took up a collection and their money filled the state treasury. Fines were imposed for everything: from not paying various taxes on time to public drunkenness. Police often raided the villages on Saturdays because small sugarbeer breweries operated there clandestinely. As men and women drank to excess their infractions became ever costlier; specifically, one pound for every container of beer found. Nowadays, even the old natives are heard to say that life was better under German rule. Then they received up to twenty-five stripes after which everything returned to normal. But now that they had to pay for even minor infractions, they were malcontent, rebellious, and lazy. Those who had no desire to work in towns or on the farms disappeared on the native reserves.

By 1921, the Herero had gathered thousands of cattle and small stock on their assigned reserves. Wealthy Herero also owned several—mostly stolen—horses. Suddenly, nearly all of the men capable of bearing arms disappeared from the larger settlements. Finding native workers became a real chore. Having organized themselves into bands stationed in the bush, they systematically practiced war games such as simulated attacks on buildings. Orders were given in German, and the detachments were modeled after the German colonial forces—even old German uniforms were worn, especially military hats with a black-white-red cockade.

Experienced farmers and several missionaries warned the government of this potentially explosive situation. A three-man police station was set up in Otjimbingwe. In their own way, the natives began warning the German settlers. The village elder, Kawahere, for example, came to Frau v. D. from his distant village and asked permission to build his pontok close to her house and dwell there because, he said, she was alone. Another old farmer was warned, "Soon we'll tell you to run, and when we do you must run fast."

Native policemen were surly and insubordinate to their superiors. Not only were the natives who lived in Windhuk restless, but also those in the areas of Okahandja, the Waterberg, Omaruru, and Gobabis. As things progressed, farmers and their families fled from the east to Windhuk, crowding into hotels. Remembering the dangers and distress of 1904, all of us felt uneasy. The natives dreamed about their future. "Soon we'll be the bosses again and the Germans will herd our cattle, but the Boers—them we'll toss into the water."

One day my servant, Max, came to me, his eyes squinting furiously—which they always did when he had something unpleasant to report. Without a greeting, he asked, "Does Madam know that one cannot see any spoors if one walks across the cliffs into the mountains? The path to Windhuk is bad. There are many pubs. You shouldn't go there." Büdi and I paid attention to his message, which in decoded high German meant, "Beware, for soon it will begin. Beware of the pub below your house, for the natives will try to plunder the stock of alcohol from the hotel. If you hear a ruckus coming from there, flee into the mountains."

That same afternoon a German settler from the Windhuk valley came to me and encouraged us to follow him. Several settler families had built a fortified camp where they thought they would be able to hold out for a time. What was I to do? Should I leave my house to chance and quickly join them? There was no one in the vicinity who could possibly help us—even the owner of the small hotel below us had already left. After some heavy drinking, the natives would incite themselves to monstrous behavior. All this I knew only too well; the settler was right. I thanked him profusely for his neighborliness and told him I would give it some thought.

Once he was gone, Büdi and I discussed the situation. "What shall we do?" I asked him. "Wait," came his undaunted reply. Still, we did make certain preparations. We fed our beloved horse, Hans, a double ration of oats, then slapped him on his flanks. He knew he was not to remain in the stable and happily trotted along the canyon that led from the back of our house into the mountains. Hans would not permit himself to be caught by anyone but us. We also took care of our cats and dogs. Max bid us farewell, saying he wanted to check on us the next morning—to see if we were still there.

Now we were completely alone. We filled our knapsacks with provisions, filled a large container with water and a second one with coffee, rolled up two blankets, and placed the browning and the parabellum (a self-loading pistol) next to them. After dressing in khaki clothes, we slipped on our hiking boots. Soon it grew dark and not a single light burned in the valley below. As all the houses were dark, we likewise switched off our lights and sat on the steps at the foot of the front door and listened. Everything was quiet and remained quiet, unusually quiet, and our nerves were on edge. Suddenly, a broad beam of light rose in the southern sky, a beam that came from search lights situated atop a small hillock. Two native villages were fully illuminated. All this meant that the South Africans had been warned at the last moment and were now on their guard.

For a while we watched the flickering lights, then I suggested to Büdi that he lie down, that I would awaken him immediately if I saw anything suspicious. Two minutes later he was fast asleep. I walked around the house and our property, after which I also lay down on the bed, still fully dressed, and listened to the night. Owls screeched, bats twittered and chased one another, crickets chirped, dogs barked in the distance. Without interruption, the light beam shone into my window. It was reassuring to know that someone out there was keeping watch over us, that we had not been abandoned!

In the morning, as usual, Max arrived on time for work; he said nothing, but started his daily routine. Good old Hans stood in front of his stable, and we drove on to school.

During the night, though, Windhuk refused to slumber because of the ongoing excitement. Someone had revealed that the artillery depot would be the first target of attack. The South Africans were using it to store weapons and munitions. Immediately, the guard had been doubled. The evening proved very cold, and a large fire had been lit in the yard. Suddenly, loud noises were heard; frightened, the guards took up arms. As from which side the attack actually came, nobody knew, but the shootings and explosions continued. Unfortunately, the fire had been ignited exactly over the spot where our German troops had hastily buried a cache of ammunition.

It was impossible to protect the old building against an attack by several hundred natives, hence, the weapons and ammunition were removed under cover of night and set in transport vehicles and taken to the train yard where they were quietly loaded onto waiting freight cars. Cannons had been placed beside the searchlight. Surrounding the villages at a respectful distance, the police kept the natives under surveillance. Between seven o'clock in the evening and seven o'clock in the morning, no native was allowed to be on the street. In the school boarding house, the older boys had taken up arms and stood on guard.

So it went for several days until a group of natives appeared before the administrator. They acted as if they had been insulted and asked the administration why such unusual measures had been taken against them. They simply did not understand and were fully unaware of having given any cause for such action. I am not privy to what the administrator answered, but the fact is that shortly thereafter great numbers of new kirris and long, homemade knives were found, primarily in Herero pontoks. The plans for insurrection seemed to have been stopped in their tracks, though the South Africans called this native agitation completely harmless.

That it was not quite as harmless as they thought became apparent in the southern part of the colony where the natives were generally unhappy. Three Hottentots had murdered a Boer woman who had refused to give them the weapons they had demanded of her. The Bondelswart captain, Abraham Morris, known from the German-Hottentot uprising, had crossed illegally—and armed—from the Union into South West and was supposed to have been arrested at Warmbad by a police sergeant. About two hundred local Hottentots intervened, all of them armed, and prevented the arrest by the threat of overwhelming force. The commissioner for native affairs and the prefect were sent to negotiate. Armed Hottentots searched their car for weapons and accompanied them to the captain. The government demanded the surrender of the guilty persons; the Bondels captain demanded exemption from punishment as well as concessions on taxation (particularly the dog tax), cattle branding, and unemployment. The negotiations ended without agreement. Volunteers were recruited from Windhuk and other areas; even some Germans registered to demonstrate that the whites were unified when it came to fighting the blacks. Naturally, they had to overcome their own reservations about the South Africans that resulted from the conditions of occupation at that time. The German volunteers, however, were not accepted. Later, the content of an interview conducted by a journalist from the *Star* with the administrator of occupied South West caused regret. “A most disappointing fact of the uprising,” according to the piece, “is the fact that German settlers did not volunteer when volunteers were sought.” The administrator was strongly encouraged to correct this false accusation.

In the end, the South African administration did not want to enlist Germans to help calm a native uprising. The Herero, however, they did want, and they worked diligently to recruit two hundred Herero to help with the situation in the south. Naturally, in this effort they utterly failed as the Herero flatly refused the offer. Instead, two hundred men, both police and volunteers, along with their newly purchased horses, were sent south.

Some of the farmers living in the area of the insurrection received unofficial warnings, but most were not given any kind of official notice. Revolutionary movements tended to be particularly successful in these areas since farms are very large and separated from one another by considerable distances. Also, the distances between police and military stations were considerable, and as telephone service was unavailable, it was impossible to call for help when danger threatened. In the meantime, the Bondels had taken weapons from the commissioner for native affairs and slaughtered the cattle of the settler farmers in the area. In consequence of this, the local military unit increased to four hundred men plus the addition of two Union Air Force planes. The administrator (whose forces consisted of a platoon of policemen, three hundred volunteers equipped with three machine guns and two mountain cannons, a signal unit, and two aeroplanes) dressed himself in a general’s uniform and personally took charge of the operation. Three engagements took place in which three policemen were killed and several volunteers wounded. The Bondels reportedly sustained heavy losses. After five days of conflict, the administrator reported that the insurrection had been quelled.

When the Bondels insurrection was debated in the Union Parliament, it was noted that the Bondleswarts and Hottentots lived along the Orange River on a native reserve assigned to them by the Germans. Furthermore, it was stated that this reserve was much too large for their needs and the natives had no reason to complain at all. As the debate wound down, they reached the conclusion that Prime Minister Smuts should ask the administrator of South West Africa, for the time being, not to collect the dog tax in these areas. “The Germans did not enforce collection of a dog tax either.” At this point, the speaker, a Mr. Pearce, was informed that the latter point was out of order.

I have deliberately presented these facts in such detail and therefore feel that further commentary is unwarranted.

It is commonly believed that we Germans are incapable of dealing properly with the natives; however, even a quick read of the history of the Kaffir Wars in the Union of South Africa reveals not a little ineptitude. When high government officials arrive and enquire after the welfare of the dear, dear natives, they are deceived, for statements of devotion have been falsely prepared. During the visit of the Governor General, for example, the Okahandja Herero read the following statement: "We assure you that we are convinced we enjoy justice and all privileges under your beneficent rule. Our old chief lived in peace under German rule until the unfortunate war of 1904, and we want to put this old dispute behind us and to let it fade from memory—now that we live under the administration of the King of England and the government of the Union of South Africa. Respectfully, we request that you provide English lessons for our children. We are satisfied with the treatment we receive from the magistrate and believe that justice is exercised amongst us, whether black men or white men." The Windhuk Herero wrote: "We are delighted to be under British rule and assure your Royal Highness of our devotion." The Berg Damara stated: "We assure Your Majesty of our obendience to the government and to the King. We are proud to be his subjects." The Hottentots expressed themselves this way: "For us there is no life if we are not free."

A while later the Rehobother Bastars became unruly. They had expected a great reward for their disloyalty toward us Germans and were most unhappy when these rewards failed to materialize. Indeed, they were a gravely dissatisfied lot and felt oppressed by heavy taxes. At the very beginning of the occupation they were allowed to keep their weapons; they then hoped for additional privileges the new masters were supposed to grant them, but nothing happened. Instead of special consideration, they were given an ultimatum: "Become citizens of the Union, and you may keep your weapons. If you refuse to become naturalized citizens, you will be obliged to give up your weapons." Brotherliness was no longer discussed; instead, the police marched in and aeroplanes circled overhead. Disloyalty defeats its own master. (I find it noteworthy that the very same natives who made life difficult for us are now defying their new masters. In view of this, how can one continue with such twaddle that we Germans are incapable of handling the natives?)

6. The House on the Mountain

When peace was finally restored in the native villages, Büdi and I felt perfectly content on our mountain. Our day began very early. We worked hard at school and enjoyed it, yet every day about noon how happy we were to reach the top of the road connecting Windhuk proper with Klein-Windhuk.

From far off we could see our house situated on the hill beyond the river. It was too much to ride our bicycles every day as the climb up the mountain was overly strenuous in the midday heat. But we felt at home up there, and when we arrived we went to the outdoor evaporative cooling room that always held a container of nice cool tea with lemon—how this satisfied our driving thirst! We were not ready to eat directly at noon; instead, we first took a seat in our airy living room, holding our very large glasses. Only then did Max set the table. After a good rest, we took a refreshing shower and then started our work again. Only when it was absolutely necessary did we return to Windhuk, for we preferred a stroll in the mountain coolness and had many adventures there.

Most dangerous were our adventures with snakes. There were many snakes in the area, both poisonous and nonpoisonous, and the rocky outcroppings there were a veritable paradise for them. We easily recognized poisonous snakes by their three-pointed heads and very short tails. Most of the time we did not bother to discuss whether or not a snake was poisonous, but aimed the gun and shot. We watched a greenish tree snake share its habitat with large nests of social weavers. When it emerged nice and plump from having fed on so many birds, fate caught up with it. One time I was about to pick up a small branch to start a fire at dusk when something glided from my hand. It was a horn viper. Another time Büdi noticed the chickens were restless, so we illuminated the hen house with our flashlight and found a mamba—a very, very dangerous snake. Our large black-and-white tomcat, which was ordinarily very shy, lay down at my feet and rolled around. As I looked more closely I noticed a whitish, jellylike foam oozing from his mouth, his head surrounded by a puddle of foam. We touched him but could not determine what was wrong, yet instinctively knew he must have been bitten by a snake. The cat walked out of the house, convulsing, and we searched every room without spotting a thing. On our second look through, we thought the gray shelf on the wall appeared rather thick; upon closer inspection we saw a gray-black mamba stretched out upon it!

Another snake incident occurred on Boxing Day. As I arranged the pillows on the armchairs and the sofa, something hissed at me: in the corner of the sofa a spitting snake uncurled itself and rose up. Our dining room leads to the veranda. In front of the wooden door hangs a gauze door that prevents flies and mosquitos from flying in. The door had shrunk a bit and no longer closed tightly at the bottom. As I stood in the dining room, I noticed the door moving softly as a yellow-and-black-striped mamba slid across the doorstep and disappeared beneath a tall cabinet in the corner. The cabinet had a glass door, behind which I displayed my goblets. I called for Bitz. At first he thought I had made a mistake, but when we shined a light beneath the cabinet, directing it to the furthest corner, we saw the coiled snake. He wanted to shoot it, but I feared the detonation would break the large panes of glass, so I took a broom handle and tried to coax it out, but to no avail. The creature had slithered up the wall behind the cabinet and suddenly appeared before us hanging from the top of the cabinet, ready to glide down towards us. Immediately, we jumped back. As it hissed and flickered its tongue, Bitz shot it. We threw the thing onto the veranda and watched to see what our dachshund, Männe, would do with it. Unsuspectingly, the dog approached it on his crooked legs, sniffed at the snake, stopped, growled, and stretched himself until he became very long. (I had never imagined a dachshund could lengthen itself to that extent). He crouched and then seized the snake behind its head and shook it again and again until the spine cracked.

When we shot smaller snakes we took them to our meerkat (mongoose) while the snakes were still wiggling a bit. It was very interesting to witness the struggle that inevitably ensued. Sometimes the snake tightly curled itself around Tuckeli (our meerkat), but Tuckeli stayed calm and simply bit through the snake's neck, devouring the entire snake except for its head.

The wide range of lizards and geckos in South West are not only harmless, but even useful as they are great consumers of insects. The natives contend that the large, colorful, red-and turquoise-blue lizards—the ones that always nod their heads before dashing away—are poisonous, but I was never bitten by one. I like to keep geckos in the washhouse or the garage, but not in the house. If they fall from the ceiling, their tails break off easily yet they are still able to crawl off by themselves. Chameleons are easily tamed and kept in the house so long as they are misted once a day with water from a flower sprayer. They are ridiculously ugly with their protruding eyes that can turn in all directions. An odd-looking sack hangs beneath their lower jaws, from which their spirally tongues unroll and shoot out at lightning speed. They prefer to change their color to yellow or green (yellow indicates they are ill or annoyed). They can also turn turquoise blue or cover themselves with dark spots. Though I have tamed dozens of them, I have never seen a chameleon blush. They eat moths, beetles, and flies from my hand, but when they become angry they swell and pinch, but cannot bite.

The foothills of the rugged Eros Mountains stretch close to our house. During a drought, there is absolutely no water in those mountains. Early in the morning, well before daybreak, and then just past dusk, a troop of fifteen baboons appeared near our house. Normally these animals are quite shy, but their punishing thirst forced them to overcome their reticence. The very strong leader of the troop sat on the highest point of the rock behind our home and watched for any signs of danger. If he saw nothing suspicious, the older males approached and sat down in a semicircle around their boss. Females and babies stayed out of sight. The first time we showed ourselves, they began screaming and railing wildly, throwing stones at us. In single file, the troop quickly retreated with their leader bringing up the rear.

Baboons are very dangerous. Should they attack, a single bite can be fatal. A dog can be easily torn to pieces—hence we worried about our four brave terriers. As soon as we noticed baboons, we locked up the dogs, concealed ourselves, and watched the creatures as they sat on the boulders. When they felt safe, the old patriarch reassured his thirsty family in baboon language. Slowly, cautiously, they climbed down into the valley. Ringed by all the mature males were two mothers carrying babies on their backs. The patriarch arrived at the water trough and drank until his thirst was gone, his front paws set on the rim of the trough. The others waited at a respectful distance. Finally, he called out to them, probably telling them, "Come, it's safe to drink." Trusting their leader, the entire troop, loudly chattering, approached the trough and drank until their thirst was quenched. While they drank, the old patriarch looked about suspiciously, occasionally surveying the windows where we hid behind the curtains. The others were more trusting and spread out the horse manure near the trough, then pulled out small bushes and chewed on the grass. When the patriarch gave the signal, the entire company started to move out, this time, though, the mothers carrying babies were in front.

For weeks we enjoyed the baboon visits, but with the start of the rainy season they were able to find water closer to their home. Büdi had no desire to shoot them since they seemed somewhat like humans, but a few years later he did shoot a female baboon in order to save our dachshund, Männe, who had picked a fight with her during mating season.

It is most pleasant to live far off from the busy life of people. From our mountaintop we can see the blue, wavelike hills of the Khomas Hochland. Above them burns the setting sun, and it appears as if the landscape were dipped in bright purple. Then the play of tropical colors begins: the clouds glow as if on fire, then the entire sky turns into a flashing sea of flames; the evening star rises and begins to shine down on the hazy valley below us; above the eastern mountaintop the moon starts to rise and the trees form black silhouettes; like yellow and red spots, light begins to shine in the valley, and where we stand, everything is bathed in a silvery light. A car rattles in the distance, a shrill train whistle sounds from the station. Dogs are barking; large, black bats rise from the nearby cliffs and fly around us. An eagle owl silently descends and perches on the slender kasuarine tree. Little owls screech. Klippdassies (hyraxes) cry. Silently we sit on the veranda and enjoy the evening. Beside us lie the dogs, and our cats rub gently against us. The world grows darker and darker until the sea of stars shines with beguiling beauty.

During the rainy season we are often afraid. Storms rage all around our home, and it sometimes sounds as if the rocks are crashing down the mountainsides. The pale, blue flashes of lightning brighten and illuminate the country around us, though below us the land has disappeared. Winds have raised glimmering clouds of dust that hide houses and gardens. A massive wall of white approaches, and the first rain drops fall, warm and soft. They hiss on the hot stones and cause steam to rise. More drops fall and make bubbles, drumming on the roof. Then the rain pours down in thick, white strings. It penetrates the closed doors and windows, and pours from the veranda. Waterfalls cascade down the rocks and into the valley, ripping deep trenches into the ground, preparing the way for the roaring, muddy masses of water that rush downwards.

As long as we can still see the Auas Mountains to the south, there is no danger of a flash flood. However, when they turn white and then disappear from view so that there appears to be nothing but a vast plain, we glance at our clock. The thick, white curtain comes closer and closer, and the houses in the valley, as well as the natives' church, are no longer visible. Forty minutes later we begin to hear a rumble, then a dull roar, then a great rushing sound along with people screaming. Down roars the great torrent of water along the dry riverbed. The river is flowing!

All of us who live on this side of the river are cut off. Bridges do not exist. If one is fortunate enough to be home, then one is safe; otherwise, one must look for shelter on the other side. Such floods rarely occur in the morning, but usually between five o'clock and seven o'clock in the afternoon/evening. By morning, the waters have receded to the point that one can safely negotiate them.

As the rains generally begin midafternoon, the natives from the Klein-Windhuk village who work in Windhuk proper cannot get home. Sometimes the water is several meters deep, so one must be patient and wait for it to recede. If the floodwaters gather slowly, wading through them can be great fun. The natives take hold of one another to form a human chain to prevent themselves from being swept away. The black women who wear long dresses act as if they are ashamed to have to pull up their skirts. The apparently obligatory screaming never fails.

I have also had to wade through the water on several occasions. Often our servant, Max, met me to help me across. Once when I saw Büdi standing wet and cold on the other side of the river, I tied a heavy rope around a big tree on my side of the river, then tied Max to the other end of the rope so he could safely enter the water and fetch Büdi. When Max got to the other side, the two tied themselves up in the rope and entered the river as I pulled them across.

Just as quickly as the storms arise they also subside. Pepper trees emit a spicy aroma, and all the flowers and shrubs smell strongly, their leaves and stems become firm and tall, for after the long drought they are able to drink their fill. Everywhere, puddles gleam in the sunlight, and the receding river glimmers like a silver ribbon. All nature breathes in a balsamic freshness. Little animals come out of their hiding places and get to work. The red tin roofs have been washed clean and delight one's eyes. People stand and drink in the damp coolness with deep, thirsty gulps. Far in the distance there is still rumbling, and great flashes of light can be seen. Two rainbows span the horizon while the evening sky burns in indescribably fantastic colors.

Soon after the rain showers abate, the landscape changes. No longer are the mountains dead and bare but are covered from top to bottom with emerald-green vegetation. Like a carpet, yellow blossoms of morning star flowers shine between the green and dot the ground that was warped by drought. Fresh, tender grasses sprout among the stones and wave their silver tufts as the soft breeze plays with them. Small red, blue, and plain yet lovely flowers appear. Vines cover the bare rocks, hummingbirds hover beside the hollow, yellow blossoms that cover the wild tobacco bushes, water turtles whistle in the puddles, and frogs croak in the warm tropical night. The plump little frogs must be Catholics since they croak in a deep voice, "Papst, Papst, Papst" (Pope). The tiny frogs are Protestants as they trill in high notes, "Luther, Luther, Luther."

The lush, verdant splendor ushered in by the rainy season does not last long. Soon enough the grass begins to turn yellow, though the plains remain very beautiful. The rain clouds disappear and the sky radiates eternal blue.

But what is that over there on the far horizon? A low, grayish cloud covers the ground, thick black smoke rises from it—a bush fire! These have various origins and are easily started. It may have been careless natives who camped there for the night and failed to extinguish their fire. Ramblers or cattle herders might have recklessly tossed a glowing cigarette end into the dry grass or emptied a smoldering pipe on the ground—all that is needed is a single spark! It also happens that sparks shooting from a locomotive are coaxed and nurtured by a single tuft of grass. Then, along comes a lone gust of wind and the entire plain is on fire. Daylight is obscured by the heavy, black smoke; fiery snakes blaze up and down the mountains; a dried-up tree burns like a torch. Over there, flames race through the yellow grass, shrubs crackle and break apart, but the hungry flames race on and on—nothing can stop them. The wind roars, howls, moans, cries, sings, and drives the black clouds before it. As quickly as possible, animals flee the dreaded flames. Man, so tiny and helpless, nevertheless begins to fight the inferno, to save the valuable grass and trees on his farmland. He starts a backfire and, with the assistance of his farmworkers, extinguishes sparks with branches and wet burlap sacks.

To help extinguish fires on the mountains around Windhuk, natives from the village are called to help and transported in lorries towards the flames.

Deaths occur sometimes: just a few days ago a well-known farmer was killed by a fire. But how beautiful are these dangerous fires at night! One observes flickering flames here and there across the landscape, then, with lightning speed, a fiery snake races down the mountain; somewhere else a flame rises high into the sky. But soon enough the brightness subsides; now, glimmering tree trunks shine red through the dark night like the lights of a distant city. Often they appear to stand in a row like the windows of a lit train carriage.

Not only have I experienced such veld fires from the protected heights of my mountain home, but also on the farm, when everyone was mobilized to protect the area surrounding the farm house from the raging fire. In the bright light of those flames, the natives appeared to be dancing—giant, grotesque movements—with their branches and cloth sacks in hand as they jumped back and forth, bending low, then stretching their bodies as they threw themselves on the ground. Against the roar of the fire, their cries went unheard; yet, when they returned home, they were deathly exhausted, sometimes burnt and wounded.

Bush fires cannot always be brought under control. If the winds are unfavorable, the fire continues to burn and an ocean of flames rages for days or even weeks across mountains and valleys, leaving in its wake a blackened wasteland. Dark, lifeless trunks rise up, whose limbs and branches, now completely burnt to ashes, lie on the ground in the guise of long, gray snakes. All this death and destruction does, however, create new life. Shyly at first, fresh green shoots rise out of the ashes and game animals eagerly come to eat the tender sprouts, which, due to the ashes, have a slightly saline taste. The game attracted by new grasses is an "El Dorado" for the Bushmen, who harvest this rich bounty.

7. Life under Mandate Rule

While we no longer feared for our lives, we did fear for the continuation of our school. Again and again, its existence was drawn into question. There was hardly enough money to cover one month's worth of salaries, so we decided to hold a bazaar for the financial well-being of the school. This was a completely novel idea; a church bazaar had been held before, but now we were going to focus a similar event on improving the school's financial situation. Whether I wanted to or not, I had to leave my mountain refuge and descend to the valley for the afternoon. All of us worked long and hard to make everything very, very nice, and it was quite touching to see everyone participate. The boys glued, carved, and hammered. Two local businesses had donated two crates filled with toys—though slightly damaged during transport—that the boys lovingly and skillfully repaired. These would be put up for sale. The girls knitted, embroidered, and crocheted. We adults supervised, mentored, and worked alongside the children. Other businesses also made donations. Our grown children danced a minuet in rococo costumes—it was the first performance of this kind and was received with great enthusiasm.

Sales at the bazaar were booming, and a number of humorous things happened. Someone desperate to wash his hands did so in the pineapple bowl and therefore paid for its entire contents. Another person sat down on half a dozen glasses, but only his wallet was damaged from the accident. By midmorning we had sold eleven hundred tickets that brought in twenty-two thousand shillings. In addition, we auctioned off twenty rams, two oxen, two cows with their calves, two drakes, six pigeons, and a riding horse from the students' kraal. Two airguns and a copper pot were yet to be auctioned. The bazaar engendered a great community spirit and willingness to sacrifice for the good of the school. Its continuation for another year was assured, though the mandate government was likely not too pleased about that.

The relationship between the Germans and administration was by no means without its complications. Difficulties arose from every quarter. The first commercial German ship chartered by the Woermann Line to return to South West was the *Baltica*, and she was permitted to make four voyages a year, but only to Cape Town for "the renewal of commerce with Germany." It was stirring to see the first passengers who came here: worried, grief-stricken women with their children who had come of age during the war years. Men also waited to greet their brides who were very much delayed in their arrival.

The Catholic Mission in South West was forbidden to bring either priests or nurses to our country. We suspect the government's intention was to cause the Elisabeth House Maternity Hospital—our stork's nest—such serious problems that it would voluntarily close down. Due to lack of funding, it was in miserable condition, besides which nurses from Germany were no longer permitted in South West as a law had been passed that only permitted nurses who had successfully passed their examinations in South Africa to work in the Protectorate. The same type of change was mandated for pharmacists and physicians. German lawyers were not allowed to appear in court. German law firms had to accept either an English or a South African partner who would represent the firm before the bar. German lawyers were no longer allowed to plead a case; now an advocate had to do that—and the advocate's services were very expensive. The legal practice has truly changed and is now founded on Roman-Dutch law, practiced without a statute book so that a judge must consider various similar cases even if they took place centuries apart. On the basis of such cases, he makes his decision. There are two courts of law: the municipal court—where offenders are fined up to fifty pounds—rules on small infractions; the superior court is responsible for more serious matters.

In 1914, thriving farms with large herds of cattle were the rule of the day; there were many wealthy merchants; artisans and workers also had good incomes; mines and diamond production were booming; really poor people or people on the dole did not exist. During the war, the thriving business sector came to a standstill. After the surrender, vague attempts were made to shore up the suffering country. The reservists of the colonial forces had been dismissed, and farmers began reconstructing their destroyed farms, gathering their cattle, cleaning out the water holes, repairing the dams, and cultivating their neglected fields.

Merchants from the Union of South Africa, laden with their wares, arrived in larger towns. Our German stores gradually changed over to the same types of goods, and ever so slowly, business began to pick up. Artisans and workers had been laid off their work, our money had been devalued, and the Seitz banknotes had been declared null and void immediately after the surrender. German imperial money was accepted as payment, though its exchange rate was heavily devalued. In the end, however, German banknotes and coins were no longer accepted as legal tender.

The political expulsions began and resulted in the loss of nearly half of the Germans in the Protectorate. Those of us who remained were in serious economic trouble because, among other things, the mandate government confiscated the German Agricultural Bank as a spoil of war. Many of our farmers had outstanding loans from that bank, and since German money no longer had any value, the debts had to be paid in South African pounds or in gold. A court of arbitration was arranged that attempted to mitigate these hardships, but these hardships were, to some extent, caused by the artificially unfavorable exchange rate.

Besides our two local German lawyers, many others arrived from the Union, wanting to take advantage of the local confusion. In November 1921, thirty-two lawyers were practicing in Windhuk alone.

The mandate government ruthlessly proceeded with the forced auction and sale of property; the result of this ill-advised course of action was that many German settlers lost their land and houses. On July 26, 1923, the one hundred twentieth German farm was forcibly sold at auction. I am not personally aware of any voluntary auctions or sales. The government requested certain information about farms that were offered voluntarily for sale on the open market and sent out questionnaires. So that the questions could more easily be answered, examples of the types of information sought by each question were included. For instance, for the question of who owned the farm and what exactly was being farmed, the following sample answers were given: "Wilhelm, German Emperor, Cultural Farmer; Michael, Small-hold Settler." A high-ranking official had signed this example by mistake. Consequently, a rather embarrassing complaint about this awkward and insulting prank was sent privately to the head of mandate governance in Pretoria.

Farmers desperately fought against their own ruination, and cattle traders seized the opportunity to hold cattle auctions, with oxen, cows, and heifers bringing an average price of thirty shillings a head, calves three to forty-five shillings, and goats and sheep one to three shillings each. Those who had not lost all their courage attempted to sell their cattle in the Union, though transport costs were so high that farmers who did so actually ended up making even less money. Besides this, the care the animals received during the long, hot journey was miserable: woe unto the cattle shipped out on the weekend, for Sunday was a day of rest in accordance with the Bible!

Farmers appealed to the government for the reduction of freight costs, and in May 1922 the tariff was "reduced" by 2.4 pounds to a figure of 2.8 pounds. Further appeals for reduction were declined even though the freight cars bringing coal for the railroad returned mostly empty. Two Germans, Herr D. and Herr B., traveled to Angola seeking markets for our cattle, but it turned out that cattle prices were even lower in Angola than in South West. Good markets existed in the Belgian Congo, but cattle simply could not be driven or transported through Angola. Besides, in Angola inoculating cattle against diseases was virtually unknown, hence, widespread cattle epidemics are common there; indeed, public health conditions generally leave much to be desired. Thus, South Africa was really the only possible market.

In the Grootfontein district, thousands and thousands of hundred-pound bags of maize lay stockpiled and unsold. Because the Union government subsidizes the cost of producing maize, South African maize is cheaper than what is locally produced. Our natives were aware of the farmers' plight and understood that farmers could not give much in cash payments, so they accepted livestock as payment. In general, natives only worked for farmers who allowed them to raise their own animals.

As an example of this, our friends, the Family D., lived near Okasise and before the war owned an ice-making machine and deep freezer for milk, cream, and butter making and storage. These products were shipped in their private deep-freeze railroad car to Swakopmund, and a large section of Swakopmund's population was supplied in this manner. The Family D. owned a car and a beautiful wagon drawn by four horses. Their entire farm resembled a German manor. And now, what has become of all their hard work? The machines are rusting away, the wagons are broken, and the horses, dead. The projected new house is still in a planning stage while the old house is in a state of ruin. Termites have eaten through the wooden ceiling and the roof beams. A third of the ceiling has collapsed onto the living room floor; everywhere else it is cracked. The doorframes have been ravaged by termites; tin or wood has been used to replace windowpanes. Because of their pigs, thorn bushes have been planted around the veranda entrance, and the gate has become useless. Tin cups sit on the table next to the old silver ones. Their porridge: maize meal mush. Money? How can one possibly earn money? Their herdsman left because he was *only* allowed to keep four thousand goats and sheep on the farm for free! The farm was almost auctioned off several times. Poor Herr D. died of a ruptured liver; his widow struggles to rear their four children by herself. They have suffered want, starvation, and slavery, but they are still on their own land. Two sons are now grown, and things are beginning to improve.

Not everyone is able to fight against decline with such tenacity. Many were exhausted from the struggle and returned to Germany shattered, penniless, and without hope. Others have become indifferent, practically giving away their animals and allowing their gardens to run down. Their farms lie as wasteland. One man slit his wrists; another man hanged himself from his favorite tree, leaving a note in his pocket: “Count me out!” Then, there was the man who took poison and wrote: “Bury me in my cupboard.” Others simply shot themselves. Gone! Finished! The end!

Yet in their place, arriving on every train from the Union, are the new “Voortrekkers.” These Afrikaners—most of whom are decidedly uncultured and uneducated bumpkins—are being systematically settled by the government on the farms appropriated from their German owners. Many of these Afrikaners had little or no capital at all but received the most beautiful farms on easy terms and subsidies of every kind. This was an expensive move and created a huge debt for our country; the Union, however, now has its *voters* in place. Some of these “Voortrekkers” were the “armen Blanken” (poor whites), descendants of the totally impoverished Boers who constitute but a troublesome lumpen proletariat in the Union. Lacking in even the most rudimentary education, though refusing to do the type of work normally done by natives, the government has transplanted them here simply to get rid of them. Special privileges are granted them, but they are incompetent farmers and their subsidies are immediately squandered. Now they are a burden on our country and, not surprisingly, according to a representative of the mandate government, more than half of the loans granted to these “Applikatie-Farmers” were never repaid. Conscientious Union officials repeatedly warned of such an outcome, but their warnings fell on deaf ears.

The situation of the Angolan Trek-Boers, or the “noble blood,” was similar. In 1876, they left the Transvaal province of South Africa and trekked through South West en route to Angola, where they settled. A few years ago, they decided they no longer liked it there and wanted to return to their old homeland. Mr. Werth, the administrator at the time, determined not to let them return to their homeland but to settle them in South West. They were given the so-called Angola-blocks, everything they needed to resettle in South West. By April 1931, not only had it cost the government thirteen million German marks to bring them here and set them up, other additional expenditures—huge expenditures—were wasted on them. Now it seems these people do not like it here and live under great economic and financial difficulty. As the promises offered them were never kept, they feel betrayed. Their situation in South West has grown worse than it was in Angola, and now they wish to return. Some have left everything behind and simply wandered off. Others swelled the ranks of unemployed persons who must be supported. The Angola schools, dormitories, and teachers cause heavy additional expenses and are a drain on the educational budget. "Why in the world were they invited to come here?" people ask. "Additional votes for the Union," is the response. As subjects of the Union of South Africa, they are allowed to vote after one year's residence.

In contrast, Germans who applied for farms, even those who had some finances, were considered undesirable settlers and their offers declined. We only gain the right to vote after five years' residence. Since no new German settlers have arrived, the influx of German money has ceased, and the effects of this have been felt.

It appears, too, that the new government needed additional natives since they arrived in droves from South Africa. Cape boys, Basutos, Zulus, and others came, all of them seeking work here. Our own natives have lost their jobs to these newcomers. Several whites have hired native drivers, leaving their well-trained white drivers without work; now these latter men labor on road works to keep themselves from starving to death.

At first the newly arrived South African settlers worked and gained a measure of success. They bought their livestock cheaply (in fact, the government gave them the money), the German equipment was still in good shape, and game animals had been preserved by their predecessors. As they had brought their extended families with them—all these *Beiwohner* (live-ins)—they had few work-related expenses. Live-ins are allowed to keep a certain number of livestock of their own, but in turn for this they must do the work that is otherwise performed by natives. Besides which, these Afrikaner Boers usually have a great many children who never go to school and are thus free to help with farm chores. Only recently has the government applied pressure to force them to send their children to school. However, since they live far away from the larger towns where schools are located, and since they have no money to pay for a spot at a boarding house, the government has no choice but to let them live in the boarding house at a substantially reduced price or simply gratis. This explains the unusually high maintenance costs for schools and boarding houses in this country; it also requires the rest of us to pay for them!

The government has allowed the land to be so densely settled that the land set aside for future farms by the German government is no longer sufficient. During German rule, farms were distributed in such a way that a piece of government land lay between three farms, and this land was to remain vacant so that during times of drought the farmers could use the land to prevent their livestock from dying of starvation. By following this logic, even in years when rain was sparse, farmers were able to maintain their livestock. The mandate government, on the other hand, gave away these stress reserves so that during the terrible seven years of drought we recently endured, animals perished by the thousands.

And how is the situation now? The agricultural bank has taken back countless farms and offers them for sale "because the owner left the country." And what do these abandoned farms look like? Houses are dilapidated; wells, kraals, fences, gardens, and farming equipment are in complete ruin. We Germans had kept everything in perfect condition. The farm my son took over from the land bank is in shockingly poor condition, yet how well do I remember how this farm used to look. The current condition of the house is simply not to be believed! Cracks in the exterior walls had not even been mended during the rainy season! The door and window frames are loose, and the windows are nailed shut with sheets of tin. In one room there is a space in the wall for a door, but no door fits that space. Fires had been made on the floor. Grain and hay were carelessly stored in an old bedroom; the kitchen and pantry are simply beyond anyone's ability to describe! So much litter and filth has accumulated all around the house that the previous owner, who had leased the farm for a short period of time, had to cart off several ox-drawn freight wagonloads of it. And this farm is hardly an isolated case. Indeed, we visited many abandoned farms and found all of them in the same dreadful condition. Farms left behind by the Boers may be purchased, but as various charges and interest payments not settled by the previous owner have been tacked on to the purchase price, they are very expensive.

Fortunately, the mandate government has recognized the distress of all the farmers and has tried to help, but it is a case of too little, too late. Dairies were established in various larger towns, which led to the necessity of transporting all this butter to the Union. Each year, about one million pounds of butter is exported to South Africa. The profits from this are minimal; they do help some people, but it is not enough. Cattle and sheep to be slaughtered and sold in the Union are purchased in large quantities by cattle merchants. Yet already countervoices in the South African parliament are to be heard. "The economy of South West threatens the welfare of the Union," was a headline in Cape Town on May 6, 1936; and just yesterday afternoon, the subject of the debate in the People's Council was a proposal by Representative Gilson (Ver. Party) for the government to take effective measures to liberate Union farmers from the burdens imposed on them. These burdens result from the competition arising from the unrestricted import of prime agricultural products from South West Africa and the sale of these in the Union. Things will only get worse if this unfair competition is allowed to continue, Gilson asserted. He also demanded that the premiums paid for the export of South West African production surpluses should be covered by the general income funds of the Union, and not the duty funds that are mainly paid by Union farmers. Gilson further reiterated that the products of South West Africa have become a very serious threat to the well-being of Union farmers.

In his reply, the agricultural minister, Mr. Reitz, urged the House to reject this proposal. He stated that it would result in starvation and ruin in South West. Payment of the butter export premium is the law of the land, and South West Africa has a right to receive this premium since it pays the same duty on butter as residents of the Union. "The people of South West Africa are our own flesh and blood, and the government must not agree to a policy that places them in a disadvantageous position," he stated. He stressed that the measures requested in the proposal would only lead to bitterness and discontent, pitting, as it were, the people of the Union against their fellow citizens in South West. Furthermore, these measures would do nothing to ameliorate the difficulties of South African dairies, and besides this, the Union had a moral obligation not to treat South West Africans as Gilson proposed. A counterproposal to return to the order of the day and not to consider Gilson's proposal was accepted by 47 votes to a mere 24. Thereupon the house adjourned. (Is this not food for thought?)

Agriculture has always been seriously neglected in this country, and the Union's unwise land policy has led to a degradation of the land. Besides the farmers, many others were affected by economic depression. Indeed, depression seemed to have encompassed the entire land. Many of our merchants went bankrupt, and white-collar employees lost their jobs. Though everyone economized as much as possible and purchased only the necessities, this led to a reduction in sales and income for merchants and employees. Government officials and those who worked in banks—people who could afford to buy things—shopped mostly in the local South African–owned stores and also imported some of their goods from the Union. They were paid in money drawn from local taxpayers and customers, yet through their actions this money did not remain in the country but left for South Africa. Since German stores also traded mostly goods from South Africa, their money was also forwarded there.

Hotels stood empty; and as no one was able to afford his beloved evening drink, public houses had little traffic. When farmers came to town, they no longer regarded their visits as recreational in the least. They shortened their stays and left their wives at home. Unlike former times, many stayed outside of Windhuk for the evening and hurriedly took care of their business the next day so they could leave town before sundown. Families were torn apart by all of this. Men somehow managed to scratch out a living; women looked for any kind of work to help pay for food and shelter, happy to be able to provide some support. A fortunate few were able to find work as typists; others sat around and ate nothing but maize meal porridge.

People went to the Administrator General of South West and asked for help. The administrator promised that construction of a railroad line from Windhuk to Gobabis would soon begin. Hope rose like a star. "Railroad construction will bring money into the country!" they thought. Construction began with five men who had to provide their own equipment: several wheelbarrows and two pickaxes. Later on, when the railroad construction moved forward in earnest, relevant local businesses hoped for income by providing materials. As it turned out, the mandate government intended to have bridges and tunnels constructed in a manner that had proven entirely unsuitable in previous railroad construction ventures in the Protectorate. In addition to this, local products such as lime and stone were not used at all; instead, all construction materials were imported from the Union. The Windhuk Chamber of Commerce vigorously objected to such action.

The hard times only grew worse. Craftsmen had little, if anything, to do and their tools lay idle. Because no construction was undertaken, our bricklayers were on holiday. And when public works projects were undertaken, the government hired mostly South African workers (who had come in droves to the "promised land"). The Union also suffered hard times. Ever-increasing unemployment, coupled with the lack of any kind of social welfare system, created a state of affairs that defies my ability to describe it. Evidence collected by employers allows glimpses into the frightening depth and extent of human misery. Although it is simply impossible to comprehend, the following is certain: Before the mandate government took over, no such conditions existed among us. Under their watchful eye, however, neither health and accident insurance nor written regulations for free hospitalization came into existence. Instead, the administration steadfastly refused to introduce such measures.

Finally, the pressure created by these dreadful conditions waxed so strong that work programs were begun. The Avis Dam was constructed, and roads throughout the land were repaired. Men of all social and economic classes were numbered among the ranks of construction workers. Merchants, craftsmen, farm supervisors, academicians, former officers, failed farmers, old people and young, were mixed together in crews. For very little renumeration, they performed the kind of hard physical labor in the biting dust, searing sun, and heat that had previously been done by natives. They received meager food rations—often less appetizing than what natives eat—and were allowed to take a short midday rest under a bush, though this was not sufficient to refresh their tired bodies. When work resumed, they suffered from thirst due to a lack of water. In the evening, completely spent, they fell onto a bed of grass under an open sky. Their calloused hands and leaden limbs were wracked with pain. When morning came, the drudgery started all over again. These men became poor, truly destitute and hopeless, in a way they had never before experienced. Ironically, among these laborers were many who had dedicated their whole lives to this country; they had loyally fought and worked for South West, and now, this was their lot in life.

The state institution that extended credit likewise contributed to our economic decline as it made accumulating debt a simple matter for just about anyone. Sometimes credit was practically forced on a person—especially for motor vehicles that could easily have been paid off after a small down payment. The result was that many gave way to the temptation to purchase a vehicle. One could, of course, save considerable time and no longer needed to spend days or even weeks on the road driving a heavy ox cart. Farm products could be delivered much more quickly and in a much fresher condition. This not only meant less time away from work on the farm, but also no need to employ an extra worker. Additionally, since droughts could last for years and the farmlands adjacent to the roadways were completely devoid of grasses, there was no need to worry about finding ox fodder along the roads. Whether working or not, an ox needs food; a motorcar, on the other hand, only requires petrol when it is in use. These were the arguments in favor of owning a car. But what about money to pay for the car, for petrol, for oil, and for spare parts? All of this money was sent abroad. It is estimated that every fifth white man owns a vehicle here now.

What happens when one is no longer able to pay one's creditors? In such cases money was collected with the utmost severity. Debtors were compelled to pay in livestock, which was gladly accepted at well under market value. Finally, if nothing could be offered as payment, the vehicle was repossessed. I know of cases when cars had been two-thirds or even three-quarters paid off, yet were repossessed.

For years and years we were exploited with high protective tariffs forced upon us merely to further the interests of industries in our neighboring country. And what benefits did we enjoy from these tariffs? Only a considerably higher cost of living! We were burdened with ridiculously high taxes to support these poor lumpen South African settlers so that they might have a livelihood and be able to educate their children. Certainly these obligations were not ours but for the residents of the Union to bear. Indeed, by forcing us to pick up the tab for the dispossessed of South Africa, the Union itself got off virtually scot-free. Citizens of the Union now residing in South West earned their living working on the railroad, as civil servants, in public works, as police officers, and as officials. We were forced to pay their wages while our own unemployed barely survived and became a burden on the country. A large number of Union citizens were fed on the interest from our debts and borrowed money. Furthermore, our own unemployed led to a new and additional debt: in 1934, South West Africa as a whole was indebted to its "guardian" in the amount of forty million pounds.

How can we fight against the worldwide depression under these conditions? For us, South West Africa is a hard country full of contrasts and contradictions, obstructions and opposition. We have always faced resistance and battles among our peoples and against Mother Nature. This seems to be our fate. Yet all our strength is needed to work hand in hand and to push forward as a united community if we are to better our existence in this hard, bitter land!

For many years, South West Africa was ignored by the colonizing nations of the world. It lay hidden behind wide bands of desert and its cruel, obstinate nature. It was a poor country, resistant to the penetration of European civilization, and remained the land of Bushmen, black Africans, and Hottentot hunters. Very late, well after the spoils of the world had been divided among the civilized nations, Germans started to battle this inhospitable land, never shrinking from the sacrifice of their possessions, indeed, their own blood, in order to develop this land into a home for German settlers. At that time, a Strong One held His hand above the land. From the sacrificial seeds, blessings began to sprout, blessings unexpected from the dead sands of the desert and from the lap of the earth to the north.

Ascent and decline, abundance and bitter need were in constant rotation, but this is the will of this country's nature: It is our destiny and it will remain so. The strong, guiding hand that raised us up as part of his own body was suddenly taken from us. Under the supervision of our Guardian, we were to walk our stony path road alone; from our rich inheritance, we were to pay for our upbringing. Along the way we had been given a rich endowment: diamonds and copper provided us a generous income that we should hold in reserve. But our Guardian had many children both rich and poor. He was willing to care for the poor ones, and the poor were to help the rich develop the new country, to gain their livelihood from the land, and, in the process, to settle it. The income generated from the new land was sufficient for that, but as more and more of the Guardian's poor flocked to the country they eventually became far more numerous than those who had arrived before them. They ruled and governed the orphan country. As its income gradually decreased, the Guardian lent the country money as long as it was deemed worthy to receive it. When the population fell on hard times, a bitterness gripping the entire world, they finally came to realize they could no longer borrow money—all their possessions belonged to the pawnshop. They stopped and asked themselves, "What shall we do now?"—the same question a leading German directed to the population as a whole.

The latter half of 1922 was a difficult time for Büdi and me. Büdi conscientiously prepared for his examinations, and my days and evenings were spent in helping him with his revisions. We were delighted when notification arrived that he had passed his exams and immediately began to prepare for a trip to Germany. Because it was both inconvenient and ridiculously expensive, we ruled out storing our furniture in some kind of storage shed, yet I had no wish to pay rent or appoint an overseer for a house that would stand empty for half a year. No one wanted to rent the house furnished, and besides that, for most people it was situated a bit too far from the city. Kindly, the estate manager of the now-deceased owner allowed us to use the three basement rooms for storing all our belongings. Two hours before our train's departure, we finished cementing the basement door so that it was tightly and securely closed. Air entered through the cracks in the walls and circulated in the rooms.

Christmas 1922 was celebrated with my parents in Berlin. Many things had changed since our removal from Germany. The economy was plagued by inflation; many relatives had passed away; and after an absence of nine years, coupled with the difficulty of the times, I found that we, Germany and I, had become estranged from each other. The Fatherland lay in bondage—I no longer knew it and could not orient myself anymore. My parents lived close to the Kurfürstendamm in the Joachim-Friedrich-Strasse. I disliked going out on the street. The Kurfürstendamm now possessed a distinctly Eastern flavor with all the Slavic and Jewish faces and people of all kinds who had taken it over; it was no longer my Germany, and it made me quite unhappy. When my father retired, I purchased a small house in the Harz Mountains where my dear old parents could spend the rest of their days. As I write this, my parents are now dead and my house has been stolen from me—as well as many other possessions of mine.

I wanted very much for my two sons to be together. My elder son had a job in Hamburg, so I took the younger one there and secured an apprenticeship in a bank for him—good training for a position promised to him in a South African bank when he returned to Windhuk. All the misery and hardship in Germany depressed me so, and I was further weighed down by a heavy heart because I was leaving my two boys, and knowing full well I would never see my parents again, I, a now lonely traveler, made my way back to South West. I had left my previous employment at a secondary school and had accepted a position at the new government school. I knew what was expected of me, and I was determined to do my duty exceptionally well. I did so for twelve and a half years.

8. Our German Schools

For twenty-one years I have served as a teacher in the schools of South West Africa, from September 1914 to December 1935, and I therefore feel I have earned the right to clearly present my point of view. In particular, I wish to counter the view of those persons who were here merely as visitors. They tell all kinds of one-sided tales and feel entitled to write long articles about our educational system. Not only are many of these articles false, they are also malicious, and therefore give German readers a mistaken impression. Furthermore, I consider those who served here as guest teachers as nothing less than incompetent—despite the fact that they consider themselves qualified in every wit to critique our educational system.

When I began teaching at the Imperial Secondary School in 1914, there were seventeen government schools in the entire country, some of which taught pupils of widely different grade levels. In Windhuk, the Imperial Secondary School was still in a developmental stage. At this school, boys and girls attended together, and at the end of 1914, the first three pupils passed their intermediate examinations. Swakopmund had a City Secondary School that offered four grade levels. The Catholic Mission in Windhuk was responsible for an upper-level girls' school. In Aroab, an area primarily settled by Boers, there was a private Boer school subsidized by the German government.

During the war (World War I) many of these schools were closed; epidemics of diphtheria, typhus, and Spanish flu also hampered their operation. When the Mandate regime took over the operation of our schools, the attempt was made to suppress German schools altogether, hence, the reason why teachers' names were placed on the deportation list. But we resisted the closing of our schools with an iron will. German school associations were founded everywhere in the country and later combined to form the National School Association. Teachers loyally supported these associations—even at great personal sacrifice. Many grueling battles had to be won, and in relating my own experiences, I have pointed out these facts on various occasions.

In 1916, the Imperial Secondary School was taken over by an English language school. We were allowed to keep our books and benches, but in place of our building, we were assigned the girls' boarding school that stood on Twentieth Avenue—on the other side of the river! This school had just been completed, but the move from one building to the other was unbelievably taxing. No wagons, no helpers, native or white, were provided us: we were simply evicted. So all of us, teachers and pupils alike, helped in the move. While the children enjoyed carrying things up and down the hilly streets, for the rest of us it was hard work. By the time the sun had set, all of us were dirty and tired, but school was ready and we began instruction the very next day. We did not remain at this location for very long. In 1918, "we were evicted from there as well and made to occupy another house," as our school song of that time amply testified. This quarter happened to be the east wing of the old German grammar school—the national museum and city library had been moved from this location. To help with finances, we combined the grammar and secondary schools, with the grammar school becoming the preparation for entering the secondary school, and added an upper class for those pupils who wished not to transfer to the secondary school. Ten German teachers taught these ten classes.

Since the Mandate Government wanted a complete discontinuance of German schools, we had many nights of heated, often bitter, discussion with the administration, but the German population argued tenaciously for the continued existence of our schools. The decrees of the first administrator, Mr. Gorges, were met with a storm of indignation. Among his ill-conceived decrees were the following:

1. The government will take possession of all German schools and boarding houses.
2. All teachers will be repatriated.
3. Replacements may only be hired from the Union.
4. German parents may choose English or Dutch, which will be introduced in first grade and from then on required in all future grades.
5. Starting in 1924, German will be regarded as a foreign language.
6. Private schools may be established, and these decrees will not be enforced in private schools.

By petitioning the administration, teachers were eventually allowed to stay on for an additional six months. Of course, all the German schools became private schools, and, by necessity, the teachers were now paid by the various school associations—a provision that increased an already significant financial burden on the German population. After General Smuts's visit to South West, the administration proposed to establish a German grammar school funded from government coffers. German lesson plans and the German language were permitted up to standard VI (U 111). The mandate government further agreed to transfer and employ German teachers who were already teaching at the time; tuition fees would not be required. Under this proposal, there would be only one private German secondary school in the entire country, and it would not be publicly but privately funded by the German population of South West. The school would have four grades so that our children could easily transfer to a school in Germany for further training. We accepted this proposal, and its acceptance marked the beginning of government-sponsored German language schools.

The Mandate government terminated the contract that allowed us to use the private secondary school building on Schulstrasse. Once again we had no idea where to go, but we found a place in the City Exhibition Hall that had been constructed in 1914 to hold the country fair. In the meantime it had served a variety of purposes such as a warehouse for animal feed as well as a stable. Now it was to become our secondary school. With great enthusiasm, though little financing, we began to furnish our new home. Unfortunately, all windows were set high on the east side of the building to allow the morning sun to shine through into the rooms, but the light and the warmth proved excessive. I canvassed the neighborhood for suitable material and was given muslin, yarn, and a variety of trim, and since everything had to be done in short order, I sewed the curtains on my own machine that evening. The director and I then hung the curtains. The severe underfunding meant that we had to economize wherever possible. We could not afford a janitor, so the girls and I scrubbed the floors and cleaned the classrooms while the boys, under the supervision of a teacher, kept the schoolyard clean.

From these humble beginnings in November 1922, I participated in the founding of today's very large university preparatory school, along with a feeder school. It is now fully organized and prepares students for the Abitur (the entrance examination for German universities). Plans are in place for students who so desire to take the English-emphasis curriculum in our school, though up to now no one has opted for it. In all, about two hundred and fifty students are enrolled. We do not yet require entrance examinations, and every applicant is accepted. The school receives a good deal of publicity, and we have made repeated attempts to persuade parents to take their children out of the government school, the "Boer's School" as we pejoratively called it, and send them to ours so they might be prepared for university and grow up to be good Germans.

Every year we received generous financial support from Germany that enabled us to enlarge the school bit by bit. A room for keeping educational materials, a library containing six thousand books, a staff room, a second building, a physics and a chemistry lab, an auditorium, an art room, and a workshop have been added over the years. The workshop was donated by the Kòsener S. C. In addition to these monies, the school received many donations in the form of furniture, books, wall decorations, teaching materials, etc., and eventually, a boys' dormitory, a girls' dormitory (called the Hedwig von Bredow Haus), and the necessities for a home economics program were added to the school. The German Colonial Society, a women's organization, takes responsibility for the care and maintenance of the buildings.

During the first two years of operation, the Mandate government supported the school on the pound for pound system, and continued to do so for an additional five years. An annual allotment of three hundred pounds was made available for the maintenance of the dormitories. Beginning in 1928, the institution received a subsidy of five pounds for every student between the ages of seven and fifteen. In 1932, the subsidy was modified so that only the first fifty children were subsidized at five pounds each, the second fifty at four pounds each, and any additional pupils at three pounds per pupil. Thus, the oft-heard assertion that the university preparatory school had to be self-supporting is not entirely correct. The government must approve of the teachers we hire and is responsible for an annual inspection. African history, geography, English, and mathematics are prerequisites—the same requirements of the German sections in government schools!

In Windhuk there is a German school for university-bound pupils, as well as a Reform-Gymnasium (university preparation school). There is also an advanced girls school run by the Catholic Mission in Windhuk, whose English language division leads to matriculation, as well as the German schools in Lüderitz, Kolmannskuppe, Karibib, and a few small farm schools.

In 1930, the government took over the private secondary school in Swakopmund and converted it into a Reform-Gymnasium. According to the curriculum in U III, grade school matriculation is feasible. Advancement to O II certifies—without an examination—that the student in possession of a Junior Certificate has passed all equivalent levels of other schools. This certificate also allows the student access to certain government positions. Matriculation follows the completion of U I and entitles a student entry to a South African university, as well as access to higher positions in the government. One year following matriculation, a student may take the German Abitur—which almost all of our students have done. A university graduate with an O I rating is paid an average salary of five hundred fifty pounds per annum by the Swakopmund School Association.

In the government high school, with an enrollment of two hundred and seventy pupils, teachers from Germany teach under the supervision of a German director. South Africans only give English and Afrikaans instruction. Achievement in the sciences is far above the average in this school. Though the German boarding schools are completely supervised by Germans, the government pays the entire cost of their upkeep and maintenance (at present there are 270 pupils enrolled). The City of Swakopmund School Association maintains an additional German boarding school, and besides this there is a secondary school for non-German children as well as Hofmeyer Hostel (also a boarding school). There are twelve German departments in regular government schools. From 1923 until 1935 I taught at the local German government school in Windhuk.

My school has been singled out repeatedly by ignorant persons who characterize it as "Boer infested" and declare its teachers to be "renegades," in absurd articles that appear in various German periodicals. "Boer infested" is only a hurtful insult directed at us, but one that also demeans the South Africans. More than half of all white South Africans are of German descent, and they cultivate their German heritage just as we do here. What is the purpose of this demeaning term? I have never heard South Africans use the term *German-infested*, though many of them speak excellent German. This entirely senseless phrase is used in these articles along with the term *Black infested*. The authors do not realize that they cause more harm than good to our German heritage by the use of such terminology. At the same time, their tactless behavior arouses a completely unnecessary bitterness directed at us and other nationals. What on earth is the purpose of this groundless and malicious agitation? They insult us as teachers, they insult our students, and they insult the parents who entrust their children to our care. They also insult the nation where we Germans are compelled to live side by side with South Africans.

These agitators also deliberately lie when they assert that our German children are taught together with children who speak other languages. Of course the Mandate Government initially made the attempt to mix our children with those from other nations, but these attempts failed. By June 1921, the administration had publicly proclaimed it would no longer require the mixing of children. Why is this proclamation never mentioned by these writers who supposedly care so deeply for German culture?

Our governmental school was attended exclusively by one hundred and eighty-five children of pure German extraction. Their parents belong to various social and economic classes. Their fathers are farmers, craftsmen, merchants, physicians, lawyers, and former German officers. Their religious orientation is Lutheran or Catholic, though some belong to the Apostolic sect, and the parents of other children still adhere to the German Belief in God (deutscher Gottesglauben) even if their children do participate in the formal religious instruction given at our school. Our students only speak German and are deliberately German; in no conceivable way are they less educated than those in the secondary school.

The school teaches eight grades, standard VI (Untertertia), and follows the instruction manual of the higher German governmental schools in Swakopmund. Strict attention is given to the school's guarantee that our pupils will graduate with a public school certificate. After successfully completing their education with us, the pupils have several options: they may leave and enter the workforce, continue their education at the high school in Swakopmund, enter Windhuk High School, or attend the local school that prepares them for entrance to a university. Seven of the teachers at our school were educated in Germany, tuition fees are not required, and the Department of Education supplies books and writing materials at reduced prices. For German instruction, we use the primer (Fibel) *Kinderland* and *Deutsches Erbe*, and for grammar, we use the books by Jahn-Witzke. For history, Kummsteller's text is used, for natural science and physics, Schmeil, for geography, Harms, and for religion, *Glaube und Werke* is the standard text. The grammar books for English instruction, Dinkler's *Learning English*, are ordered from Germany. Mathematics texts have been commissioned by the department and compiled by German teachers, and are also in use at the University Preparatory School in Windhuk. Besides these readers, *Lesehefte für deutsche Schulen in Südwest* are used from standard III (Sexta) to standard VI (Untertertia). German teachers have assembled these readers from reports, travel descriptions, and stories by German-African authors, and they depict various aspects of our schoolchildren's second home, South West.

Naturally, we must give particular attention to the second language here, English. Initially, when South African teachers taught this subject they had little success, but this stemmed from the fact that none of them spoke German and all were quite unable to give proper explanations to the children. Beginning in 1924 and continuing until December 1935, I taught all English classes. Instruction begins in the third grade, and the children receive three hours per week of English lessons. After the children overcome the initial fear of a new language, they participate wholeheartedly. In the other grades, English is taught six or seven hours per week, and besides the German-English grammar books, we use the English readers, *Alers and Charter*, and *Young South Africa*, which supply a good deal of material for conversation. In standard VI, I supplement the assigned reading material with newspaper articles, selected stories, and chapters from modern English books. Beginning in fourth grade or standard II, essays are required that become increasingly challenging in each successive grade. In standard VI, the child must be able to master grammar and idiomatic English to such a degree that he can write a fairly grammatically and orthographically correct essay.

Judging by the average level of competency in a class, I sometimes reached this goal quite easily, other times only with difficulty. I must say, however, that the requirements for English are very challenging and that I have to work very hard with the children for them to meet the standard. My education in England has been a great help to me, and I am certain that teachers arriving directly from Germany experience a fair degree of difficulty in teaching English. Even though I had the responsibility for standard III (Sexta) class, I taught exclusively in English in the classroom. My entire time is devoted to English instruction.

Afrikaans is taught twice a week as a second foreign language, but only in the upper grades, from standard VI on, and it is the only instruction given in our school by a non-German (we employ an Afrikaans speaking South African).

From the very beginning, instruction in mathematics is most challenging in all the grades. Since England, along with her dominions and colonies, has not yet adopted the decimal and metric systems, children and teachers alike have a difficult time learning one-twelfth, florins, half-crowns, inches, yards, chains, furlongs, bushels, muids, cape hundred weights, etc. German instructors have compiled lessons for this country because the children must know how to think when learning these tricky lessons.

A comprehensive knowledge of history is necessary for an appreciation of the children's second homeland, and for this reason we differentiate between general and African history. Only ignorant or ill-willed people would assert that our German children in South West are not sufficiently initiated into German history. It is impossible for me to list the entire curriculum, but such a list would easily put to rest any notion that our children are not learning with great enthusiasm, as children in Germany do, of the heroic deeds of our nation and our great men. African history is closely connected to European history, and many South Africans have German blood coursing through their veins. Twenty-four of the first thirty-one settlers in South Africa were, in fact, Germans. The German mercenaries Wintervogel, Müller, Darkert, Wrede, and von Meerhof were among the first explorers. Wrede compiled the first dictionary of the Hottentotten language. Gneisenau served as an officer in Cape Town. From 1786 to 1808, the Württemberg Regiment was stationed in the Cape Colony with the consequence that the two forward-facing watchtowers of the old fortress in Cape Town are called Hessen (after the German soldiers from the German state of Hessen) and Katzenellenbogen (cat's elbow) respectively. After the Crimean War, twenty-nine hundred members of the German legion immigrated to the Cape and settled between the Fish River and Tugela River. In 1858, an additional two thousand Germans settled there. Thus it is not only desirable but essential that our children be taught the history of South Africa, for South West is our second homeland. It is proper that our children learn how their forefathers fought and worked in South Africa, just as their parents do here in South West. I consider this a strengthening, not weakening, of our Germanic consciousness. One thing is for certain: our children are inspired by history and have developed an insatiable appetite for it.

Only experienced instructors teach the natural sciences. Sometimes our classroom becomes a meadow when specimens of the "beloved" twenty-four species of grass lie on the lectern. A piece of a bat's wing is scrutinized under the microscope for parasites; anopheles and various other mosquito larva are cultivated in Petri dishes; diagrams of refraction are on the blackboard; the essentials of first aid are taught, along with many details about health and well-being.

The classroom of standard V is richly decorated with pictures of Frederick the Great, Paul von Hindenburg, the Führer (Adolf Hitler), and others. The latest newspaper photos are pinned to the board in the back of the classroom. The walls of the classroom are neatly arrayed with maps so that pupils can make trips through every part of Germany, by water, by train, or by plane. In a large sandbox on the school grounds, South West is represented in miniature—complete with railroad lines and roads; even dams and rivers are marked with colored chalk powder.

After the death of von Hindenburg, the children asked permission to take up a collection in order to purchase a picture of the immortalized Reichspräsident. In no time at all they had collected the necessary funds and together we looked for just the right spot to hang it. We decided the best place was in the back of the classroom above a small cupboard. I intentionally had nothing to do with the collection but allowed the children to do it on their own. But I have provided blooming pelargonies and placed them under our beloved picture. Each day, girls lovingly brought vases filled with fresh flowers. One morning, a strip of black, white, and red cloth (the former imperial colors) appeared beneath the picture; the flowerpots were also draped in these colors. This particular class was especially lively, often very loud, yet under the picture's influence they became models of self-discipline. I was pleasantly surprised, and when I praised them, they pointed at the picture and said, "He is watching us, therefore we must be good."

Are these children who have lost their love and devotion for Germany and the German language? I wish the mendacious reporters might come and see for themselves. They need only listen to our choir singing our beloved German songs, led by our choir director, S., or simply observe the way the children do their round and folk dances.

Our school also has two shop rooms, one for cardboard and wood, the other for metal. The Department of Education supplies the materials, though since the government must deal with the effects of the depression, it recommends frugality because funding for these materials is limited. We ourselves must "dig deep" to scout up the materials we need. Much of what the students do in our shop courses helps the school. Maps are framed and books are bound; shelves, bicycle stands, podiums, tables, and benches are constructed; doors, door locks, and keys are repaired; the fences round the school property are tightened; even the wrought iron gates are produced by our students.

Our art teacher, Herr F., is also certified to teach cardboard and woodworking. From our stocks of cardboard, pupils make small boxes, bowls, flower boxes, and calendars. They produce sheets of colored paper, bind notebooks, regular books, sheet music, etc. When it comes to woodworking, the boys saw, carve, plane, and turn it. They make electric floor lamps, hanging lamps whose lighting fixtures are made from ostrich eggs cut in half, small tables with carved African animals, small carts, radio sets, kitchen utensils, and deck chairs. Last, but not least, entire sets of furniture—the so-called Shell furniture—are constructed from Shell's wooden gasoline crates. In the midst of all this activity and noise, the teacher hurries back and forth, clothed in his white coat (soiled with spots of oil and paint) advising here, lending a hand over there.

Now to the blacksmith's shop! The metal shop teacher, Herr W., who, only minutes ago a natural science and physics teacher, stands before us. He is tall, and in his blue work apron with his shirt sleeves turned up, appears massive. The bellows hiss, the fire smolders, the boys hammer on the anvils, and sparks fly here and there. Metal is filed, hardened, soldered, and riveted. Often I stood by and watched—both astonished and amazed! What would South West be without tin cans (that range in size from four-gallon gasoline canisters to the daintiest tins of specialty foods)? And what is made from these discarded tins? The inventive genius of the shop teacher is truly astounding, and under his tutelage they create luggage, lunch pails, milking pails, baskets, flower pots, kitchen lamps, cups with exchangeable handles, even depression kitchen utensils. They also manufacture wire, meat forks, meat hooks, grates, soup ladles, skimmers, garden utensils, fire tongs, and many, many more things. Indeed, innumerable are the things they make.

After visiting our shop exhibit, a slightly jaded observer stated, "I have no idea why so much fuss is made when new tins are made out of old ones. Planting a tree is more important than this!" Of course our children are fully capable of doing that; however, it is good form to appreciate the transformation of old things into new—even if they are small things. We also note with satisfaction that even our university preparatory school has begun to practice this form of manual arts. I am also reminded that Bodenstedt consoled us thus:

> Don't be angry, my friends, when mockers laugh at you—
> Meet their mockery with a smile and know:
> The wit of the mocker can't make despicable
> Anything that is not truly so!

We also have new materials such as iron, copper, and brass sheets in various thicknesses. From these, charming little engraved boxes and smoking needs have been produced. We have made writing items from cartridge cases; ashtrays, candlesticks, and tobacco boxes from the heads of wild animals, some with initials soldered on top. Cigarette and cigar boxes lined with curing materials and invisibly soldered together have been produced by us. The boys who help to manufacture these items do so with such enthusiasm they have no wish to stop working in the afternoon. How proudly they display their works of art! And when they eventually return to their farms, how capable they will be of helping themselves now that they have acquired a knowledge of how to work with all manner of tools. Yet beyond all the joy and enthusiasm they have experienced, a serious work ethic and the discipline of learning have also been instilled. Tools must be maintained, knives and scissors must be sharpened, and saw blades straightened. They have also learned how to make new handles and how to oil and maintain the machines.

Our young girls are not neglected; indeed, the department gave us two sewing machines and scissors, as well as crochet hooks and knitting needles. Since we are permitted to order materials free of charge every year, we do so and receive thread and yarn for sewing, knitting, and stitching, and wool and other necessary items. Before I became completely occupied with teaching English, I supervised needlework instruction in the senior class. My colleague, Frau S., and I set out to elevate needlework to such heights that it was noticed across the entire country; happily, the teachers who took over from us have continued to uphold this tradition. Now Fräulein Sp., a capable and talented woman, expertly supervises the needlework instruction so that when our girls graduate, they know how to cut cloth, make bed linens and dresses, sew, crotchet, knit, stitch, and weave. Instruction in needlework goes hand in hand with instruction in drawing. Here, the children prepare designs for handkerchiefs, pictures to be stitched on blankets, pillows, and bed linens. The designs are then used in their needlework classes. Drawing classes also support the work done in the shop classes by teaching precision drawing to scale and design. In all three subjects a sense of color matching and good taste is instilled to avoid the production of unattractive goods.

The schoolgirls participated in service activities by using their newly developed manual arts skills. While I was a member of the board of trustees of the Elisabeth Haus (a maternity home for German women), I had the girls sew baby clothing, as well as design and sew bedding, pillows, and kitchen towels for the facility—an activity they enthusiastically continued for several years.

In this year's home-industry exhibit, our pupils received many, many individual prizes, and for the third time, were collectively awarded the trophy for the best work projects of all the schools. Three times our pupils contended for this trophy; three times they were successful! Needless to say, we teachers are proud that our youth are so well prepared for life.

In addition to all of this, we instruct our pupils in gymnastics, swimming, and other sports, and nearly all our children belong to the scouting and maidens youth groups. We likewise prepare our own work exhibitions, slide lectures, and parent evenings with musical presentations and theatrical performances, for which all scenery, costumes, and theater props are made in our own atelier. It is our own tried–and-true director, Herr Kr., who assumes responsibility for all school activities with his steadfast composure and prudence.

Two student dormitories, one for boys and one for girls, are part of the school complex. The boys' dormitory has been supervised for many years by our faithful director and his wife, along with one other teacher. One of our longest-serving teachers, Herr W., who has been in South West since 1914, together with his young wife, oversees the girls' dormitory. They carefully see to the girls' personal hygiene as well as their physical and intellectual development and are further supported in this by one of our female teachers.

The annual fee for boarding in the dormitory is forty pounds per child per annum. Parents who cannot afford the fee are subsidized by the government. Some German parents pay as little as ten pounds per child per annum. Both school and dormitory support each other in successfully educating our youngsters. One has only to look at the children's happy faces in our bright, airy rooms and watch them at play to realize how well our German community spirit and comradery are nurtured here in South West.

Radios are the private property of the teachers but may be used in both dormitories. Not long ago, I was invited to the boys' dorm to listen to a speech by our Führer, Adolf Hitler, broadcast by the Deutschlandsender (the official Nazi broadcast service). Sitting close together in the director's living room, the boys on the floor shoulder to shoulder, eyes shining, we listened with enthusiasm to the distant voices emanating from our old homeland. In complete silence, each of us listened to the first whistling and cracking of the radio signal, then, after a pause, the voice of the reporter, "The Führer disembarks from the airplane and walks to the automobile." We hear the roaring applause and the shouts of "Heil Hitler" by the people. Another pause, then his speech. I regret that I am not able to fully describe the expressions on the boys' faces, their little red cheeks, eyes filled with tears. Snuggled against my knees sat a little blond boy, "Great, great!" he shouted for joy. "Man, that was great!"

This is how a German government school operates under the direction of "renegade" teachers! German newspaper hacks have written insidious articles about these schools, but we have met with visitors to this country and requested that justice be done. I do not know whether it is more difficult to work in a public school like ours or in the German private school. The middle school receives generous financial aid from Germany, but it needs to be stated that the Mandate government subsidizes it as well. We guard this outpost of Germany with finesse and sensitivity and rear our children to become the same kind of Germans as those in the Fatherland; we preserve our Germanity and pass it on to our children. But simply because we perform these functions quietly and thoughtfully so as not to raise the ire of our current masters, we are treated with hostility. There was a time when the teachers of the university preparatory school would not greet us and mocked and ridiculed our pupils. We certainly did not deserve this kind of treatment; it hurt us. There were many tears, and how often did we have to console youngsters to remove bitterness from their hearts. Often the children fought one another.

How gladly would we have received gifts from Germany, as did other schools! Our pupils need this support as much as the others. Our library is relatively small, and we finance it with a lottery as well as contributions from parents; the state subsidies it on the "pound for pound" system. Only once did friendly donors from Germany send us a hundred German marks for cultural purposes, and that allowed us to enrich our library holdings with ten pounds worth of books. It seems that no one ever thinks that we, the stepchildren of German culture, in our difficult outpost in South West, might be in need of a bit of help. Our German pupils would also like to have received a portion of joy from home. The officials here are fully aware of the hostility shown toward us. Several times they have intervened and helped us, even by compelling those in this country who could be reached and who had spread false reports to rescind those reports. But we march forward without hesitation. Though I am no longer a teacher, I still oppose—all the more vehemently now—all of those malicious and intentionally false reports given against us.

9. My Sons

After a year of pressuring me, I finally gave in to my sons' desires and let them come home to South West. It had been especially difficult for my younger son, who suffered greatly from homesickness. I had been able to save enough money to pay for their passage, and it was a joyous time when at last the three of us were together again. Both of them were able to find situations. The elder son wanted to become a lawyer; the younger, a book dealer—he had always been interested in books, book dealing, and everything else connected with the world of books. Each of us, though, in the back of his mind, had been nurturing the desire to buy a farm as soon as we could afford it. A city in Africa, like Windhuk, though it is small, is still a city, and all of us longed for nature, for hunting, for animals, for freedom, for solitude. We wanted to work in the kind of freedom the countryside affords, to be close to the soil—which we loved—and with joy to wrestle from that land everything necessary for life. But as before, we lived in Klein-Windhuk, renting a house below the one in which we had previously lived.

Hurrah, we have a car! So that the boys' commute to work twice a day might be easier, we purchased a small, four-cylinder Chevrolet and were delighted with it. The boys were already good horsemen and hunters; now they became skilled, intelligent drivers. There was hardly a Saturday or Sunday during the hunting season that the boys were not outside. Over longer weekends, I accompanied them.

We were doing very well and were happy to be together again; indeed, it was a season we were able to enjoy without any serious cares. My colleague, Frau S., decided to take a vacation. She was not only a teacher but also supervised one of the governmental school's girls' dormitories. The school had appointed a substitute teacher, but no one was available to take over responsibility for the dormitory. In order to make it possible for her to take her vacation, I agreed to substitute for her as dormitory supervisor for the entire four months. Since this was bound to interfere in the relationship with my boys, I did not agree readily, yet was able to rent my house furnished—except for the boys' bedroom—for the entire four-month period. Because the renters were friends of ours, it worked out well, especially as the boys were able to breakfast with them. The Education Department consented to allow my sons to have their noon and evening meals with me. We did have to pay for the meals, yet we could eat together and visit twice a day. It was only after dinner that they drove home.

In time, my younger son began to appear pale and hardly touched his food. When I enquired after his health, he always answered that everything was fine—he even laughed at me when I asked him to visit a doctor. One afternoon, though, he became so ill that I had a second bed placed in my room so I could care for him until he recovered. During the night, his fever rose above forty degrees centigrade. As soon as it was morning, I sent for the doctor. He had typhus and was immediately taken to the hospital where I was not allowed to care for him myself. I was, however, permitted to stand at the door and witness him fall ever deeper into this dreaded illness. I had responsibility for the children in the dorm and looked after them as my son continued to suffer for four long weeks, often in a state of unconsciousness. The faithful doctor, our dear friend, did his very best; tirelessly, the nurses cared for him round the clock. I, his mother, could only stand at the door and watch. He left us. Africa had taken Büdi from me.

After my four months of supervising the dorm, I was still unable to return home as my colleague had been transferred to another school. Thus I had to to stay on and oversee the dormitory another four years. With this I was content, and because of the many things I had to do, and the fact that so many children depended on me, I was denied the self-indulgence of brooding and self-absorbed sorrow. Mornings found me in the classroom teaching; from noon on, I worked in the boarding house. Most of the pupils were girls, though a few little boys slept in one of the bedrooms. I learned to love my little ones very much, and to this day many of them come to visit me and remain attached to me. With all these duties I hardly had a moment to myself—day or night. A housekeeper, who was responsible for the kitchen, and another teacher who helped supervise, supported me during this time. The bookkeeping had to be exact, almost pedantically so. Every worn-out broom, every damaged cup, every torn bed sheet had to be retained until shown to a commissioner.

Our children here in the tropics are prone to disease: measles, whooping cough, and a number of less serious illnesses were weekly occurrences. I maintained a correspondence with the parents, looked after visitors, negotiated with merchants, and had my troubles with natives. Except during the holidays, I had no time for myself at all—only the sad, sleepless nights were mine.

During this period of time, my elder son was informed that he would be unable to become a lawyer as he had not taken the English form of Abitur, the Matric. He had wanted to become a lawyer and had been promised that his two years of working in a lawyer's office would be recognized as part of his apprenticeship. In South Africa, one "learns," that is, one works as an apprentice, in order to become a lawyer. I had been assured that he would reach his goal without any difficulties—especially since he was the first German born in this country to be prepared for this profession. The board of attorneys had promised to help me, and I had firmly counted on this promise. My son had worked hard so he would be able pass the required examination. Now, in addition to all my other troubles, this huge disappointment was set about my neck like a millstone. Neither of us had anticipated such a blow! My son pursued everything that could possibly secure a firm position for himself. With my continuing sorrow for my youngest, now I was worried about the future of my eldest.

He worked as a substitute in various positions, but our life was one of unfulfilled expectations and uneasiness. During this period of time, we responded to two hundred calls for applications, sending them to various cities in the Union, as well as overseas. Since he was an excellent driver, he agreed to guide an Anglo-Franco hunting party to Livingstone, Northern Rhodesia, through the Caprivi area in the northeastern part of South West. He was not, however, paid for his labors. In fact, I had to wire him money for his return trip to Windhuk! Later on he guided a big game hunter and sportsman from England who wished to collect record-breaking trophy animals. While it was a physically stimulating time, it certainly did not help him get ahead. He prospected for gold in the area around Rehoboth, illegally hunted for diamonds, and was ever in danger of losing his life in the desert. The end result of all this activity was the two hundred pounds I had to pay for his troubles—without him ever having found anything of value. My life was worry, unrest, disappointment, and fear of what would become of my son when I became unable to work.

Finally, he had a bit of good fortune. With limited means, he purchased the farm, Narachaams, which lies about one hundred fifty kilometers from Outjo in the Franzfontain area. Happily, he also married a very fine young woman. Now, he concentrates his efforts on the breeding of karakul sheep, and I dearly hope that he will make a success of this. Alas, being a mother often leads one to despair.

10. Hunting in the African Bush and a Return to our Home in Okombahe

During my holiday breaks, I felt in dire need of relaxation and a change of scenery, which I found in hunting safaris that took us deep into the veld, as well as to the great savannahs of South West. We had exchanged the little Chevrolet for a larger Chrysler motorcar. It was definitely past its prime: the paint was faded, the doors creaked and groaned, occasionally a cable broke, the poor thing was blind in one eye, and its only "good" eye peered but dismally into the darkness of night. A chocolate-colored piece of canvas covered the top; the springs in the seats nipped at one uncomfortably; the horn was reluctant to blow. The engine, however, was in splendid condition and propelled the car through water and sand, pulled us up steep hills and across rocky terrain, and led us safely downhill. Its radiator had a hole we patched by pouring a bit of cement into it to seal the leak, after which we attached a second radiator, in form of a "desert kettle," to the right side of the car near the bonnet. A complicated system of tubes allowed the water to circulate through both radiators, and though the car moaned a bit when driving through deep sand, it always pressed forward. Together with Bitz and me, this little car has experienced a great deal as it has taken us to impassible areas of South West—staggering through aardvark holes, running over thick stands of thorn bushes, it bravely pressed forward, covering three hundred thousand kilometers and more.

When we undertake an extensive journey, everything must have its proper place. To the left and the right of the bonnet are two rifles; the third is in my arms. Between the bonnet and the mudguards sits the folding table with tent poles tied to the front of it. A railing has been attached to the running boards so that containers of oil and water, canisters of gasoline, a crate of kitchenware, and various other provisions can be stored there. The back seats have been removed to make room for luggage, more gasoline, tools, reserve hoses, field bottles, field beds, chairs, and snacks of all kinds. Rolled blankets, the tent, picks and spades, and so forth are strapped to the luggage rack in back. Thus every item has its own firm place, and the weight is properly balanced.

By traveling and camping in this manner, I have been able to revisit all the areas familiar to me when we lived in Okombahe—all the places that captured my heart. From the village of Karibib we drove along the Erongo River. Again and again, I was moved by the extraordinary size of the boulders, the rocks, the huge pinnacles (as if constructed by a giant's hand) that tower steeply and defiantly into the heavens, and the immense weathered spheres of stone that seem to rest precariously atop their steep, rocky pedestals. When next a powerful flash flood roars through the area, these enormous balls of stone might roll down the polished hillsides, following the fate of many other huge orbs. At the foot of the mountains, between rubble and lesser stones, mighty colossi of stone reach up to the blue infinity of the sky, standing there as lonely guards of this great expanse of desert.

In days of yore, Bushmen lived in this mysterious wilderness. We found their signs and drawings on the overhanging rock walls and wondered who the great artists were from so long ago. Footprints have been hewn into the rock; holes both small and large have been bored into the rock as well. Their drawings have been interpreted: the small holes represent food, the large ones water sources; the footprints are to warn the intruder that once he sees this he must turn around, for the berries and water belong to us, the "us" being a warrior painted in red, who stands there with bow and arrow. Beside him one sees pictures of game animals, which also belong to him, and are a warning that the animals belong to the people of this area and they will defend these animals, thus the intruder must go elsewhere to hunt. I have often stood here both amazed and astonished as I attempted to unravel the riddles of this strange people.

We drove through the Namib Desert toward the glowing, reddish conical peaks of the Spitzkoppies. In front of us, individual gemsbok and hundreds of springboks walked and grazed, and my son shot a splendid trophy specimen from the herd. We rested for a time, then pressed on. When the broad, blue shadows fell across the Namib, we had reached the mountains.

I searched for that old water hole but did not find it. The old farmhouse that had once stood there with pride was now in ruins—the garden, the kraals, and the well had completely disappeared. Of the dam, only a tiny ruin remained. I continued searching for the graves of those who had been murdered in 1904, but darkness fell and we had to make camp. Wood was not difficult to find, and before long we had built a mighty fire. Hyenas and jackals howled, hyraxes came down from the rocks and gnawed, unafraid, on the bark of trees close to our field beds. I lay awake the entire night listening to the voices of the wilderness. The wind blew into the fire causing sparks to dance upon the rocks. I reminisced about our first life in South West and about how much life had changed; I thought about the many deceased persons I had known from those times. Early in the morning, I got up and found the remnants of graves I could still identify.

We began our trek to the Bushman's paradise, a difficult and wearisome climb. How marvelous it was to note the differences in the vegetation of that high valley perched between massive boulders, and I eagerly anticipated showing my son the famous Bushman drawings. Alas, it turned out to be a great disappointment as they had been so wantonly destroyed. We also found places where people had dug for gemstones—the Spitzkoppies and environs are rich in helodores, tourmalines, aquamarines, sapphires, and rock crystal. Feldspar is also found there.

To the west of the Spitzkoppies extends the long back of the Okombahe Mountain. We drove there, my heart aching as we approached the huge ana trees growing along the river. We carried on through the narrow gully and stopped in front of our dear home. Two former acquaintances of ours had purchased it and were now trading with the natives. Okombahe had shrunk; the population had dwindled. Most business owners had moved away since they could no longer make a living there. Many people worked in the surrounding tin mines; others had settled with their cattle near some lonesome water hole.

Our house had changed. The exterior had been painted white, the paintings on the gable had been covered over, yet, "Caelum non animum mutant qui trans mare currunt" (For those who cross the sea, the sky above changes but the heart does not; Horace, Epistle I, xi.27). One room had been added, another enlarged. Now I sat in our former dining room, kindly hosted and entertained by the current occupants of the house. I did, of course, participate in the conversation, but the room was filled with memories—living shadows too. The old couple still remembered me, greeted me with affection, and squeezed my hands. The old woman wept when she saw Bitz and tenderly stroked him as she said, "You, you are my child, for when your mama was away, I secretly came to you, and here," she said, touching her bosom, "here you drank from me." The old woman stroked Bitz again, but he felt rather awkward, rather embarrassed.

My old friend, Captain Josua, was not there that evening, but the next morning at dawn I heard his shuffling and the tapping of his cane. There before me stood Josua, now an ancient man. Immediately, he fell to his knees, pulled my hand towards his mouth, and kissed it again and again. Before I knew what was happening, a wooly, white-haired head lay on my shoulder;

Josua sobbed like a little child. His embrace did not relax, and I—I also sobbed. I could not let this joyous occasion pass without giving dear old Josua a gift, so I gave him money for sugar as well as an entire springbok. I felt relief at being able to provide him this kindness. We departed.

The old fortress, on whose battlements I once stood guard, had disintegrated. The Protestant mission had been abandoned, but a kindly, long-bearded Catholic priest attended the congregation and dressed the garden. He gave us oranges and papayas. On the opposite side of the river, we briefly stopped in at the home of our former neighbors, the Merkers. Both Herr and Frau Merker were now seventy years old, and we celebrated our reunion with wine they themselves had pressed.

Now, at last, Okombahe lay behind me; my life there was finished and done.

11. From the Brandberg to Franzfontein

One of our trips took us to the Brandberg. En route we passed the large tin mine at Uis, located in a desolate area of extended plains of black, broken stone, devoid of all vegetation. It was our intention to hunt springbok on these plains that lead to the foot of the Brandberg, and then drive on toward the Ugab River. The rainy season was passed and the grass in the Namib had grown so tall it was impossible to follow the track, so in Uis we hired a native to guide us through this great expanse of waving green. As it turned out, however, this "guide" was barely able to lead us in the right direction.

At the Ugab River, we came across a village of old Okombahe natives. From those earlier times, I still knew the village elder, Galeb, and our reunion was happy, though not without certain expenses. A shriveled little woman stood in front of Bitz, her toothless mouth smiling, and said, "I am Helene—you know, your old Eia." She patted Bitz.

Bitz led Galeb to the car so the two of them could hunt for a springbok and an ostrich to bring back to the village. As opportunity now presented itself, I prepared to take a bath in the Ugab River. Firstly, I had to set up a canvas screen to afford me some privacy, for the chance to watch a white woman bathe had about the same interest for the women and children as going to the cinema. In the meantime, I hired women to scrub the rusted pots and to clean our dishes. In no time at all, they had scratched out a hole in the ground with their bare hands, filled it with water, and used it as a basin to wash the dishes. Soon, all the shining pots and pans were drying in the sun. I wished to pay them for the milk they offered me, but Galeb took this as an insult and declined the money. "That is my child," he said, directing his eyes at Bitz, "I am his father; what I own he owns." Bitz replied, "My goodness, I had no idea how many mothers and fathers I really have."

We were a bit apprehensive about crossing the Ugab River, as it had been years since a car had driven across it. Just in case of difficulty, Galeb kept some oxen at the ready to pull us out, though happily, we encountered no difficulties at all. On the opposite bank, near the water's edge, we discovered lion tracks in the damp soil; but since lions can cover a lot of ground very quickly, it did not mean much. With no trace of a track, we simply enjoyed the drive. Galeb had given us a piece of advice: "You must lie flat on the ground and look through the grass. Wherever the grass is a bit shorter and firmer, you can be certain it was caused by the ox-drawn army carts that were driven across the plain during the uprising in 1904. At those places, you will see the path." Now we knew how to navigate!

Bitz drove the car while I practiced gymnastics while being flat on my stomach. A deep, narrow gulch slowed us down and even bent the luggage rack a bit as the car descended into it, but the old jalopy bravely pulled us through. It would not have been so bad had we brought along someone from an auto repair place for the next thirty or forty miles of driving. But instead of a repair shop, there was silence and solitude; as far as the eye could see, there was not a soul in sight to help us. In countryside like this one must rely entirely on oneself; yet it is for this precise reason that a journey through the bush is so alluring. Maps are of little use, especially since the old routes do not appear in them.

A main track crossed the one we were looking for, so we continued on with a fair degree of confidence. This main path had been created about a year and a half ago when a government employee drove his car along it while searching for a shorter route the Angola-Boer Transports could use. Such a route is called a main path. A piece of wood, hanging crooked and loose on a tree, had something now illegible written on it. We knew that where bushes cluster thickly around tall trees on the riverbank, there had to be a water hole. We did find it, but it was dry. On all of our excursions to the bush, we always drive from water hole to water hole—for we never know whether it will be just to replenish our own supply or if someone else is in need.

Shortly before sundown, we heard the snapping of twigs and the breaking of branches just before we spied a herd of zebra noisily moving through the dry brush. Bitz shot a stallion that got away, though it left a trace of its blood behind. We found that trace and searched for the animal. Soon, however, there was no more blood to be found. The sun went down, and we had no choice but to return to the car and make camp. Where there are zebra, there are also lions, for zebra are their favorite bounty. But these zebra were so shy they must have been chased from their cover into the open. This is an area where stacks of thick, rocky slabs, called sokibank, slope ever downward. As evening settled in, a cold wind began to blow across the plain. It was tempting to pitch camp under the protection of these rocky slabs, but in countryside where lions roam about freely it is far better to have a clear view. We had to decline the comfortable site and set up camp beside the car. Some distance from the car blazed our huge campfire. We had quickly gathered stacks of wood to keep the fire burning brightly during the entire night, if necessary. We fried fatkukkies, a kind of pancake, on the glowing coals and felt quite comfortable and at home despite the heavy wind. Sometime during the night the wind turned into a storm. To prepare for the gusts of wind, we firmly tucked the blankets about our bodies and weighted them at the foot of our beds with stones. The raging storm drowned out all other sounds—even the anticipated roaring of lions.

The next morning we quickly shook the sand and dust from our camping equipment and began our search for the wounded zebra. We did actually feel it was wounded, perhaps even dead. Not more than two hundred meters from camp we found the animal's remains, encircled by the large tracks of two lions that had had themselves a delicious meal. This, of course, explained the silence of the night, for lions are taught while still cubs to be silent during meals. We returned and loaded the car, and after extinguishing our fire—the unwritten law of the savannah—continued on a northeasterly course. Countless game paths and fresh dung indicated that many kinds of animals were nearby. Lifting the fieldglasses to our eyes, there before us, on a small meadow, stood two gemsbok grazing peacefully. They scraped roots out of the sand, then pulled up bundles of grass, ate, and slowly moved ahead of us.

Suddenly, a grating shriek ripped through the morning silence. Right beside us, a guinea hen flew up, warning anyone within hearing distance. Her call sounds like the noise of a child's rattle—the kind sold at a German Christmas market. The hen was frozen in place, as though suspended in midair, cackling and cackling. Other guinea fowl immediately joined the call of warning, all of them cackling and grating. This silently dreaming plain burst into life, with antelope in complete stillness suddenly stamping their feet, raising a mighty cloud of dust, then disappearing. The meerkats sat upright on their tails in front of their holes, their front paws pressed into their bodies as they curiously pondered the reason behind all this ruckus. One of them ducked, turned, and disappeared—a sign that all of them should hide.

Hunting in Africa is not a simple affair. Game animals seem to have a protective spirit. Did something move up there on the rim of the dune? Quickly, the fieldglasses! Ah, the strap is tangled up with the case, and by the time it is untangled, whatever it was is gone. Now the engine is running quietly, and we look about carefully. Back there are gemsbok bulls! The car puffs and jostles about, then bumps against a root or a stone, making a truly unwelcome noise. Another time, my finger on the trigger, some insect flew about my face and would not be driven off. Meanwhile, the antelope that stood there as if rooted to that spot is gone, darting off to one side. We scoured the area with our fieldglasses, but the little creature was much amused as it stood directly behind us.

Late that afternoon, we reached Franzfontein—a reserve for the Swartbooi Hottentots. A dear old friend of ours, J., lived there. His house had been built of red, unfired clay blocks and looked welcoming and peaceful. We were invited to enter and sat comfortably in the high-ceilinged living room amid a surfeit of animal skins, horns, and other hunting trophies. Yes, this was the home of an experienced hunter; ninety-six leopards he had shot, and a large lion skin decorated the wall, his mighty rifle standing in the corner against the wall. He had been accustomed to shooting the leopards that had been caught in his traps. An entire pack of large dogs circled us again and again, and I worried about Männe, our reddish-brown dachshund. Should even one of these dogs so much as yawn, our Männe would be swallowed up whole. But he seemed to understand the situation and stayed in the car, not even daring to growl from his secure corner, but made himself at home in the car and kept as quiet as a mouse.

In earlier days, Franzfontein was larger than it is now. Scattered across the ground are countless little white limestones, and here and there a mopane tree spreads its delicate crown. A strong spring rises in the Franzfontein Mountains and forms a brook that flows through the village, with thick, green grass covering the banks of the stream. This is precisely how it looks as it flows through my son's farm. Narachaams is ten miles from Franzfontein, and Bitz was fortunate indeed to have a spring all to himself.

At one time Franzfontein was a mission station, but it was abandoned long ago. The mission house, in a state of ruin and decay, is still visible; a small church still stands on its original foundation. Every Sunday the Hottentot schoolmaster conducts a church service; weekdays, the chapel becomes a school and he is the schoolmaster. The small bell tolls busily on Sunday, and the Hottentot men, dressed in black coats, at least those fortunate enough to own one, and the women in colorful European dresses, hair covered in a scarf, stroll to church—and stroll they do, never do they "walk." The schoolmaster plays the harmonium, and the Hymn of Namaqua is performed with certain notes being held far longer than their due. The schoolmaster rises and delivers the sermon in his nasally voice. Just outside the front door, small, yellowish-skinned Hottentot children scrap and play, and among them, their young kids bleat and frolic. The Swartbooi-Hottentots of the area own both large and small stock, and cultivate grain, pumpkins, and tobacco. Some labor on neighboring farms, even in the town of Outjo; thus, the tribe always has some limited funds. The spring is truly their source of life, and the water is divided every few hours between the police station, our friends, and the Hottentots. A ringing bell indicates that the flow of water is being redirected.

The people of Franzfontein are well known for their savage hunting expeditions. That last time we visited, about thirty Hottentots—and almost as many dogs—had cornered a lion and shot it. For this kind of a hunting party the people are issued several guns and ammunition by the police at the local police station. At this particular time, the police station was a one-man operation. The Hottentots involved sold the hide and divided the fifty-shilling profit among themselves. This lion had killed a cow very close to Franzfontein, and at the site of the kill, the spoors of three other lions were plain to see. At first, the beasts had walked off together, but after a certain distance had been covered, the spoors of the largest and strongest lion suggested the alpha male had parted ways with the others. It was this one the men decided to pursue, leaving well before sunrise. The spoors led across the limestone cliffs, and during the course of the pursuit—in the heat of the day—the stony ground had torn open the lion's paws. The spots of blood that dropped upon the rocks grew ever larger. Eventually, the lion slowed down, then finally, surrounded by Hottentots and their dogs, it surrendered. The poor beast was torn asunder by the continued shooting, and it pains me that such a noble animal should meet its end in a manner as pathetic and dishonorable as this.

Our friends, the Js., owned a small store at Franzfontein and traded in cattle and animal skins; they also raised cattle on a nearby farm. Produce from their garden and orchards provided a very good income as well. On average they shipped out sixty thousand oranges a year; under their attentive care, grapes also grew very well. Many an evening found us sitting in the mellow light of a hanging lantern with a large teapot sitting on the table, filled with a delicious smelling punch prepared from their own homemade wine. Then we listened to the dear old gentleman's hunting stories or played a game of skat. His very able wife is exemplary of all that is good in a German woman.

The door to the Js' living room was open, and the soft, cool evening air entered. A few moths, attracted by the light, circled the lamp. From the distant Hottentot village, we heard the laughter and singing of the natives. Their small fires were lit; some of their dogs were howling. Now and again, we heard cattle lowing in their kraals, goats were also bleating, the small brook lightly splashed its way past the house, the limestone cliffs glowed white. In the garden, the crowns of the date palms swayed, and moonlight, in wide stripes, flowed across them. Occasionally, one of the big dogs got up and looked out the door, then stretched itself and tenderly set its head in my lap. Before I knew it, it was bedtime. How I loved this house and its inhabitants!

While in Franzfontein, an old Hottentot man visited me and conversed with me in perfect German. He wished to know the latest comings and goings in

Windhuk and proudly told me he had been employed there as an imperial document hanger—the person who pasted official notices on all the notice boards about town—when the Germans were still in charge. It was his greatest wish that the Germans return and run the country, and should that happen, he would immediately reapply for his previous job. Since he was elderly, he hoped that day would come soon—but until then he would wait. The day would come, it was just a matter of time.

12. Korichaams

Together with our friends, we decided to go hunting at Korichaams, a place so small and remote that even on a large map it can hardly be found. However, if one enquires of an old South West soldier, one who had been stationed in the north of our former colony, a wistful smile will appear on his weathered face, and he will sigh, "Ah, yes, the old Korichaams." It is a Namaqua name, and the final syllable, *aams*, means spring.

Korichaams is nothing but a limestone pan in which there are three tiny springs, the only source of water in a very dry place. Midmorning, we loaded into our car and headed out. At dawn, we had sent off the Js' Hottentot teamster with an ox cart loaded with provisions and bedding. At a moderate speed, our car followed a path called "a trail along the mountains." I was very much looking forward to seeing Korichaams and hoping I might overcome my distress, my sadness, in the beauty, solitude, and wonderment of nature. On and on we drove, skirting along hills and mountains, across wide, desolate plains torn asunder by the brilliant, unrelenting sun. We passed thorn bushes, mopane trees with their branches swaying in the breeze, the odd native trekking across the landscape, a narrow winding swamp whose reeds softly rustled, then more mountains, hills, another plain, a river, and patches of sand.

Before long, the area became increasingly desolate. Isolated trees stood in a soft, hilly area; gnarled dwarf acacia trees spread their umbrella-like crowns. We had difficulty finding the track followed by many an oxcart in days of yore. Then in the distance we noticed something dark against the gleaming sunlight. Ostriches! To the right, a bird started calling; then, without warning, dry branches began cracking and we heard a great pounding of the ground as a herd of zebra rushed by, followed by a splendid stallion. We did not trouble them as we intended to stalk the ostriches. Hunting ostrich has become more common, even popular, since the Boers began hunting them. There used to be so many ostriches that they were truly a plague. But even though ostrich feathers are no longer very fashionable, they are still appreciated, and traders pay seventeen shillings or more for an ostrich skin. According to last year's statistics, about five hundred thousand ostrich skins were exported! Not only is the skin used but also the dried meat, called biltong, which is fed to natives working on the farms, or to chickens and pigs. Ostrich fat can be boiled and made into soap, though the soap has a peculiar fatty smell that transfers to clothing laundered with it.

We slowed down and came to a halt. Our friend and my son got out, but the ostriches had already gotten wind of us. Nonetheless, two shots were fired and two enormous birds lay dying: man had arrived in this perfect world, leaving a trail of agony in his wake. We covered the carcasses with branches to protect them from vultures until they could be loaded into the ox cart.

To me, these vultures are the strangest, most mysterious animals of all. I looked up at the bright, clear blue sky. Nothing at all could be seen in that glimmering, impenetrable vault that only a moment before had been robbed of its silence. The sound of a stone being moved or a branch cracking under someone's foot was all that was needed. I walked a little way beyond the car, ducked behind a rock, crept into the sparse shadow of a bush, and watched the sky. I noticed a black spot that grew larger and larger, then another spot and another until at last there were many. Like a missile, something fell from the sky and landed in the nearby tree. A vulture! Then another and another until a large flock of these ugly creatures perched all around me. Slowly, awkwardly, one of them flew closer to the fallen animal while the others, eyes filled with greed, carefully watched its movements. Wobbling ever nearer, the vulture struck its hooked beak into the animal's belly and pulled out the intestines. There was no holding back the others, and they rushed into the fray. At first, baring their beaks and their claws, they pulled and picked at one another, then finally came to some agreement and devoured a gruesome meal. After an hour or two, only cleanly picked bones remained. These creatures are truly the health services of the savannah, and they do excellent work!

On we drove and swayed from side to side as the car passed between sharp, porous rocks. Encountering fewer and fewer trees, eventually there were no bushes, then, not even a single blade of grass. The car bumped its way across the rocks, and we finally arrived at Korichaams—a wide hollow covered with boulders both large and small, a few isolated shrubs, and between them, a small tree. The stones completely cover the ground, and so round and smooth are they that one must walk gingerly to avoid slipping and falling. What is the cause of their smoothness? Over the centuries, the hooves of game animals walking across the stones every morning and evening to drink at the fountain have polished them smooth and round. In those few places where dust and sand cover the ground, visitors to the fountain are recognized by their tracks—zebra, ostriches, guinea fowl, oryx, jackals, and hyenas. In addition to these, and hardly to be measured with two human hands, were the enormous spoors, fresh spoors, of lions. This was precisely why we had come here. At last my son would be able to hear the Lord of the Wilderness roar, and then to hunt him down in his own kingdom. Following the wisdom of old hunters, "where there are zebra, there are lions," we had already heard small groups of zebra breaking through the thicket on our way to this place.

I was overcome by elation and a feeling of solitude in this vast African grassland. A canopy of blue, cloudless sky stretched above us. A few little deep green trees and bushes stood on the plain, with a covering of bright limestone boulders that stretched as far as the eye could see. The water hole was miniscule. One of its openings was so small we could easily have covered it with a single flat stone, and this we did, leaving the other two uncovered. About fifty steps from the fountain stood a small tree with a semicircle of fairly large stones stacked around it—remnants of a shooting pontok, a hide in which hunters could wait and shoot without being seen by the animals. We settled down in it while our friends made camp for the night in a similar pontok near the water hole on the west side. Truly, our pontok was nothing less than enchanting.

It was just a few weeks earlier that a policeman had spent the night here without spotting any lion spoors. He had made himself quite at home in the pontok and, in anticipation of an adventure, ate the muttonchops he had brought with him, then without concern, he tossed the bones in front of the thorn bush barrier. The pontok's stone perimeter wall was about waist high, and the policeman had hastily constructed a roof for the dwelling from branches and brush set against the boughes of the stunted tree growing inside the pontok, leaving open a tiny slit just wide enough for him to see the spring. He made an additional opening in the thornbush portion of the pontok, one barely wide enough for him to creep through should he need to flee his lair. Then, with his rifle tightly secured under his arm, he waited and peered through the slit at the fountain now bathed in pale moonlight.

Suddenly, he heard a crackling in the narrow passageway. Looking with great intensity, he saw a shadow fall across the opening, and in the heavy darkness, two iridescent green eyes. A lion, doubtless attracted by the smell of the discarded bones, had found the entrance to the pontok and was now so close that the policeman could not possibly lean back far enough to take aim. Keeping the rifle under his arm, he closed his eyes and squeezed the trigger. The rifle engaged with a mighty clap of thunder, then there was silence. His eyes still tightly closed, he waited for the lion to pounce on him—yet nothing happened. Nervously, he waited a moment longer, then looked into the opening and saw a body lying about ten steps from the opening. Hesitating a bit, he tentatively ventured out. Before him lay the lion, dead, a gunshot wound in its side. This story is not simply a hunter's tale! No, our good friend, J., related it to me himself, otherwise I would not have believed it. A few days after this kill, he saw the lion's skin himself and heard the story from the policeman's own lips.

Needless to say, we proceeded cautiously and parked the car far away behind a hill so the animals would not be spooked by the strange smell. In the meantime, the ox cart had finally arrived, bearing all of our goods as well as the two ostriches. It is truly marvelous how the natives can detect and read the smallest signs along the way. We had placed two ostrich feathers, held in place by small stones, and had made a pointer with a few other stones to direct the teamster toward the ostriches. The Hottentots had found both birds without much trouble or thought.

As the cart rumbled toward us, both the driver and the rope guide shouted, then quickly unloaded all the bedding. The oxen were urged onward and trekked for quite a stretch. There the ostriches were offloaded and cut open. Next, the oxen had to be cared for as they were exhausted from the long journey. The beasts were led to water and drank their fill, then driven to the east on the broad plain where they found yellow clumps of grass between the stones. Meanwhile, the teamster searched for wood and carried large, dry trunks and branches back to the cart. Because of danger posed by the wild animals, the oxen were to remain tied to the yoke during the night and protected by six bright fires. In order not to lose time, we two ladies warmed up the meal we had brought along and quickly made tea so we could retire to our pontoks before the sun went down.

Our men had the entrails from the two ostriches tied to a stick that the natives carried to the spring and back, a distance of perhaps three or four hundred paces, all the while scattering small pieces of meat to attract predators to the spring. We quickly cut down a few bushes thickly overgrown with parasitic creeper and covered the top of our huts with them. Inside I spread skins, blankets, and pillows, placed the night glasses next to me, and made certain the guns and shells were ready.

The sun set, sending streaks of yellow, red, violet, and green across the sky. We remained silent as my son lay facing the slight opening that allowed him to view the fountain. On the other side lay my friend, Frau M., who was charged with the difficult task of keeping Männe, our dachshund, quiet. This little reddish-brown monster had to come along everywhere, of course, and always barked at the most inopportune moments. How we would like to have left him in the car, but our friends' great dane was there to protect the oxen, and just the thought of that Great Dane sent poor Männe into a panic. Left alone in the car he would have howled all night, so he got the softest skin to keep him happy and was lovingly covered with a coat.

A full, silver moon was rising above the hills in the east, and black silhouettes of trees and bushes appeared on the hilltops. All of nature fell completely silent—no screeching birds, no clicking insects. Without making the slightest sound, bats fluttered around us as moonlight washed over the plain and turned the smooth stones gleaming white. Silvery light dripped from the bushes and glistened on the spring waters that had risen steadily during the course of the day.

We waited together but soon felt tired and decided to take turns keeping watch. Staying awake is quite easy for me, and I have passed many a night without sleep. Thus, sitting there with no one but myself for company, I looked through the branches at the magic of a moonlit African sky. The solitude and the beauty carried me away.

I had spent many nights under the stars, and they had always given me strength. I lay back gazing into the sky, watching myriads of stars twinkle. I searched for my friend, bright Jupiter, in the white ribbon of the Milky Way. Later, Kanopus, the dog, rose with a greenish light gleaming in his eyes—like the mischievous eyes of a beast of prey. Then I watched the Southern Cross as it gradually set. Shooting stars fell along the sky, then disappeared into oblivion. I sat and listened. From far away, I heard a slight noise: the crackling, the rustling, and then the pounding of many hooves in the darkness at the edge of the bush. The first herd of zebra approached the spring. The animals wished to move closer but dared not do so. Suddenly, the stamping and snorting ceased, replaced by an angry neighing: the dominant stallion had detected a hostile scent. It seemed as though the herd had turned round and walked in a circle, for soon they appeared on the other side of the spring. This time they came close enough for me to see their shadows. They stopped and silently froze in place. A silly little colt dared to move closer, but his mother followed and tried to give him some kind advice, but to no avail.

I saw a quick turn of her head and heard a high-pitched scream: the mother had bitten her son. But he remained obstinate and seemed to say, "You can't stop me!" Stamping his hooves, the young zebra walked closer still. The others followed cautiously. The same stallion, recognizing his duty, threw back his beautiful head, whinnied, and approached the spring. Ten steps from our pontok, he stopped; the mother and her son stood so close I could hear the colt smacking his lips and snort with delight as he drank from her. Step by step, over slippery stones, they dared to approach the water's edge. Other herds, encouraged by this one having suffered no ill effects, also approached, and to my delight I saw about two hundred zebra on the plain. A dozen or so animals could drink at one time, always keeping a suspicious eye on the dark spot: our pontok. Männe awoke once, shook his not-too-long dachshund ears, and the entire herd of zebra vanished. About a quarter of an hour later, they returned. After all of them had quenched their thirst, they began flirting and playing. Some females were passionately courted, but woe to that creature when the dominant stallion found him out. He pounced on the competitor, and both fought with hooves and teeth until the would-be lover had been chased away. Whinnying at the top of his lungs, the victor lifted his mighty striped head. This kind of play continued until about two o'clock in the morning when a sudden harrowing scream penetrated the night—an obvious cry of warning. The pounding, rolling, and cracking that ensued is indescribable, but all were gone, the water hole deserted.

I looked through the fieldglasses and noticed a small shadow coming from the east. It was a jackal following the trails of meat. Soon another came, and then a third. They enjoyed the meat and appeared sociable, even sharing their meal. Their howls penetrated the night and summoned their fellows. As we had set an additional three traps with ostrich meat, we did not mind the frivolity. Already one of them had been caught. He howled loudly and angrily at first, then his cries of pain were long and drawn out.

Silence followed, only to be torn asunder by a scream so loud and terrifying that I actually shuddered. My son sat up and reached for his rifle. There he was again, the lord of the water and the wilderness! I grew cold. Männe growled. Strangely, the scream turned into a gruesome sort of screech, a kind of laughter—it was a hyena! It had also come to drink, but only after chasing away the jackals. The hyena ate, then drank. The wind carried the pungent scent of a beast of prey to my nose. The hyena disappeared, but in its place came a pair of animals stepping ever so softly. "Well," I thought, "it must be Herr and Frau Leopard." I was mistaken—it was just a pair of wild dogs. But somewhere close to me a wild kitten meowed.

The sun gradually rose in the cold morning air. For a few moments I had diverted my gaze from the small opening, but when I looked again, I saw two animals standing in the shadows. They were not darkly colored, and I was able to make out a long tail with a tassel at its end. Sleek flanks grew into broad chests, their heads bowed low over the water—but more I could not see. Poking my son, I whispered, "There they are, perhaps." He rose and aimed his rifle. I leaned toward him and whispered, "Go after the smaller one first, she is the lioness." The lioness is more dangerous than the lion, for she will always attack and pursue should something happen to her companion, whereas the lion is a cowardly beast and will leave his lioness to her own devices and run away. Finally, they had drunk their fill and lifted their heads. Despite our disappointment, we laughed, for they were only gemsbok and we could clearly see their horns as they slowly walked towards the bushland.

It grew lighter, and birds began to awaken. Quail were the first to appear. They chirped, beat their wings in the air, circled the water hole for a time, then disappeared. A while later, we heard more robust chirping and a strong whirring—a second patrol of about a dozen birds that circled the water to make certain it was safe, then flew off. After that, the entire flock drifted in like a cloud in the silvery air. The cloud descended, and the birds took possession of the water hole. I find that quail taste very nice, and if carefully prepared, are similar to Krammetsvögel (a juniper thrush). Often when I am in the bush early in the morning, I enjoy watching flocks of birds in flight, and it seems to me that all of them fly toward the same destination, the water. It is helpful to know this, especially if one is lost or if one is in an area devoid of human beings (for water is rare in these areas). Well-worn animal trails that seem to converge into a single path often lead to water as well. Natives are experts in reading and interpreting such signs and trails, and it is perfectly useless to consider oneself superior to the natives and to pretend to know such things better than they. We always let them lead the way, and they have never failed us.

Behind the quail strolls a philosopher, a crowned heron that must have developed a thirst on its wanderings. Carefully, and slowing its pace nearly to a halt, it strutted ever closer. Minute after minute passed as it stood on its long legs, first on the one, then on the other, until finally the heron made up its mind and dipped its beak deep into the water and slowly lifted it up again. Half an hour must have passed before it made the decision to move on. By then, full daylight shone across the land, and we were just about to walk to camp for breakfast when a flock of guinea fowl arrived, walking single file toward the water. What a comical, gabby, and quarrelsome company they were. Tripping about like a bunch of prim old maids from days past, they stopped, tidied themselves, gossiped, quarreled, called one another names, were separated by the presiding rooster, took offense at such treatment, and finally—finally—began to drink. I counted more than eighty of them.

By this time, breakfast was very enticing to us, and we crept out of our hide never having fired a single shot. We walked to camp and found that our friends from the western side of the fountain had already arrived and were just as disappointed as we about the absence of lions. Because all of us had expected His majesty to make an appearance, none of us had dared fire at any other animal. His Majesty, however, had decided not to drink.

The natives had already checked the traps and gathered in three jackals and a hyena. We Africans are tough and obstinate, so we decided to stay another night. At ten o'clock in the morning ostriches appeared, around noontime a few springbok, after that, a good many gemsbok. We did not bother to hunt these animals as we had plenty of meat. Smaller herds of zebra came all day long. Quite early on, we retired to our pontoks, and things went pretty much as they had the previous evening. The zebra were not quite as shy, and we enjoyed watching them as they approached our pontok, some coming within ten or twelve feet of us. No lions had yet come. The night scenery was magnificent. A blood-red moon rose above the eastern hills, beneath which spread an endless, dangerous, thirsty wilderness. The great lion had to come from that direction, and we waited to hear the echo of his hungry roar. Oh, how fervently I wished to hear him roar! He need not even appear—all I wanted was to hear his roar again. But my hope proved to be of no effect. Jackals and hyenas started their nightly howls as the zebra calmly drank, followed by the chamois, the quail, and the guinea fowl. With disappointment, we left our hides. However, we discovered that the lion had been present—though he had not come from the direction of the mysterious hazy blue mountains, but from the opposite side. A native, who was supposed to lead the oxen into the veld to graze, excitedly came after us. About three hundred yards from the fountain, fresh spoors from that great cat could be seen circling the water. No doubt warned by the strange, unfamiliar scent, he had not dared to come any closer. We discovered the very spot where he had crouched to assess the situation, but something must have given him a feeling of unease, so he departed from here and headed east, back toward the blue mountains. We lost his spoors between the cliffs.

13. Etoscha and the Waterberg

During other holidays, and on our way to the Etoscha Pan, we drove past the copper mining town, Tsumeb, sixteen kilometers southeast of Lake Otjikoto. The journey to the lake is not too difficult, and our car took us through dense thickets of brush, past deciduous forests and the occasional palm tree. As we neared the lake, the only thing to be seen was an unsightly pumping station situated just a bit higher than the lake that used to be the primary means of supplying Tsumeb with water. Unfortunately, it desecrates the landscape. Otjikoto is a crater lake with calm, deep-green waters well embedded in a rocky basin that reaches a depth, according to local knowledge, of one hundred meters. The limestone cliffs that tower around it are overgrown with shrubs, tropical trees, and aloe plants, all of which are reflected in the deep, clear water. A single, narrow trail—access to which is not easy—leads down to the lake. Cattle and small stock, as well as wild animals, use this trail, and the lake supervisor told us he sets his traps along the path at night and catches large predators. He actually took us inside a shed and showed us six leopard pelts he had spread along the walls to dry.

The waters of the lake are deceptively still and dreamy, yet they have much to tell us—many eerie legends and stories. Otjikoto has yet to be explored scientifically, but it is generally believed that a subterranean karst connects the lake with another lake, Lake Guinea, and a second karst connects it directly with Tsumeb. It is apparently unwise to swim in the lake as a subterranean whirlpool is said to pull the swimmer into the depths of the lake so that the body never reappears. At various times both white men and natives are alleged to have vanished there; cattle have also fallen in and their bodies never floated ashore. Natives fear these strange waters where evil spirits roam. Indeed, the many Vambos who must pass the lake on their way from Vamboland to Tsumeb prefer to endure thirst and arrive in Tsumeb dehydrated and exhausted rather than draw water from Otjikoto.

Our German troops, not wanting to surrender their cannons and rifle ammunition to the enemy, sent all of it into the lake for safe keeping. Someone trusted a woman with this secret, but she sold the information to the enemy and collected her contemptible money. The enemy searched for the materials and was successful in fishing out the cannons, but they were completely useless; several crates of ammunition were also recovered and the shells eventually sold to the Germans, though most of them were ruined: this is a fact. It was later said that the lake was reluctant to return the goods that had been entrusted to its keeping. One of the most eager salvage fishermen slipped from his boat, sank into the deep water, and never again saw the light of day. That was his punishment, people said, the lake took its revenge! It is a pity, though, that the lake was unable to claim the life of the woman who betrayed her Fatherland.

The Etoscha Pan is a game reserve, but accessing it is far from easy. One must apply well ahead of time for a permit to enter the protected area, and one must state the purpose of the visit. No arms are allowed. We were granted permission to enter and stored our rifles in Tsumeb. Driving in a northerly direction from Lake Otjikoto, at first through savannah, mopane forests, and groves of palms, we followed a well-marked track, indeed, the one used by Ovambos coming south to look for work. They typically march in groups as a protection against wild animals and carry their few belongings on bamboo sticks, along with a bundle of provisions and a calabash with water. Some of them had never before seen a motorcar and stared in both amazement and fright at the hissing monster. Horrified, they jumped aside when we blew the horn, though the braver among them absolutely shook with laughter.

We occasionally encountered Bushmen—treacherous riffraff. Just an instant ago, we saw a dark, dwarf-like figure at the side of the trail; now it is gone, swallowed up in the wilderness. It is impossible for him to be far away; the savannah is not a place to hide. We strained, looking in the direction of his disappearance, and indeed, there he stood, paralyzed, as motionless as a tree trunk, completely one with his surroundings: the bushveld. I have often been told that the motionless stance is a well-known trick of the Bushmen and the Hottentots when they are hunting or when they face danger.

In the early afternoon, we drove through extensive bush forests, of which there are many in the Etoscha Pan. Ages ago, the Etoscha Pan was a large, salty lake that dried out over time. I am not sure if my impressions were accurate or if I simply imagined it, but these forests looked different. The mopane trees had almost no leaves, the dense scrub here was even more impenetrable than the scrub in other places, everything was covered with thorns, the palm trees were all stunted in their growth, and the greenness of the vegetation appeared faded and far less verdant. We soon left the forest. Pointed, thorny bunches of ugly, sour grasses rustled harshly; a few single, diminutive, reddish-brown prickly scrubs stood in the limestone rubble. Suddenly, the Etoscha Pan lay before us, a wide, gleaming, glittering, endless white and desolate pan. Sky and earth melted together in the glimmering, bluish-gray air. But where was the sky? Where was the earth? Where did the one end and the other begin? The white salt desert stretched endlessly, farther than the eye could see, and I was horrified by all this desolation.

Finally, my searching eyes found some rest as they focused on a clump of trees—the game warden's station. Because we had already been officially registered, we encountered no difficulties; indeed, the stationmaster, a South African major, assigned an armed policeman as our guide. He came and took a seat in our car, and we began our journey across the pan, passing large herds of gemsbok, wildebeests, hartebeests, zebra, and occasionally ostriches. At the center of this great expanse stood a few small islands amid the marshes, and a small forest. Huts that had served as lodging for the Governor General of the Union and his entourage were located on the higher ground, and the police officer had been instructed to accommodate us in these huts. We would certainly hear the roar of the lion, we were told, but probably not see him; but at dawn we would have ample opportunity to observe the wild animals.

I prefer not to sleep in abandoned pontoks, as I am afraid of snakes lying in wait, and much prefer to sleep in the open. We sat at the fire for a very long time and listened to the officer's tales. When I entered the pontok, I held my lamp high and shone the light everywhere. As far as I could determine, everything was in order. A little mouse and a gecko rustled in the dry leaves; a few stars shone through the branches that collectively were the roof. In the distance, I heard a bird call but I did not recognize its call; it sounded to me like a child crying in its sleep. Sometimes the call was softer, then it grew louder, but was continually a part of the night. A jackal walked around the hut howling. A hyena screamed. The lord of the wilderness, however, did not favor us with his voice. A cold wind blew across the pan and shook our huts. Etoscha slept, glittering in the moonlight.

Another time, another holiday, we went to the Waterberg Plateau via Hamakari. There I got out of the car and made my way through dense thorn bushes to a former battlefield of 1905 (the German-Herero War). I paused before the old warriors' graves, standing there for a long time, those giants now resting quietly in our hard earth. They had triumphed; they found life's deepest meaning in their readiness to do battle, in their willingness to sacrifice their lives for our Fatherland. A breeze blew softly through the high grass. A tiny bird sweetly sang its song from a thorn bush. I stood there in place of the people who had to bury their loved ones in this very spot; how well I knew their feelings, and though I was not personally acquainted with them, my thoughts were with them and I honored them.

Our emperor had a stone prepared and placed there with the following inscription: "The country into which the German eagle has struck its talons is German and will remain German." German eagle, you are flying once again!

We drove around the massif of the Waterberg Plateau, whose canyons are so wide that farms have been established in them. We climbed to its tablelike top, wandered along it, and enjoyed the panoramic view of our African country. Strangely, we did not encounter any beasts of prey there. Visitors, who often stay only for a few weeks, report that we old Africans have run out of luck. A dear, long-time friend of ours owns a farm in one of the Waterberg's canyons; his wife and children had climbed for years without guides and without weapons. They laughed at me when I asked them if they had ever been afraid to do so.

The final day of our trip had come and we were, as usual, in quite a rush to get home. As we had done on so many other occasions, we decided to drive through the night and brave the long, dark, threatening shadows of bushes and trees that flew past us in the blackness. The headlamps of our car shone with unnatural brightness along the lonely tracks. Frightened rabbits, blinded by the lights, zigzagged in front of us with oddly placed jumps; but in their terror, they simply could not find their way off the tracks. Along the sides of the road we saw flickering green lights—reflections from the jackrabbits' eyes. Here and there, tiny flames seemed to flare as small groups of jackrabbits raced back and forth.

Jackrabbits have short front legs and long hind legs. At night, they break into gardens and fields, causing a good deal of damage. They are nocturnal animals and are rarely seen during the day when they sleep in their subterranean burrows. At dusk they appear, and wherever they are found they should be shot. They are edible, but when they are skinned, one often finds a multitude of thick worms just beneath their fur, worms that have eaten their way into the flesh. Hence, I do not favor eating these wild rabbits. Our car's headlamps also caused owls and eagle owls to fly off in fear and a jackal ran for its life. Jackals are much despised in South West, for they are scavengers, mere beggars, among the beasts of prey. In broad daylight, they jump at the udders of sheep and goats and tear them off; they take lambs and kids by the throat and strangle them; they are a scourge for rabbits and game animals that have lost their health. Following the tracks of lions, they gorge themselves on what the big cats have left behind. But they are tenacious—if a jackal is caught in the jaws of a trap, it will bite off its foot to free itself.

14. 1934: The Year of the Torrential Rains

South West had experienced seven years of severe drought. During the time of year when the rain typically falls (October–April), the sky was relentlessly blue, and when even the smallest clouds dared to appear, a hostile wind immediately drove them away. The riverbeds were dry, water holes had to be dug ever deeper by both native and white, and still they eventually ran dry. Not even a trace of moisture remained in the ground. Water in the swamps had receded considerably, and now even they were dry. Large cracks tore the clay earth; bushes and trees were mostly naked; beautiful wild flowers no longer sprouted though their seeds slumbered in the hard earth. Grass had become pale and yellow, and the wind broke off the dry blades. Powerless stumps stood in the parched ground. Animals in the bush suffered tremendously: springbok and chamois that normally require very little water now used the hooves to dig out moist roots from ground as hard as stone; kudu jumped fences and ventured ever closer to human dwellings. Twice, on my way to school, I met a completely emaciated kudu doe. Skin and bone ostriches got caught in the fences that surrounded our houses; young ostriches were nowhere to be seen. Female game either bore their offspring too early or did not bear at all. Guinea fowl did not mate and their flocks grew ever smaller, and when a hen was shot, it turned out to be very old (their ages are determined by their combs) and nothing but skin and bone.

Reservoirs and springs on farms all over the country were completely dry. Gardens and pasturelands were no more; consequently, cattle and small stock died by the thousands. Even the reserve pastures could no longer support them. A bit of grass and water was still available in the eastern part of the country, but the animals were far too weak to be driven overland and had to be loaded into freight cars and transported by railroad. When the livestock arrived at their destination, all the expired animals had to be offloaded, and a terrible pestilence-like stench rose from those carriages.

The situation was even worse in the southern part of the country. Thousands upon thousands of sheep died, and it was not unusual for a single farmer to lose between six hundred and a thousand animals. Farmers had become so poor they were starving themselves. Coffee was made from the wild berries of the witkat tree; their meals consisted of wild foods they collected from the bush and starving livestock they judiciously slaughtered. Only the better heeled could still afford to purchase maize meal. Milk and butter were simply unavailable, as the few drops of milk the cows still produced had to be used to preserve the starving calves. Littered along the animal paths that crisscrossed these huge farms were dead beasts, and the wind carried the dreadful stench of rotting carcasses in every direction.

The government had asked that prayers be offered in every church and by all denominations. It was December 1933, and every day people looked into the sky and saw that clouds were forming in the east. They became increasingly blacker, ominously so, as well as heavier and ever more massive. The mountains appeared to be closer than before. Then, finally, it began to rain. Christmas Eve saw rains of such force that the river truly roared through the Klein-Windhuk valley. It became impossible to cross, and provisional quarters were organized for those who lived on the other side of the river and could not return to their homes. It rained for days on end, and we wondered whether it would ever cease. But South West is, after all, a land of extremes.

Rivers flowed wildly, and the rain turned the dusty roads and paths into impassable swamps. On the day after Christmas, forty-eight cars were stuck in the mud; trains arrived late or not at all; the railroad dam had been breached in sixty-four different places; and the railroad bridges were severely damaged or in such danger of outright collapse that trains were not allowed to cross them. Water roared down from the mountains. Streams grew into brooks, brooks into rivers, and a wide, yellow wall of clay and water rolled into every omuramba (stream or riverbed) seeking to join a larger river that would eventually lead to the sea. So great was the force of the flood that huge boulders were torn from their foundations, yet in all of this roaring tumult one could still distinguish the rumbling of stones beneath the water.

Huge clods of earth were torn from the riverbanks and dammed them up so that the water had to create new channels. The massive clods also caused flooding, tore apart gravel roads, and carved deep rifts across the countryside. Many large trees that had been ripped from the riverbanks now drifted in the floodwaters, then shifted across the rivers only to get caught on the banks. This caused disastrous backup flooding, and the savannahs were turned into a morass of mud and debris. Water flowed into earthen dams until they weakened and burst, a circumstance partially exacerbated by mice that, during the drought, had burrowed tunnels into the dam walls that effectively destabilized them, and this, coupled with the work of termites, caused such damage that these dams collapsed under the pressure of these multiple strains. Needless to say, the water released in this way resulted in even greater damage.

The Swakop River received so many in-flows that it ran all the way to Swakopmund and into the Atlantic Ocean with such force that technical emergency personnel had to be ready day and night. Neighborhoods were flooded, houses were destroyed, trains could not run, and food and clean drinking water were in short supply. The bridge across the Swakop River had been torn away so transport to and from Walfisch Bay was only and irregularly maintained by air flights. When the floods finally subsided, the citizens of Swakopmund constructed a primitive kind of cableway across the river—though later on passengers from Walfisch Bay were carried across the river by natives.

The Avis Dam, which supplies Windhuk with water, was in great danger of bursting. Day and night, prisoners and the unemployed worked to prevent a break. The overflow gates were too small, and on both sides of the dam the buildup of water caused a wide, natural stream of release that now threatened the dam. Sandbags and bundles of twigs bound with string were pressed into the openings, yet still the water roared over them, opening huge holes in the soil as it went its way. The neighborhoods in the Klein-Windhuk valley were in grave danger.

School started, though many of the schoolchildren did not arrive because the roads were still impassable. Teachers from many of the schools who had spent their holidays in the Union arrived late, often coming by aeroplane. Others had spent many days on trains, in motorcars, in carts of various kinds, and on special trains that only ran between two stations because the rails had been damaged in more than eighty different places.

Every morning we walked down to the foot of our mountain and found the riverbed filled to overflowing. When the water was too high for me to cross, Bitz had to carry me. If it was low, I waded through water that often reached up above my knees. Besides the books in my satchel, I always carried shoes, stockings, and a towel so that when I reached the other side I could change my clothes. During these weeks of raging rivers, I had to leave my house early enough to arrive at the school with sufficient time to change my clothing before the bell rang. I did not mind the hem of my dress being wet, but often much more than that was water soaked. Normally, I stood or walked about my classroom, but one day I sat on top of a child's desk in front of the class and did not move. After class, a curious pupil asked me, "Frau von Eckenbrecher, why did you sit still today?"

"Well," I answered, "I'll tell you. My underwear was wet and clammy, and I couldn't bear to walk around in it, so I sat down to be more comfortable."

"I thought it was something like that," he said.

The next morning he came up to me and asked, "Do you have dry underwear on today?"

When the rains did not stop, I began to keep dry clothing at the school and simply changed there. Even though by this time I had become a rough-and-ready African, getting wet morning after morning for eight weeks was just a bit too much.

Because of the rain, we were unable to journey into the deep countryside on the second Easter holiday. It had rained moderately during the afternoon, and though the river was flowing, it had not exceeded the capacity of its bed. Around six o'clock in the evening, the Auas Mountains were hidden in a thick, murky cloud. A neighbor friend of ours, the person who lived below our house and very close to the river, asked Bitz for help. Her valuable race-horse had been injured. I accompanied Bitz.

The sky was covered with clouds, and darkness fell quickly. We sat in the living room under the light of an electric lamp when the wife of another friend, a surveyor, who lived even closer to the river, arrived, anxious and wet, carrying a large bundle under her arm. She had just received a telephone message informing her that the river was rising quickly. She feared her property would be flooded and had gathered a few belongings to bring to Frau St. for safekeeping. The telephone rang. Because news of the rising river had circulated through town, Frau St.'s son-in-law enquired if all was well. Bitz calmed him and assured him that all was well, and just at that moment he noticed a small stream of water—like a glittering snake—flowing beneath the closed door. Bitz had just enough time to shout into the telephone, "The water is coming!" before the line went dead. Immediately the light went out and we sat there in the dark. All around us roared the water, and when I opened the dining room door, it came rushing in toward me. A house fully wired for electricity has neither candles nor kerosene lamps, so Bitz struck a match, and as far as the light reached, there was nothing but water—water that splashed, lapped, and roared. It flowed across the threshold and into every room in the house; there was not even time enough to lift the rugs.

We rushed to open doors and windows so the water could escape, then gathered, as quickly as we could, important papers, jewelry, silverware, weapons, and blankets, and splashed our way through the water and up the hill. The surveyor came running and asked Bitz to drive his three cars to safety, but the water had risen so high that the engines would not start. He left the motorcars and swung open the stable door and headed toward the stall where the purebred horse was kept. Cutting open the fence around the stall, Bitz ran to safety. Frau St.'s beautiful racing horse stood up to his belly in water but could not be coaxed to leave the stable. Unable to firmly step on the ground, he remained there and suffered intense pain. Our dogs paddled and happily swam through the floodwaters. We tried to reach a third neighbor, a

South African, to warn and help him, but his house was dark; the door was open with no one inside.

The bright headlamps of a car shone from the direction of the mountain. It could only be Bitz. Because the batteries in our flashlights were spent, I carried with me a kerosene stable lamp to provide a bit of light. Completely soaked and badly shaken, all our neighbors came with their bundles and found refuge with me. A pile of wet clothing became a feature in every room.

Silently we stood at the windows and in the doorway listening to the roar of the flood and the din of the storm. The houses were left to fate alone. Throughout the night, we heard pleas for help coming from the opposite side of the river, but we could not cross the torrent to help those poor souls. We did, however, notice cars coming over the pass, and a chain of lights soon appeared. The blaring car horns actually gave us a sense of comfort as we now knew that the people living on the opposite side of the river were finally receiving desperately needed assistance. Lights and searchlights flashed up and down the river. Suddenly, we heard a long, dark, drawn-out groaning.

"What was that?" someone cried.

"A collapsing house," replied the Ds.

We were unable to see anything, but wondered whose house it might be. Quickly, I prepared hot tea with cognac—which everyone enjoyed very much—and when the rain subsided, we took our lanterns and trudged down to see whose houses were still standing. Frau St.'s home was still in place! The next day we had a better look round and noted that the water had rushed through all the rooms at a height of nearly one meter, lifting the rugs and toppling the furniture as it did so. Had the water gone a centimeter or so higher, it would have reached the keys of her precious piano.

Everywhere one looked the destruction was truly indescribable. Half of the Ds.' house and all their furniture had disappeared in the flood—no doubt floating down the river somewhere.

I did my best to provide sleeping quarters for all my guests; after that, we simply had to wait until dawn.

At daybreak we began to see the full extent of the destruction. On the opposite side of the river, one house and all of its outbuildings had completely disappeared; the only remaining thing was part of the house's foundation. Another house, located some distance from the river, was still under water up to its windows. The soil around it, the trees, the fountain, and the windmill water-pump had all washed away. The river had powered its way through a small eucalyptus forest, felled nearly four hundred strong trees and washed them downstream; the house that stood to the side of this little forest had collapsed. It was destruction and desolation as far as the eye could see. Around eight o'clock in the morning, the volunteer fire fighters arrived to help those affected by the flood. Tying the men to secured ropes so they would not be swept away by the water, they began their work. A growing chain of cars parked on the other side of the river was evidence of many curious onlookers as well as those who had come to prospect for spoils.

A piece from the roof of a truck was barely visible in the sand; a child's bed lay stuck in the bushes; we eventually identified a peculiar-looking object as the motor from a windmill; and a piece of metal that had once been a basin but was now flat as a pancake came into view. The Ds searched for the remains of their belongings down river and did find the top of their desk and its legs, along with the upper part of a sewing machine and pieces of their beds. A beautiful, brown, silken evening gown, now completely dry, had hung itself on an uprooted bush and flapped in the wind. In the distance, the door to a wardrobe was found; just beyond lay its mate. And thus did things continue.

Fortunately, no lives had been lost. At the very last moment, a number of courageous men had been able to save nine children from a collapsing house by strapping them to the backs of horses. When the waters finally subsided, we helped Frau St. dry out her belongings. We did the same for the Ds. Under the current government, everyone was responsible for his own personal damages. Who else should shoulder the burden of replacing lost or destroyed property? I think the tax collectors should have had enough sensitivity to forego gathering taxes for half a year. "But why did you build so close to the river?" was their standard line. For as long as those houses had stood in Klein-Windhuk—some fifty years—there had never been such a flood. Later the cause of this catastrophe was determined with a high degree of certainty. The Avis dam had not ruptured, but further up river, near the Gobabis Railroad, a break in a dam had occurred. The hole was temporarily sealed with debris, but when the pressure of the water became too great, the floodwaters burst through and, according to one witness, caused a tremendous surge of water to roar its way downstream. The flow continued until the end of April, though never again with its original force.

The citizens of Klein-Windhuk assembled and planned to petition the government to undertake some projects to prevent such disasters in the future. To this day, however, it has not been decided who actually owns the river—the city or the state government. One entity points its finger at the other and assigns it full responsibility for the damages; consequently, nothing ever gets done despite the fact that both city and state collect taxes.

After the rainy season, the landscape was beautiful. Everywhere there was green foliage and blooming flowers. Yellow morning stars climbed all the way up the mountainside, and varieties of flowers we had never before seen bloomed wondrously. Many bushes were covered with red blooming creepers, and the countryside offered so much natural food that many natives—

always children of the moment—left their places of employment. They had enough to eat, so why work? Plants and animals breathed new life. Innumerable tortoises of all sizes crept along the roads and trails. In the swamps, along puddles—indeed, wherever there was some water—small, medium, and large frogs croaked. At many watering holes, hundreds and hundreds of little tadpoles twirled in the water. Later these fully grown creatures hopped through flower and vegetable beds, relaxed on all walkways, and without invitation entered the yard, the rain barrel, and even the kitchen.

Millions of plump beetles, yellow fire bugs, stink bugs, and vinegar bugs found joy in their existence. By lantern light we collected beetles and in a single evening easily filled a large clothesbasket to the brim and fed them to the chickens. Billions of flies buzzed about, ranging from the thick, disgusting, greenish gleaming blowfly to the humble housefly. As soon as a door opened they swooped into the house. Hoards of them buzzed around the cooling chamber in the garden because they could smell the meat inside it. Leftover foods could not be saved. Bread molded quickly; after just a day the rice was musty, fruit spoiled, butter was spotty, milk soured, and despite every possible precaution, after a single day of storage meat was full of white maggots. Bed bugs and fleas appeared, cockroaches celebrated with great orgies in the kitchen, moths and silverfish ate anything and everything in the closets, and crickets chirped and screeched outside. Bulging-eyed dragonflies buzzed everywhere; thick swarms of male mosquitoes danced in the evening light.

The plague of mosquitoes brought with it that dreaded disease, malaria. A hitherto unknown strain of malaria tropica raged throughout the country and caused many deaths. Physicians were summoned to distant farms, and since comparatively few cars were in working condition, aeroplanes were used. The doctor arrived in an aeroplane that somehow landed in the vicinity of the farm. More than once he crash landed, often walking for an hour or so before finally reaching the farm and its inhabitants. There were times when he arrived too late. Many of the ill were airlifted to Windhuk where the hospitals were already overcrowded. Despite one's best efforts, many a farmer's call for help had gone unanswered. On some farms the entire family lay sick in bed, but no one else knew of it and no one came by because the roads were impassable.

Typhus was widespread, and outbreaks of scarlet fever, measles, diphtheria, and whooping cough were a result of the great year of rain. The plague of rats and mice was unbearable. Every evening we set twenty-seven traps around the house; twice each evening they were full plus an additional trapping early in the morning. The terrier went wild with all the rodents, and the cats were fat and refused the food handed them. I personally identified six varieties of mice on the mountain and compared them with Brehm (a standard

German biology text). There was one variety I could not identify, for it had grayish-white spots, the tail of a squirrel, and large ears. I was inclined to believe it belonged to the mongoose family. They jump and climb exceptionally well, and can squeeze through the narrowest openings. Although we kept our screen doors closed, several came into my bedroom at night and jumped on my face, and when I reached to brush them aside, they emitted a strange, trumpet-like sound. I set four traps in my room, and when they were caught, they proved to be most tenacious and terribly angry, biting at everything. After a great deal of effort, I was finally able to get rid of my gray, nocturnal friends—to see one now, thank goodness, is a rarity.

The large gray cliff rats damaged plant life, especially the valuable cacti and succulents. Rats and mice also climbed the fig trees and devoured all the figs, leaving none for us. Locusts were the worst of the plagues. On the far horizon was a dark cloud that moved with the wind and came ever closer. At first, just a few appeared, then more and more until, darkening the sky, they flew as thick as snowflakes. It made me dizzy when they buzzed close around me. The fertile Klein-Windhuk Valley, with its many private and public gardens, seemed most attractive to them. Rarely did these creatures fly past the gardens without causing damage; indeed, they often stayed on for days, and whirled about by winds that blew hither and thither between the mountains—it was not easy for them to find their way out of the valley. When a swarm approached, all throughout the valley one heard a loud outcry, then the deafening racket began. Each person tried his best to protect his garden from being devoured, and together, people howled, screeched, and squeaked until they were hoarse while waving sacks and bed linens, beating pot lids together, blowing into old cow horns and whistles, beating drums, and filling large gas canisters with rocks and rattling them. Others beat sticks against empty tin cans or simply rattled them. Far above all these people, the sun's light was blocked out by the swarm, yet when they flew low, grown-ups and children jumped contortedly and beat the nasty things however and wherever they could.

To create a heavy smoke, we quickly wet down bushes and grass, then lit them, and in the midst of the terrible racket the locusts made, heavy clouds of smoke rolled through the valley. The locusts, however, did not appear to be affected but flew about us, hopped all over the ground, hung onto twigs, and were absolutely glued to flowers. Despite all our efforts, luck was not on our side and we surrendered, exhausted from the struggle. Not long afterwards, the trees and bushes were no longer green but a rusty brown; flowers and stems were bent down from the weight of the locusts hanging on them; the ground was also brown. Yet the wretched things continued whirling round us. With every step I took I crushed them beneath my feet. The locusts flew into my face, my hair, and hung on my dress. Those that flew over my head dropped their green waste upon me as a kind of calling card. I simply do not have the capacity to describe how disgusting a plague this is!

Once the locusts settle in, everything is lost. The morning wind will drive them off, but I was never quite prepared to walk through our garden after the locusts had visited us. My large, beloved casuarina bush looked like a straw broom, the palm trees were skeletons and never recovered—eventually I had them removed. Where before flowers had bloomed, now the garden was empty, here a bare stalk, there a nacked one. A few locusts still jumped about in all this devastation. Strangely, I began to notice that all locusts do not look alike, that some have almost human faces, and we were able to detect facial similarities between them and certain people we knew and even gave them their names!

When we drove along the dirt roads we frequently encountered great swarms of locusts and had to stop the car because we could no longer see. On these occasions we saw their first developmental stage, the pedestrian stage. The tires of our car squashed their way through these blankets of walking locusts, as deep as the length of an index finger, continuing on until the tires had gone completely green and so coated in dead locust glue that they no longer turned. Even trains were delayed by locusts, sometimes for hours, and the effort involved in clearing the rail lines was massive.

The government went to great lengths to destroy the next generation of locusts and required farmers to inspect their properties and to report the findings. If larvae were found, they were sprayed with poison by specially trained workers. If the young locusts were at the walking stage, they were also sprayed. The government even attempted to kill the swarms by spreading the poison with aeroplanes and many were killed in this manner; unfortunately, the poison settled on trees, bushes, and even the grasses of the veld, and when the cattle ate these plants, they also died. Natives who gathered wild foods from the bush became ill as well.

A not-ineffective method of controlling the locusts is to dig trenches so that when the walkers approach and wander into the trenches, they can be poisoned or destroyed by fire. But do such small things truly help against the huge masses? Nature itself always strives toward a balance. Often a particular species of fly accompanies these swarms and lays eggs in the bodies of the locusts. The worms that develop cause their host to die. In some places piles of dead locusts were found with these worms in their bodies. The natives also like to eat locusts. If a swarm descends near their village, at daybreak women and children march with sacks and gather the locusts—which are stiff and lifeless from the cold of the night—stripping them from bushes and stems. This kind of work cannot be done, of course, without loud screaming and cackling. The natives start a fire, and as they fill their many sacks, they eat piles and piles of locusts. Many times I observed them gathering these creatures, and the filled sacks moved like waves because the locusts, rustling and buzzing inside the burlap, are taken alive. Natives eat them roasted or ground into flour and made into mush or baked into a sort of bread. Locusts are said to be sweet to the taste and quite filling.

> We are present as insects,
> And have tiny, sharp claws;
> To worship Satan, our Lord,
> And show him our respect. (Goethe)

Indeed, this is what they do!

In this year of the great rains, the natives found an abundance of wild foods, more than they had found in all the past drought years combined. Wild berries and fruits are rare in this hard country, but the natives have developed skills to find them. They dig up unkies, a type of mealy onion, which is also a favorite food of baboons. Unkies are roasted or cooked and mashed. Natives also find truffles and set them on hot ashes and sell them to the whites. We prepare them as we do French truffles—though for me they are not a particular delicacy. Ozonbanies taste very nice, a kind of nut that grows on long vines, and in many a farm household they actually replace almonds. These nuts are very good when fried in oil and salted. One can find wild cucumbers, but one must be very selective, for many of them are bitter. Wild korintjies, small red berries that grow on bushes, are very popular, and while still green, are used by natives in brewing sugarbeer.

Tsamas are wild melons that have saved many a person from dying of thirst. In places all over the landscape, hundreds of them lie scattered across the ground like green bowling balls. The melons that are not bitter actually have no taste at all but are meaty, juicy, and quench one's thirst. Men, horses, cattle, and wild animals all appreciate the tsama melon. Their seeds are roasted and eaten. The more frugal South African housewife mixes tsama with fruit mush to make it stretch, and the tsama flesh takes on the flavor of the fruit and thus helps to supplement the household supply of food.

The nara, a fist-sized melon that grows at the edge of the Namib Desert, is also popular and is the mainstay of the Topnaer-Hottentots who eat it raw or dried. When dried, it resembles flattened cattle dung—though it is supposed to be a delicacy. Natives tell about a mysterious bush that grows at the edge of the Namib, and after a man eats the berries of this bush, he is either quite drunk or "runs to find a woman." Strange stories constantly circulate about the exploits of white men who accidentally eat these berries, though I have not been able to determine which of these tales, if any, is true.

Now the Windhuk natives are called "city coats," and their village, a conglomeration of huts made of corrugated tin, has developed its individual culture—if one can properly employ that word in this instance. All their houses are pretty much the same size and constructed of sheets of tin. The streets run perfectly straight; they operate a hotel for natives, as well as a café and dance hall; they have their own bands for concerts; and at their local pub a special kind of beer is served under government control. Both males and females love to parade about in beautiful clothing; some even own a car. The older natives still prefer to speak German. I once had occasion to hear a high South African official's wife speak the most excellent German, and when I complimented her, she explained that she had learned German from her servant, who adamantly declined to speak English and who would only stay on with her if she spoke German. So she learned to speak German very, very well. The Windhuk natives seem to have forgotten their traditional ways and customs, and now one must sit beside a campfire, way, way out in the bush, with a hired, uncultivated native guide, if one is to learn about the old, traditional ways. Such natives are typically very shy and suspicious, and only once one has gained their trust, which is a rather slow process, do they begin to speak of such matters.

15. Of Diamonds, Gold, and Other Treasures

For gold contend,
On gold depend—
All things and men.
Faust I, 2800

The very first man who dares to venture in a new land is a special kind of person: the prospector. It matters little to him whether the new land is buried under a thick sheet of ice or bakes under a relentless tropical sun. Prospectors do not come from any one country, neither do they belong to a particular social class; rather, they come from all parts of the world, belong to every conceivable social class, and speak a plethora of languages. What they do have in common is the dream of finding gold or precious stones and becoming rich. Prospectors were mining in South West long before Adolph Lüderitz hoisted the German colors on the shore of what is now called Lüderitz Bay. For this type of prospector, no landscape is too immense, no mountain too steep, no canyon too wild; they will suffer hunger and thirst until they go mad; it matters not at all if the sun dries out their bodies or the cold nights stiffen their limbs. They penetrate the lonely solitude of the wilderness, the ocean, and the endless sand dunes, driven by a single passion: gold and precious stones.

Another species of prospector never reveals the secret of why he wishes to prospect, but it is certainly not an addiction to riches. For long, long ago these sorts gave up everything that, in the eyes of others, makes life worth living. They have no home, no family, they neither write nor receive letters. Were they ever young? Are they now old? Their faces look like furrowed leather, their beards are unkempt, their hair is as colorless as the khaki clothing they wear.

There is another category, perhaps a third category, of prospectors, nay, adventurers, whose ranks are comprised of rascals, spies, deceivers, and con artists, and whose principle aim is to separate people from their money and salt the fields to entice naive people to purchase land from them.

All of these sorts have one thing in common: they love the bush and the savannah, and feel perfectly at home there, and are equally repulsed and suffocated by the confinement of towns or even a single room. Hence, they are driven by a deep longing for something else, something different. How many of these men have disappeared? Somewhere beside a lonesome path lies a nameless grave; somewhere in the sand of a dune lie their bodies, drying up. Few enjoy what their prospecting gives them, for as soon as they find gold they squander it. Mining diamonds has been profitable, but not necessarily for the original prospectors, many of whom have not been able to enjoy the fruits of their labors. But for those clever men who rode on the backs of these prospectors, it has been enormously profitable.

South West enjoyed quite an economic boom when the diamonds were first discovered. The origin of these precious stones is still not known, but they are found loosely strewn in the desert sands. After South West became a mandate of the Union, all the German diamond companies were obliged to sell their rights to the large South African mining concern, Consolidated Diamond Mining Company. C.D.M. has so severely restricted the mining and trading of diamonds in South West in favor of their South African operations that unemployment has taken on frightful dimensions and contributes to the economic depression in South West. Lüderitz Bay, once preeminent in diamond mining and trading, has lost much of its importance.

The Mandate government also has declared certain diamond fields off limits, though time and again daredevils and desperate men venture into those lonely, desolate sand dunes to prospect. Occasionally a short note appears in a local newspaper: in some bleak spot a camel patrol found the bodies of white men or natives who died of thirst. Next to their dead bodies lay sand sieves. Maps and drawings may be secretly purchased from men who claim to have been in these places and know whereof they speak. The police, naturally, keep a sharp lookout for these charlatans and employ both white and native stoolpigeons to find them. A host of them now sit in jail with plenty of time to think about the consequences of breaking this particular law. But despite all of this, there is an active illegal "diamond exchange" wherein uncut diamonds trade hands. Certain German and South African nationals are suspected of dealing in illegal diamonds.

Woe to those people whose cars are stopped by a patrol at the border of the restricted area or even in the forbidden zones (which are not even clearly marked), for the officers mercilessly slice open car seats, coats, and tires, and demolish the radiator and the gas tank. Shootouts have even been reported, and the wildest stories imaginable are told, some of which I personally know to be true. One man, for instance, had a large photograph of himself nicely mounted and framed with the intention of sending it via airmail to his bride. Diamonds, however, had been carefully set into slits between the picture and the backing. Another man threw diamonds into the bottom of his canteen and poured plaster on top of them. A woman had baked this man a loaf of "diamond" bread, but felt she had not been sufficiently compensated for her efforts and went to the police. Someone once drilled a hole into the leg of a chair and set diamonds in the cavity. Customs officials noticed that he handled this chair with unusual care and became suspicious; next came a farewell voyage to Germany! A certain woman, who did not appear the least bit suspicious, had diamonds hidden on her person. Her stupid husband divulged this to some of his close friends, who then decided to rob her. Two of them went aboard ship to wish acquaintances of theirs a "pleasant journey." The "friends" planned to wait for the woman and rob her once she entered her cabin. Unfortunately for them, this woman had other plans and stayed on deck at the railing until the ship sailed. Unbeknownst to her, their plan was foiled and the stones were safe.

A particular man who regularly traveled to Swakopmund boarded a certain steamship just before its departure time. He strolled around the deck for a while, but not, according to his accomplice, because he longed to sail home. Instead, he had packed diamonds in the extreme ends of the horns of wild animals and was shipping them home. Diamonds were sometimes hidden in a small bag tied to the leg of a passenger, and when a hint of danger arose, the string could easily be loosened and the diamonds inconspicuously deposited on the ground. X-raying the body to find illegal diamonds is of no use as they cannot be detected by x-rays. Young smugglers often choose a rather unpleasant human orifice for stashing diamonds before boarding the ship, purportedly with great success. It is the same cavity used by natives who smuggle diamonds by asking to leave their workplace in the diamond fields for just a short break. Now they are forced to bend their knees and, in difficult cases, castor oil is applied!

It is taken for granted that deceptions of all sorts are part and parcel of diamond trading. Rock crystals, tourmalines, and aquamarines are sold as diamonds to naive customers. The temptation to cheat is enormous, and the cheated person cannot complain since he or she has also broken the law. Late at night, stones are sometimes offered for sale in hotel bars; most of the time, however, police informants try to trap illegal buyers. Native informants have also been trained to stop victims on the street and offer diamonds to people who might be prospective buyers. Most of the time the intended customer has been pointed out to the informer in advance. The stones that are being "sold" are marked and the property of the secret service. This so-called system of trapping has been used on other occasions as well and is just as despicable as the system of turning state's evidence. It must be shameful for the *trapper*, and bitter and spiteful for the *trapped*!

The mining of precious and nonprecious metal is regulated, and miners are required to obtain a mining permit at a cost of ten pounds, with an additional monthly fee of two to six shillings paid to the district, and a five-shilling fee to the government of South West. Mining claims are restricted to certain sizes: those for gold and silver measure one hundred by two hundred meters, all others, three hundred by six hundred meters. A stake must be raised at each corner of the claim, with an additional stake at the center. A notice must be attached to each stake on which, in a completely clear and legible manner, the last and first name of the miner, the mining permit number, and the exact date and time when the stake was set in place are written. Stakes may only be raised during daylight hours.

Taxes for a mining claim are to be paid six months in advance at a rate of one to ten pounds for precious metals and fifteen shillings for all other metals. From this point on, every month five shillings must be paid for a gold claim, two to six shillings for additional claims. In my experience, most of the men who try their hands at mining are poor fellows, which makes things very difficult, for if a miner fails to pay his fees on time, his mining permit and claim are terminated—no matter how long the miner had been in possession of the claim. When a claim and permit are terminated, another person may immediately plant his stake next to the stake of the one who can no longer pay. In the language of the prospectors, this is called "jumping a claim." People pay close attention to the status of very promising claims, and it is not unknown that a man's best friend will jump his claim. Needless to say that such things undermine a man's character and sow the seeds of mistrust.

The goldfields near Rehoboth, south of Windhuk, are presently worked by small operators. It is rumored, however, that a big South African mining concern intends to purchase the fields. How typical this is! Once the initial work has been done, through the toil and sacrifice of small-time individuals, then big capital arrives, exploits the work already done, and turns a profit. How sweetly do the shareholders smile as the small-timers are forced to leave and seek for gold elsewhere. Small-time operations have yielded some gold near Omaruru, but these operations are insignificant.

Our country possesses other mineral resources as well: copper, tin, vanadium, iron, zinc, fluorspar, lithium, and marble, along with all types of precious and semiprecious stones. So far, little has been done to investigate the feasibility and profitability of mining these minerals. Copper is presently mined in Tsumeb, and tin is gathered in open pit mines. Much of the time prospecting for metals or finding precious stones is simply a matter of luck; many search for years on end without any success, and others simply fall into it.

What our farmers produce, faithfully and diligently, day after day, year in and year out, has far greater certainty and value. These are our *black* diamonds and come in the form of karakul sheep. These have the greatest economic value, as their little pelts and their wool fetch consistently good prices. At this moment in time, as well as looking into the future, not only the German farmers but settlers from other nations living in this country are successful in the battle for survival. It is not the Mandate Government but the courageous individual farmers who must be credited for this, men who are not frightened off by economic depression or lack of immediate success, but who, through thick and thin, persevered and continued working until they finally began to reap their reward.

When the breeding of karakul sheep began to succeed, the spirit of enterprise and courage to begin new ventures started to increase in our country. A short article printed in our *Allgemeine Zeitung* (the local German daily) on May 13, 1936, serves as evidence of progress and new flourishing enterprises:

> On April 30 of this year, the steamer, *Ibena*, sailed from Walfisch Bay with a large load of freight: 16500 crates of butter bound for London; 500 tonnes of vanadium ore going to Hamburg, 1000 bales of wool, most of it karakul wool, primarily intended for German consumption. This was the largest amount of wool ever shipped from Walfisch Bay. Because prices for karakul wool are very favorable, it is hoped the wool trade will expand considerably. The export of ore is good news for the potential revival of mining in South West. Every month, small but regular deliveries of tin, tantal-metal, and wolfram are shipped to England. A few days ago, the Italian freighter, *Duchessa*, departed for Marseille with 235 tons of lithium ore. One expects these types of ore shipments to be regular from now on.
>
> The season for snoek (a delicious fish) fishing is now over and eight snoek boats have sailed for Cape Town to be renovated for the nest season which begins in December. A total of 621,000 snoek, weighing some 1500 metric tonnes,

were shipped from Walfisch Bay to Cape Town. In addition to this, fishing boats sailing to Cape Town carry 150,000 other varieties of fish. The largest catch was made by the snoek-boat, *Snoeker*, with 113,000 fish. The snoek catch netted the port authority approximately 4000 pounds in port and storage fees.

A Concluding and Final Word

I am now a bit fatigued from this endeavor. Tentatively, I lay down my pen, acknowledging that much more could have been said about the things I have written, while other matters have not been mentioned at all. My intention is not to vainly criticize the Mandate Government; indeed I have done my very best to present events and conditions as objectively as possible, to present things as I experienced and understood them from my own limited perspective. No one should do injury his own blood and his own people. I have never done so and never wish to do so. I have happily, proudly, and as an undisputed matter of course, stood by my Germanity. But I also fully appreciate other nationalities, as I have occasion to rub shoulders with them, and have no desire to harm anyone. At the same time, it is clear to us Germans that certain things appear to us in a different light, that South Africans do not always see what we see. We are tolerated in a country that we conquered with our own blood and strength, and though we are permitted to raise our colors here, it is simply not enough. In each of our hearts burns our greatest desire: to become part of our Fatherland once again. O Fatherland, our great and beloved Fatherland, do stretch forth Thine arms and embrace our thirsty and difficult land, a land so dear to our hearts, and restore us to the strength of Thy bosom!

THE END

Postscript

I: ESTABLISHING ROOTS IN SOUTH WEST AFRICA

Margarethe von Eckenbrecher had a genuine, deep, abiding love of South West Africa. Twice in her life she made the decision to relocate there, the second time under particularly difficult circumstances. South West Africa became her preferred place of residence; indeed, the days of her life spent there eclipsed that of the days she lived in Germany. The pursuit of her ideal, her Idyll, of a farming life in South West Africa was the principle undertaking of Frau von Eckenbrecher's first stay in the Protectorate. Though it entailed wading through considerable inconvenience, difficulty, occasional heartache, and true danger, it was also, as she chronicles it, an extraordinary period of her life. Taming a wild and foreign landscape, wrestling with the land to make a living, exploring a strange, often inhospitable countryside, bringing a child into the world, enjoying the times she and her husband spent with their favorite neighbor, Old Ertmann, engaging with the native peoples whose ways she often found incomprehensible, learning to accept the untimely deaths of persons dear to her (both natives and settlers), and taking a "heroic" part in the defense of Okombahe during the beginning stages of the Nama/Herero Uprising are but a few of the many experiences that ultimately distilled into deeply etched memories—memories whose savor remained a melancholic mix of joy and pain.

With her 1914 return to South West Africa as a divorced woman with two young sons, Frau von Eckenbrecher had no immediate intention of pursuing a rural ideal, though she always maintained a personal preference for rural over urban. Prior to her removal from Germany, she corresponded with close friends in Windhoek, inquiring after the possibility of finding work so that she could support herself and her sons. With affirmative responses, she arranged for the passage. After a time she found work and then looked to secure a house with sufficient property that her sons might have a sense of freedom, a feeling for the wider landscapes of South West Africa. For Frau von Eckenbrecher, the city (though Windhoek was diminutive in the extreme compared to Berlin) was a completely different experience from Okombahe. Ever adaptable, she made a good life for herself and her sons.

Though the personal qualities that characterized Frau von Eckenbrecher's approach to her rural life in Okombahe were likewise manifested in her Windhoek life, there is an indefinable, peculiar strangeness, perhaps a telling obscurity, about the relationship of her life in Okombahe to her life in Windhoek. Indeed, Part II of her memoir is written in a manner that might suggest little if any deep connection between these two pieces of her life in South West Africa. Given her love of the years she passed in Okombahe, it is difficult to understand why she waited until her son, Bitz (who was about eleven years of age when she returned to South West Africa), was a full grown man and owner of a motorcar before she organized a return visit to Okombahe—the very first one since her hasty 1904 departure. It was a decidedly brief visit. There she met with friends and acquaintances, both natives and settlers. Together they reminisced and they wept. When her capacity to retrace the footsteps of her former life in South West was spent, she took her leave. Commenting on this very moment, this revisitation of a set of bittersweet memories, Frau von Eckenbrecher noted that now Okombahe was behind her, that her life there was finished. What, precisely, she had in mind when she penned these telling phrases will never be completely known.

The general tone of Part II of her memoir is noticeably different from the tone of Part I. Casually, it might appear that Frau von Eckenbrecher's joie de vivre, or perhaps better stated, her *Lebensfreude*, was somewhat in decline. One must bear in mind, however, that upon her return to South West Africa her immediate domestic and parental concerns and duties were daunting and demanding. In addition to this, not only had the circumstances of her life greatly changed, but she herself, the Margarethe von Eckenbrecher of 1914 (let alone the deeply retrospective woman who finished writing Part II in 1936), was not the Margarethe von Eckenbrecher of 1904. Frau von Eckenbrecher had added another decade of experience to her tally, some of it very difficult indeed: a child who died at birth[1] and the dissolution of her eleven-year marriage. Through the combined effects of her personal happiness and woes during that ten-year interlude in Germany, together with the natural process of aging (from twenty-nine to thirty-nine) and the maturity, perhaps hard-won wisdom, that sometimes accompanies years of moderate adversity, the *perspective* from which Frau von Eckenbrecher engaged the world had changed. Her practical orientation in life had shifted. She was hardly a sullen, morose, and despondent person; rather, quite to the contrary. Yet now she had to consider the future on very different terms. Not only did she have concerns about her own prospects, her ability to earn a living, and to provide her two sons with a proper *Erziehung* (nurturing, training, rearing, and educating both mind and spirit), now she had to consider her sons' long-term possibilities, their education, their future employment, and the potential for their ultimate success and happiness in South West Africa. She had chosen South West Africa as the place she wanted to live, but was it the correct choice for her sons? This is a question she revisited again and again over the next two decades.

It was altogether appropriate that Margarethe von Eckenbrecher considered her sons' futures, and her experience proves that she was wise to have done so. Her younger son, Büdi, wished to own and operate a bookstore in Windhoek but died of typhus at the age of twenty-two. "Africa took him," is how Frau von Eckenbrecher described it. Bitz had decided to become a lawyer, but the changing circumstances of South West Africa (from German Protectorate to League of Nations Mandate and the de facto annexation by South Africa as its fifth province) made it impossible for Bitz to pursue this goal. Indeed, Bitz became a longer-term worry for Frau von Eckenbrecher as he struggled to find employment in a country now dominated by English- and Afrikaans-speaking South African transplantees. He worked as a driver and a hunter, he drifted from one thing to another, even illegally prospecting for gold. Since Margarethe von Eckenbrecher had spent eleven years of her life with a husband who had unsteady employment, her concerns for Bitz were not unfounded. Bitz, however, married well and eventually took over a cattle farm in the northwestern part of the country. This he husbanded with the occasional modicum of success, though he and his family ultimately removed to Cape Town because of financial struggles. It was there that Bitz found some financial security. Frau von Eckenbrecher lived the rest of her life in Africa, and all but the final year of her life in Windhoek.

The von Eckenbrecher story does not end with the matriarch's death. Of the African branch of the von Eckenbrecher family established by Themis and Margarethe von Eckenbrecher, none, at the time of this writing, resides in Namibia. However, third-, fourth-, and fifth-generation African von Eckenbrechers do live in neighboring South Africa, mostly in the Western Cape. Though South West Africa was Margarethe von Eckenbrecher's land of promise, the land of her dreams, an ideal worth striving for, and the place she loved more than any other, she could not possibly have foreseen the longer-term consequences of adopting South West Africa as her homeland. Insecurity in her day was both political and economic,[2] and in her own case was partially a consequence of the aftermath of World War I and the South African dispensation in South West Africa. She complained about this, about being a second-class citizen in her own land, not seeming to detect any irony in connection with the Imperial German displacement of African natives. She likewise struggled with economic uncertainty for a time before eventually obtaining a secure position as a schoolteacher. It is likely that Frau von Eckenbrecher would never have entertained the proposition that thirty-five years after her death South West Africa would become Namibia, an independent and majority rule country, or that South Africa would follow suit four years after Namibia.

Though in her book Frau von Eckenbrecher professes loyalty to the Fatherland and desires that South West Africa become a German possession once again, her surviving son, Bitz, likely felt no such loyalty to Germany. Bitz's two children, sons, had even less connection to Germany, and after four generations of African-born von Eckenbrechers, they are no longer Germans who live in Africa, but Africans of German descent. The very same issues of political and economic stability and security are at play in the lives of Margarethe von Eckenbrecher's great grandchildren, as several of them have immigrated to other countries—perhaps in search of what their great grandmother had found in German South West Africa.

II: THE VON ECKENBRECHERS TODAY: ROOTS AND BRANCHES

Margarete von Eckenbrecher (1875–1955)

Margarethe von Eckenbrecher (neè Hopfer) was born in Bernburg,[3] Saxony-Anhalt, Germany, on September 30, 1875. In February of 1902, she became engaged to her cousin, Themistokles (II) Hans Robert von Eckenbrecher, for the second time. They married six weeks later in Charlottenburg. Just days after their wedding, they sailed for South West Africa. In 1903, at their farm in Okombahe, South West Africa, Themistokles (III) Heinrich Adolf Curt von Eckenbrecher (known as Bitz) was born on the eve of a native uprising. In 1905, in Weimar, Germany, Themistockles and Margarethe had another son born to them. The von Eckenbrechers divorced in 1913, in the city of Braunschweig, Germany. Frau von Eckenbrecher never remarried. She returned to South West Africa with her two sons in 1914 and lived in Windhoek until 1954 when she moved to Cape Town, South Africa, due to failing health, and died there on July 18, 1955. She was buried in Windhoek.

Themistokles (II) Hans Robert von Eckenbrecher (1876–1935)

Themistokles von Echenbrecher was born in Düsseldorf in 1876. Sometime after the divorce, he immigrated to South America, eventually settling in Paraguay. Themis appears to have spent much of his life as a wanderer and "ne'er-do-well," though not for lack of talent or capacity.[4] In Paraguay he married and had children. He died of cancer in Asunción in 1935. In 2004, H-H. F. von Eckenbrecher, of Cape Town, South Africa, was contacted by a Señior von Eckenbrecher of Argentina, wishing to know if they were somehow related (personal communication with H-H. F. von Eckenbrecher). They believe they have a common grandfather—Themistokles II.

Heinrich Adolf Curt Themistokles III (Bitz) von Eckenbrecher (1903–1976)

Bitz attended school in Germany and in Windhoek. In 1935, he married Constance Juliete Bridgman in Cape Town, South Africa. The couple had two sons, Themis (IV) Henry von Eckenbrecher, and Hans-Henning Friedemar von Eckenbrecher. They settled on a remote farm in north-central Namibia and bred karakul sheep and Afrikander cattle. While most German-speaking men in Namibia were incarcerated in internment camps in South Africa during 1939 to 1945, Bitz was not because of his expressed anti-Nazi sentiment. In 1950, they sold their farm and relocated to Cape Town, where Bitz entered the business world as an accountant/bookkeeper until his death in 1976 at the age of seventy-three.

Constance Juliete Bridgman von Eckenbrecher (1908–1992)

Constance was the daughter of Dr. H. M. Bridgman who was educated in the United States during 1882 to 1891, and the granddaughter of the Rev. H. M. Bridgman, who was sent by the American Board of Commissioners for Foreign Missions to preach the Christian faith among the Zulus from 1861 until his death in 1896. Constance was born in 1908 in Cape Town, South Africa, but was partly schooled in America. She was a graduate of the University of Cape Town. Before her marriage to Bitz, she taught English at the German School in Swakopmund, Namibia. She was fluent in German, which was the language that Bitz, Constance, and their two sons spoke at home. She died in Cape Town in 1992, at the age of eighty-four.

Hans-Henning (Büdi) von Eckenbrecher (1905–1927)

Büdi was the second son of Themis and Margarethe von Eckenbrecher, and died in Windhoek of typhus at the age of twenty-two.

Themistokles IV Henry (Tim) von Eckenbrecher (1936–)

Tim is the eldest son of Bitz and Constance von Eckenbrecher and was born in 1936 in Swakopmund, Namibia. He is a businessman in Cape Town, South Africa, and is married to Bridget Abery. They have two children, a son, Themis (V) Neil von Eckenbrecher (married in 2001 to Nikki), and a daughter, Mary Ann von Eckenbrecher.

Hans-Henning Friedemar von Eckenbrecher (1938–)

Hans is the second son of Bitz and Constance and was also born in Swakopmund, Namibia. He is a biochemist and resides in Cape Town, South Africa. Hans married Annamarie Zoutendyk in Cape Town in 1966. They divorced in 1976. They are the parents of Juliette von Eckenbrecher (born in Cape Town in 1966, married to Eugene Theron in 1994, currently residing in the United States); Heidi von Eckenbrecher (born in Cape Town in 1969, married in 2000 in Piet Retief, South Africa, to Andrè Maree, currently residing in South Africa); and Lisa von Eckenbrecher (born in Cape Town in 1971, married in 1994 to Marius van Zuydam, and currently residing in the United States).

III: A REMEMBRANCE OF MY GRANDMOTHER, MARGARETHE VON ECKENBRECHER

I apologise if my presentation seems to lack any real substance, but I write of my grandmother (my father's mother) as I knew her during the years 1945–1947, when my brother and I had to leave our family farm in north-central Namibia in order to enter school in Windhoek. During the years we lived with my grandmother, our days were largely uneventful, following, as they did, a simple, unvarying routine. I regret that I am unable to comment on my grandmother's inner thoughts and views about life and the world, but when I lived with her I was a child and she treated me as such. Never did she discuss adult topics with me.

My grandmother was actually my teacher at the Deutsche Realschule in 1946. Our relationship was that of teacher/pupil, not grandmother/grandson, and remained as such during that period of time. Grandmother was a disciplinarian who lived a Spartan lifestyle—not by choice but of necessity. She was a serious-minded person who had little time for frivolity. By birth she was an aristocrat and behaved as such her entire life. I believe she may well have accepted the idea of the divine right of kings and the correctness of manorial/feudal system, in which birthright or the circumstances of one's birth established one's status and station in life.

We lived under austere conditions even relative to those times. My brother and I dwelt in an annex adjacent to her home built on the slope of a mountain overlooking Klein-Windhoek. Her very large property was bounded on one side by the von Eckenbrecher Strasse, which nowadays forks with Mandela Street. During those years we did not enjoy the luxury of electricity to power lights, stove, oven, refrigerator, hot water heater, washing machine, vacuum cleaner, or radio. We had no typewriter, no wind-up gramophone, no piano, no telephone, no flushing toilet, no motorized form of transport, and thus had little opportunity to socialise except when occasional visitors dropped by for afternoon coffee and cake. During that era, children were seen but not heard, and though we may have listened to adult conversations, I doubt we understood much of what went on.

We did enjoy having clean, piped, running water, and I remember the pipes were bound with sacking material to prevent them from bursting should a winter night's temperature drop below freezing. Grandmother employed a male domestic worker, called Phillipus, who actually came from our family farm in the northern half of the country. He performed the heavy-duty domestic chores such as gathering and chopping firewood to fuel the manual washing machine and to fuel the "geyser," which provided hot bathwater. He also ironed our clothes using an old-fashioned iron into which one set glowing hot coals. He watered the garden, cleaned and maintained the main house, and polished the silver, copper, and brass with wood ash. My brother and I made our own beds and cleaned our own bedroom.

Grandmother's house was tastefully furnished with Persian carpets, a comfortable lounge suite, a dining room suite, and a grandfather clock. Walls were adorned with many beautiful paintings painted by our great-grandfather, Themistokles I, whose paintings, at the time of this writing, are still being eagerly sought in the marketplace based on information I have gathered through the Internet. Our grandmother had some of the finest porcelain china dinnerware ever produced in Meissen, Lower Saxony, as well as precious silver flatware and cutlery embossed with our family crest—all heirlooms acquired from our opulent, aristocratic family in Germany.

On many an evening, we sat on the upper terrace, which overlooked the Klein-Windhoek valley well below us. There, we enjoyed the cool breeze, listened to the noises of the night, and watched the twinkling lights in the valley—simple entertainment, to be sure. During the dark nights, Grandmother often pointed out the major stars of the various constellations. Never once did our grandmother "go out" at night for pleasure or duty, and leave us alone at home. We left for school at the same time each morning, and returned home within an hour of one another.

Grandmother was a highly intellectual, cerebral woman, which made it difficult for us to communicate with her at that time simply because of our intellectual disparities. There was seldom a meeting of the minds. Her home abounded with classical German books, mostly printed in Gothic type, English books, and French books, which I was incapable of reading, let alone understanding. We had no children's books, no access to a public library.

While we kept no cats or dogs as pets, we did care for a tame owl, a monitor lizard, and a hornbill bird. Toys such as jigsaw puzzles, legos, mecannos, etc., we did not have, though we did have a collection of leaden soldier figurines and animals, and played with them in the dirt. As our school possessed no sports field, after-school activities such as cricket, rugby, field hockey, or soccer were not offered. Our school also had no tennis court, no swimming pool, no athletics track, and no gymnastics hall. Drama, choral, and speech clubs were also unavailable. Ironically, our old school now houses the School of Music in Windhoek.

Initially, we walked to and from school—which seemed like miles and miles away—come frost, rain, or searing heat. In course of time, however, we began cycling to school, both of us on a single bicycle. En route to Grandmother's home, a number of Afrikaaner children occasionally waited for us to pass a certain point from which they hurled stones at us. Their intent was to hurt us, but not to kill us. Anti-German sentiment was rife during those years.

After school, our grandmother performed the necessary light-duty house chores, which included cooking meals, and then marked class papers, read the *Allgemeine Zeitung*, or literature, until the light failed. We ate simple food and I always felt hungry. Our garden provided us with oranges, lemons, pomegranates, and prickly pears. We had cypress trees growing along the terrace and I shall never forget the eerie sound made by the wind as it blew through the cypress trees. The soil was dry, infertile, and unforgiving.

Once my older brother, Tim, and I owned our own bicycles, we became good endurance cyclists. Saturdays, Tim and I cycled to town to purchase incidental groceries; we helped with the weekly washing of the linen using the manual rotating, wood-fired washing machine; we also helped dust the Persian carpets and mats by suspending them from a line and beating them with a modified stick. Tim and I helped water the garden, and the two of us cleaned the bedroom we shared.

I cannot remember having much time to play, or even an incentive to play. As the climate was often too harsh, we were compelled to rest on our beds during the afternoons until 16h00 because of the heat of the day. Businesses even "shut shop" between 12h00 and 15h00 for this very same reason. It was seldom that we ever entertained a friend, mostly due to logistics. We witnessed no dinner parties, no theatre, no ballet, no shows, no musicals (*Oklahoma*, *Annie Get Your Gun*), but once, perhaps twice, we attended the big tent circus. I also remember going to the cinema to view a film called *Lassie Come Home*.

Early to bed, early to rise was our basic routine. Several mornings, we woke to discover a herd of Kudu grazing on our terraces. Mountainous regions were their habitat as they were sure-footed creatures. Our pastime was riding our bicycles from nowhere to nowhere, and using a pellet gun with improvised pellets—which always caused indeterminate, variable trajectories. Seldom, if ever, did we hit our target.

Occasionally, Tim and I visited our family farm during our school holidays. We sometimes travelled by train, pulled by a steam powered locomotive to the small farming settlement of Outjo where our parents fetched us by truck. Petrol was rationed during World War II, and tyres were unprocurable due to short supply. The net result of this was that our normal mode of travel was by donkey cart. Only if it were absolutely necessary did our parents drive their truck. I always longed for the company of my parents and craved the sanctuary of our farm.

At no time during my life did Grandmother take a vacation to Germany, to Cape Town, to the local seaside resort of Swakopmund or, for that matter, to any other location of interest, other than visiting us on the farm once, maybe twice. She pretty much remained housebound. Her extensive property was mostly bounded by a mountainous no-man's-land, owned by the government, except for a lower part of her property, situated in the valley, that adjoined a property owned by her best friend, Frau Starke, the wife of Dr. Starke, a lawyer. Another acquaintance who lived adjacent to her property was Herr Berker, whose son became the first Judge President of Namibia. Most of Grandmother's acquaintances were from academic circles and professionals such as doctors, lawyers, and senior government administrators.

My grandmother taught classes at the Realschule until the end of 1948, at which time she had reached the age of 73. After her long teaching career, she busied herself during her leisure time by writing a sequel to her book, though I do not know what became of the manuscript. Once Tim and I left for boarding school, the well-known Namibian landscape artist, Adolf Jensch, boarded in the annex we had occupied. I can never recall my grandmother ill in bed, or complaining of arthritis, influenza, heart palpitations, headaches, etc.

In the course of time, a man who called himself "Count Halwill" came to occupy the annex. He exploited my grandmother's memory lapses by fraudulently "relieving" her, bit by bit, of her hard earned savings, which ultimately caused her financial ruin. My parents had long since removed to Cape Town, where my father took work as a bookkeeper/accountant. In 1954, my brother, Tim, and I paid Grandmother a visit and noticed that her memory was failing. We reported our observation to my father who decided to bring Grandmother to Cape Town to live with us. Her large property overlooking the Klein-Windhoek valley was sold in 1954 for a "song" due to her financial constraints (and since then the property has been sub-divided into many, many plots on which beautiful mansions and houses have been erected).

Grandmother lived with us in our home in Claremont, a suburb of Cape Town. As both my parents worked, a household servant looked after her during the day, but Grandmother repeatedly tried to leave our house and wander off. Good, constant care was what she needed and my parents decided to place her in a nursing home for the elderly where, in 1955, she died peacefully in her sleep. Her body was returned to Windhoek where she was buried beside her younger son, Büdi, in the Windhoek City Cemetery.

I truly regret that I was unable to relate to my grandmother on an adult basis. Today, I greatly appreciate her intellectual prowess, her high moral standards, her single-mindedness, her courage, etc. I cannot recall her wasting a single minute of her time on frivolities. While my grandmother was not a Mother Theresa, she was, nevertheless, a truly remarkable person in her own right, a person who accomplished a great deal during her long life.

Hans-Henning Friedemar von Eckenbrecher
Claremont, Cape Town, South Africa
10 May, 2012

Notes

A READER'S INTRODUCTION

1. In undertaking a translation of von Eckenbrecher's memoirs into English, we were very fortunate to find an original 1911 edition of the book in the collections of the Harold B. Lee Library at Brigham Young University. To this, we joined the additional chapters von Eckenbrecher added to the book for the 1940 edition that updates the reader from her May 1914 return until 1936. The edition of the book we used for translating the latter portion is a 2008 reprint of the 1940 edition, published under the auspices of Peter's Antiques, Swakopmund, Namibia.

2. See, for example, the multivolume work produced by the Feminist Press at CUNY: *Women Writing Africa* (2003–2007).

3. Frau Irma Schimansky Hübner was born in German South West Africa on January 4, 1914, and died in Namibia on January 13, 2006, at the age of ninety-two. She married Alfred Hübner, a master watchmaker from Lübeck, Germany, on August 8, 1944. Herr Hübner worked for Meyer Jewellers, Watchmakers, and Optometrists in Windhoek. The Hübners eventually took over the business and fostered its further success. Their son, Klaus Hübner, continues to operate the optical portion of the business. An exceptionally well-known person in the community, particularly the German-speaking community, Frau Hübner was committed to her wide circle of friends and acquaintances. She was also an invaluable bridge between old South West Africa and modern Namibia. All references to Frau Hübner are taken from private conversations with her in Windhoek in 2003.

4. Today Bernburg is a provincial town of thirty-five thousand located at the Saale River in the State of Sachsen-Anhalt in the Federal Republic of Germany.

5. Von Eckenbrecher's family owned property in the picturesque mountainous Harz region where the extended family occasionally gathered and discussed philosophy, literature, and the political and economic situation in Germany and the world.

6. Both of these were distinguished theologians who lectured at several German universities. Wilhelm von Humboldt, the eminent German scholar and linguist,

founded the university in 1810. It was later named Humboldt University after him and became a model for much of the European and American university systems and became one of the most prominent of the great German research institutions.

7. See Friedrike Eigler, "Engendering German Nationalism: Gender and Race in Frieda von Bülow's Colonial Writings" in *The Imperialist Imagination. German Colonialism and Its Legacy*, eds. Sara Friedrichsmeyer, Sara Lennox, and Susanne Zantop (Ann Arbor: University of Michigan Press, 2001), 69–86; Lora Wildenthal, "'When Men Are Weak': The Imperial Feminism of Frieda von Bülow," *Gender & History* 10, no. 1 (1998): 53–77. See also Joachim Warmbold, *Germania in Africa: Germany's Colonial Literature* (New York: Lang, 1989).

8. Margarethe von Eckenbrecher, *Was Afrika mir gab und nahm. Erlebniss einer deutschen in Südwestafrika 1902–1936* (Berlin: Mittler, 1940).

9. Homi K. Bhabha, *The Location of Culture* (New York: Routledge, 1994).

10. Paul E. Kerry, *Enlightenment Thought in the Writings of Goethe: A Contribution to the History of Ideas* (Rochester, NY: Camden House, 2001).

11. The von Eckenbrechers did not actually own a farm but were looking to purchase one. Indeed the status of the small holding they occupied in Okombahe is unclear. Captain Cornelius, a leader of the local Damara, had given Themistockles permission to use some land, not only for building a house and for penning his livestock but also access to pasturelands and tracts along the banks of the Omaruru River where the von Eckenbrechers cultivated extensive gardens.

12. For this biographical information see Katharina Gerstenberger, *Truth to Tell: German Women's Autobiographies and Turn-of-the-Century Culture* (Michigan: University of Michigan Press, 2000), 69–77.

13. Daniel J. Walther, *Creating Germans Abroad: Cultural Policies and National Identity in Namibia* (Athens, OH: Ohio University Press, 2002), 2.

14. Roger Chickering, "'Casting Their Gaze More Broadly': Women's Patriotic Activism in Imperial Germany," *Past and Present* 118 (1988): 156–85, see 177.

15. Susanne Zantop, *Colonial Fantasies: Conquest, Family, and Nation in Precolonial Germany 1770–1870* (Durham: Duke University Press, 1997), 43–75.

16. Carol Aisha Blackshire-Belay, "German Imperialism in Africa: The Distorted Image of Cameroon, Namibia, Tanzania, and Togo," *Journal of Black Studies. Special Issue: Image of Africa in German Society* 23, no. 2 (1992): 236–45.

17. Gerstenberger, *Truth to Tell*, 99 and 92.

18. Ibid.

19. Lora Wildenthal, *German Women for Empire, 1884–1945* (Durham: Duke University Press, 2001), 152–53. Daniel J. Walther asserts that femininity cannot be analyzed without reference to masculinity, as they are both needed to define each other. "Gender Construction and Settler Colonialism in German Southwest Africa, 1894–1914," *The Historian* 66, no. 1 (2004): 1–18.

20. Lora Wildenthal, "'She Is the Victor': Bourgeois Women, Nationalist Identities, and the Ideal of the Independent Woman Farmer in German Southwest Africa," *Social Analyses. Special Issue: Nations, Colonies and Metropoles* 33 (1993): 68–88, see 81.

21. Nancy Reagin, "The Imagined Hausfrau: National Identity, Domesticity, and Colonialism in Imperial Germany," *The Journal of Modern History* 73, no. 1 (2001): 54–86, see 85–86. See Nancy Reagin, *Sweeping the German Nation: Domesticity and National Identity in Germany, 1870–1945* (Cambridge: Cambridge University Press, 2007). Cf. Benedict Anderson, *Imagined Communities: Reflections on the Origin and Spread of Nationalism*, rev. ed. (London: Verso, 2006).

22. See Megan Vaughan for an analysis of the complexities and tensions of the colonial medical discourse and practice in Africa: *Curing Their Ills: Colonial Power and African Illness* (Stanford: Stanford University Press, 1991).

23. Reagin, "The Imagined Hausfrau," 82, fn. 79; Helmut Bley, *Namibia under German Rule* (Hamburg: Lit Verlag, 1996), 279.

24. See Niall Ferguson, *Empire: The Rise and Demise of the British World Order and the Lessons for Global Power* (London: Allen Lane, 2002).

25. Johann Wolfgang von Goethe, *Faust Part Two*, ed. and trans. David Luke (Oxford: Oxford University Press, 1994), 223. By permission of Oxford University Press.

26. Wildenthal, for example, asserts that von Eckenbrecher in a later edition of her book "altered the original text to suit Nazi racial sensibilities." *German Women for Empire*, 153.

27. J. H. Esterhuyse, *South West Africa 1880–1894* (Cape Town: C. Struik, 1968).

28. For an excellent overview of Britain's exploration of the Namibia coast circa 1795 to 1895, see Jill Kinahan, *By Command of Their Lordships* (Cape Town: CTP Book Printers, 1992).

29. Esterhuyse, *South West Africa*, 7.

30. Ibid., 8.

31. Heinrich Vedder, *South West Africa in Early Times* (London: Oxford University Press, 1966), 145.

32. Jon M. Bridgman, *The Revolt of the Hereros* (Berkeley and Los Angeles: University of California Press, 1981), 32–34.

33. Ibid., 32.

34. Ibid., 32–33.

35. Ibid., 33.

36. It is important to bear in mind that although one writes of Herero or Nama or Orlam as if they were unified peoples, this was certainly not historically accurate. It is fundamentally mistaken to write of any of the "native" peoples of Namibia as if they possessed a centralized political and economic system that governed all of the Herero or all of the Nama or all of the Ovambo, and so on, because such is not the case. These peoples were and are internally subdivided into groupings on the basis of kinship—whether patrilineal, matrilineal, or a combination of the two. Though the elder men of a kinship-based grouping took decisions on matters brought before them, it was not the case that they necessarily had the means to enforce their decisions. It is far better to think of each of these peoples as comprised of autonomous, self-governing, property-owning kinship groups that could temporarily confederate with other autonomous groups of the same "tribe" for specific purposes, especially in the case of the Herero. It was and is also the case that great leaders could arise,

such as Jonker Afrikaaner, who, through raiding the property of other kinship groups, or outsiders, were able to gather about them a great deal of productive wealth (land, livestock, etc.) sufficient to attract others to them who wished to use that productive property to make a living. The concept of a paramount chief who spoke for all the Nama, Herero, Ovambo, and others was imposed by colonial governments that desired to converse, trade, consult, and negotiate with just one person, rather than the scores of men who headed sundry kinship groups. The result of this colonial imposition was that while missionaries, traders, or government officials congratulated themselves on having concluded a peace pact with the Herero "led" by a particular great leader, the pact was not recognized by any other group of Herero, perhaps not even by the great leader's underlings. Under most circumstances, individual kinship groupings—whether Herero, Nama, or Orlam—consistently sought to maintain their autonomy. See D. P. Crandall, *The Place of Stunted Ironwood Trees: A Year in the Life of the Cattle-Herding Himba of Namibia* (New York: Continuum, 2000).

37. Bridgman, *The Revolt of the Hereros*, 30.

38. A good example of this is Missionary Henri Junod (1863–1934) and his well-known, two-volume work on the Thonga, *The Life of a South African Tribe* (New York: University Books, 1962).

39. *The Letters of Emma Sarah Hahn*, ed. Dorothy Guedes (Windhoek, Namibia: Namibian Scientific Society, 1992), 69.

40. *Charles John Andersson, Trade and Politics in Central Namibia 1860–1864, Charles John Andersson Papers, Vol. 2*, ed. Brigitte Lau (Windhoek, Namibia: SWA/Namibia Archives Services Division, 1989), 168–69.

41. Brigitte Lau, *Southern and Central Namibia in Jonker Afrikaaner's Time* (Windhoek, Namibia: National Archives of Namibia, 1987), 31–33; 52–56.

42. Ibid., 110.

43. *The Letters of Emma Sarah Hahn*, 282–83.

44. Bridgman, *The Revolt of the Hereros*, 35–37.

45. Ibid., 37.

46. Ibid., 37–38.

47. Ibid., 38.

48. Ibid., 41.

49. For an overview of Germany's colonial ambitions, see Woodruff D. Smith, *The German Colonial Empire* (Chapel Hill: University of North Carolina Press, 1978).

50. Dr. H. E. Göring was father of Hermann Göring, the infamous Nazi Reichsmarschall.

51. Bridgman, *The Revolt of the Hereros*, 41–42.

52. It is interesting to note that both von François and Manasse refer to each other in the letter as "friend," yet while von François always addresses Manasse in the familiar form of you (*Du*), Manesse always addresses von François in the formal you (*Sie*).

53. Curt von François, *Deutsch-Südwest-Afrika* (Berlin: Dietrich Reimer, 1899), 146–47.

54. Bridgman, *The Revolt of the Hereros*, 44.

55. Samuel Maharero was son to Kamaharero, also known as Maharero, and attended the local Lutheran schools, where he was seen as a potential priest. When his father died in 1890, he gained the "chieftainship" of Okahandja, although much of his father's wealth, especially wealth in cattle, went to the latter's matrilineal relatives. Samuel's early adult life was hardly exemplary, but he did manage to become a leader of the failed uprising against German rule, and he fled into exile in Botswana. His remains were eventually repatriated to Okahandja, Namibia, and he remains a great hero of the Herero. For a more sympathetic portrayal of Samuel Maharero, see Jan-Bart Gewald, *Herero Heroes: A Socio-Political History of the Herero of Namibia, 1890–1923* (Oxford: Ohio State University Press, 1999), especially chapter 3.

56. *The Hendrik Witbooi Papers*, trans. A. Heywood and E. Maasdorp with annotations by Brigitte Lau (Windhoek, Namibia: National Archives of Namibia, 1989), i.

57. Ibid., 87–94.

58. Bridgman, *The Revolt of the Hereros*, 46.

59. Ibid., 48.

60. Ibid., 48–50.

61. Ibid.

62. Gerhard Pool, *Samuel Maherero* (Windhoek, Namibia: Gamsberg Macmillan, 1991), 144.

63. Ibid., 143–45.

64. Bridgman, *The Revolt of the Hereros*, 48–50.

65. John H. Wellington, *South West Africa and Its Human Issues* (Oxford: Oxford University Press, 1967), 190–92.

66. Ibid.

67. Bridgman, *The Revolt of the Hereros*, 50.

68. Lau, *Southern and Central Namibia in Jonker Afrikaaner's Time*, 109, footnote 9.

69. Pool, *Samuel Maherero*, 180.

70. Ibid., 175.

71. Annemarie Heywood, Brigitte Lau, and Rajmund Ohly, *Warriors, Leaders, Sages, and Outcasts in the Namibian Past: Narratives Collected from Herero Sources for the Michael Scott Oral Records Project (MSORP) 1985–6* (Windhoek, Namibia: MSORP, 1992), 162–63.

72. Wellington, *South West Africa and Its Human Issues*, 194.

73. Ibid.

74. Ibid., 195.

75. Theodor Leutwein, *Elf Jahre Gouverneur in Deutsch-Südwestafrika* (Berlin: E. S. Mittler und Sohn, 1906), 305–6. Leutwein's memoir is an effort to rehabilitate his efforts and recast the specific German colonial gambit as a part of a global effort, including the United States. See: Jens-Uwe Guettel, "From the Frontier to German South-West Africa: German Colonialism, Indians, and American Westward Expansion," *Modern Intellectual History* 7, no. 3 (2010): 523–52, see 551.

76. Ibid., 196.

77. Pool, *Samuel Maherero*, 174–84.

78. Ibid.

79. Jürgen Zimmerer and Joachim Zeller, *Genocide in German South-West Africa: The Colonial War (1904–1908) in Namibia and Its Aftermath*, trans. Edward Neather (Pontypool, Torfaen: Merlin Press, 2008). Jeremy Sarkin, *Germany's Genocide of the Herero: Kaiser Wilhelm II, His General, His Settlers, His Soldiers* (Rochester, NY: James Currey, 2011). Jeremy Silvester, Jan-Bart Gewald, eds., *Words Cannot Be Found: German Colonial Rule in Namibia: An Annotated Reprint of the 1918 Blue Book* (Leiden: Brill, 2003). Jon Bridgman, *The Revolt of the Hereros.*

80. From BBC News, http//news.bbc.co.uk/2/hi/Africa/7033042.stm, October 7, 2007.

81. See Marion Wallace and John Kinahan, *A History of Namibia: From the Beginning to 1990* (New York: Oxford University Press, 2011) for a complete history; see Patricia Hayes et al., *Namibia under South African Rule: Mobility and Containment, 1915–46* (Oxford: James Curry, 1998) and Wade C. Pendleton, *Katatura, A Place Where We Stay: Life in a Post-Apartheid Township in Namibia: Katutura Before and Now* (Windhoek, Namibia: Gamsberg Macmillan, 1996) for accounts of native peoples during the time period.

82. www.worldbank.org.

83. For an interesting account of the transition from South African governance to independence, see David Lush, *Last Steps to Uhuru: An Eye-Witness Account of Namibia's Transition to Independence* (Windhoek, Namibia: New Namibia Books, 1993).

POSTSCRIPT

1. Gehrecke suggests that after 1905 the von Eckenbrechers had a third son who died at birth, though Gehrecke offers no conclusive evidence for it. Margarethe von Eckenbrecher herself mentions nothing about a third son in her book. See Siegfried Gehrecke, *Themistockles von Eckenbrecher, 1842–1921* (Glosar: Museumsverein Goslar, 1985), 18–19.

2. At the end of World War I there was also insecurity about residency as the South Africans deported many, many German settlers. Frau von Eckenbrecher also experienced this but was eventually granted permanent residency by the South African government.

3. Today Bernburg is known for its psychiatric hospital that was used during the Third Reich to euthanize the mentally ill.

4. Gehrecke comments that Themistockles' father, Themistockles I (a well-known painter), found his son to be gifted but also possessed of an unsteady character. He was an incessant traveler and someone who had an uncanny ability to invest and lose great sums of money in business ventures others would instantly have recognized as unsound and unsavory. See Siegfried Gehrecke, *Themistockles von Eckenbrecher.*

Index